Ngoc Thanh Nguyen, Bogdan Trawiński, and Jason J. Jung (Eds.)

New Challenges for Intelligent Information and Database Systems

T0137888

Studies in Computational Intelligence, Volume 351

Editor-in-Chief

Prof. Janusz Kacprzyk
Systems Research Institute
Polish Academy of Sciences
ul. Newelska 6
01-447 Warsaw
Poland
E-mail: kacprzyk@ibspan.waw.pl

Further volumes of this series can be found on our homepage:
springer.com

Vol. 329. Santi Caballé, Fatos Xhafa, and Ajith Abraham (Eds.)
Intelligent Networking, Collaborative Systems and Applications,
2010
ISBN 978-3-642-16792-8

Vol. 330. Steffen Rendle
Context-Aware Ranking with Factorization Models, 2010
ISBN 978-3-642-16897-0

Vol. 331. Athena Vakali and Lakhmi C. Jain (Eds.)
New Directions in Web Data Management 1, 2011
ISBN 978-3-642-17550-3

Vol. 332. Jianguo Zhang, Ling Shao, Lei Zhang, and
Graeme A. Jones (Eds.)
Intelligent Video Event Analysis and Understanding, 2011
ISBN 978-3-642-17553-4

Vol. 333. Fedja Hadzic, Henry Tan, and Tharam S. Dillon
Mining of Data with Complex Structures, 2011
ISBN 978-3-642-17556-5

Vol. 334. Álvaro Herrero and Emilio Corchado (Eds.)
Mobile Hybrid Intrusion Detection, 2011
ISBN 978-3-642-18298-3

Vol. 335. Radomir S. Stankovic and Radomir S. Stankovic
From Boolean Logic to Switching Circuits and Automata, 2011
ISBN 978-3-642-11681-0

Vol. 336. Paolo Remagnino, Dorothy N. Monekosso, and Lakhmi
C. Jain (Eds.)
Innovations in Defence Support Systems – 3, 2011
ISBN 978-3-642-18277-8

Vol. 337. Sheryl Brahnam and Lakhmi C. Jain (Eds.)
*Advanced Computational Intelligence Paradigms in
Healthcare 6,* 2011
ISBN 978-3-642-17823-8

Vol. 338. Lakhmi C. Jain, Eugene V. Aidman, and
Canicious Abeynayake (Eds.)
Innovations in Defence Support Systems – 2, 2011
ISBN 978-3-642-17763-7

Vol. 339. Halina Kwasnicka, Lakhmi C. Jain (Eds.)
Innovations in Intelligent Image Analysis, 2010
ISBN 978-3-642-17933-4

Vol. 340. Heinrich Hussmann, Gerrit Meixner, and
Detlef Zuehlke (Eds.)
Model-Driven Development of Advanced User Interfaces, 2011
ISBN 978-3-642-14561-2

Vol. 341. Stéphane Doncieux, Nicolas Bredeche, and
Jean-Baptiste Mouret(Eds.)
New Horizons in Evolutionary Robotics, 2011
ISBN 978-3-642-18271-6

Vol. 342. Federico Montesino Pouzols, Diego R. Lopez, and
Angel Barriga Barros
*Mining and Control of Network Traffic by Computational
Intelligence,* 2011
ISBN 978-3-642-18083-5

Vol. 343. XXX

Vol. 344. Atilla Elçi, Mamadou Tadiou Koné, and
Mehmet A. Orgun (Eds.)
Semantic Agent Systems, 2011
ISBN 978-3-642-18307-2

Vol. 345. Shi Yu, Léon-Charles Tranchevent,
Bart De Moor, and Yves Moreau
Kernel-based Data Fusion for Machine Learning, 2011
ISBN 978-3-642-19405-4

Vol. 346. Weisi Lin, Dacheng Tao, Janusz Kacprzyk, Zhu Li,
Ebroul Izquierdo, and Haohong Wang (Eds.)
Multimedia Analysis, Processing and Communications, 2011
ISBN 978-3-642-19550-1

Vol. 347. Sven Helmer, Alexandra Poulovassilis, and Fatos Xhafa
Reasoning in Event-Based Distributed Systems, 2011
ISBN 978-3-642-19723-9

Vol. 348. Beniamino Murgante, Giuseppe Borruso, and
Alessandra Lapucci (Eds.)
Geocomputation, Sustainability and Environmental Planning,
2011
ISBN 978-3-642-19732-1

Vol. 349. Vitor R. Carvalho
Modeling Intention in Email, 2011
ISBN 978-3-642-19955-4

Vol. 350. Thanasis Daradoumis, Stavros N. Demetriadis, and
Fatos Xhafa (Eds.)
*Intelligent Adaptation and Personalization Techniques in
Computer-Supported Collaborative Learning,* 2011
ISBN 978-3-642-19813-7

Vol. 351. Ngoc Thanh Nguyen, Bogdan Trawiński, and
Jason J. Jung (Eds.)
*New Challenges for Intelligent Information and Database
Systems,* 2011
ISBN 978-3-642-19952-3

Ngoc Thanh Nguyen, Bogdan Trawiński, and
Jason J. Jung (Eds.)

New Challenges for Intelligent Information and Database Systems

 Springer

Ngoc Thanh Nguyen
Wrocław University of Technology
Institute of Informatics
Wybrzeże Wyspiańskiego 27
50-370 Wrocław
Poland
E-mail: Ngoc-Thanh.Nguyen@pwr.edu.pl

Jason J. Jung
Yeungnam University
Department of Computer Engineering
Dae-Dong
712-749 Gyeungsan
Korea
E-mail: j2jung@gmail.com

Bogdan Trawiński
Wrocław University of Technology
Institute of Informatics
Wybrzeże Wyspiańskiego 27
50-370 Wrocław
Poland
E-mail: Bogdan.Trawinski@pwr.wroc.pl

ISBN 978-3-642-26779-6   ISBN 978-3-642-19953-0 (eBook)

DOI 10.1007/978-3-642-19953-0

Studies in Computational Intelligence   ISSN 1860-949X

Typeset & Cover Design: Scientific Publishing Services Pvt. Ltd., Chennai, India.

Printed on acid-free paper

9 8 7 6 5 4 3 2 1

springer.com

Preface

Intelligent information and database systems are two closely related and well-established subfields of modern computer science. They focus on the integration of artificial intelligence and classic database technologies in order to create the class of next generation information systems. The major target of this new generation of systems is to provide end-users with intelligent behavior: simple and/or advanced learning, problem solving, uncertain and certain reasoning, self-organization, cooperation, etc. Such intelligent abilities are implemented in classic information systems to make them autonomous and user oriented, in particular when advanced problems of multimedia information and knowledge discovery, access, retrieval and manipulation are to be solved in the context of large, distributed and heterogeneous environments. It means that intelligent knowledge-based information and database systems are used to solve basic problems of large collections management, carry out knowledge discovery from large data collections, reason about information under uncertain conditions, support users in their formulation of complex queries etc.

Topics discussed in this volume include but are not limited to the foundations and principles of data, information, and knowledge models, methodologies for intelligent information and database systems analysis, design, implementation, validation, maintenance and evolution. They cover a relatively broad spectrum of detailed research and design topics: user models, intelligent and cooperative query languages and interfaces, knowledge representation, integration, fusion, interchange and evolution, foundations and principles of data, information, and knowledge management, methodologies for intelligent information systems analysis, design, implementation, validation, maintenance and evolution, intelligent databases, intelligent information retrieval, digital libraries, and networked information retrieval, distributed multimedia and hypermedia information space design, implementation and navigation, multimedia interfaces, machine learning, knowledge discovery, and data mining, uncertainty management and reasoning under uncertainty.

The book consists of 35 extended chapters based on original works presented during a poster session organized within the 3rd Asian Conference on Intelligent Information and Database Systems (20-22 April 2011 in Daegu, Korea). The book is divided into four parts. The first part is titled "Information Retrieval and Management" and consists of eight chapters that concentrate on many issues related to the way information can be retrieved and managed in the context of modern distributed and multimedia database systems. The second part of the book is titled "Data Mining and Computational Intelligence" and consists of 13 chapters. In the majority of them their authors present and discuss new developments in data mining strategies and algo-rithms, and present examples of their application to support effective information retrieval, management and discovery as well as show how chosen computational intel-ligence technologies can be used to solve many optimization problems related to in-telligent information retrieval and management. The third part of the book is titled "Service Composition and User-Centered Approach", and consists of six papers de-voted to user centered information environments design and implementation. In some of these chapters detailed problems of effective autonomous user model creation and automation of the creation of user centered interfaces are discussed. The fourth part of this volume consists of eight chapters published under the title "Intelligent Manage-ment and e-Business". Its chapters were worked out on the basis of the works pre-sented within the Third International Workshop on Intelligent Management and e-Business (IMeB 2011).

The editors hope that this book can be useful for graduate and PhD students in com-puter science as well as for mature academics, researchers and practitioners interested in merging of artificial intelligence technologies and database technologies in order to create new class of intelligent information systems.

We wish to express our great attitude to Prof. Janusz Kacprzyk, the editor of this series, and Dr. Thomas Ditzinger from Springer for their interest and support for our project.

The last but not least we wish to express our great attitude to all authors who contributed to the content of this volume.

January 2011   Ngoc Thanh Nguyen
                                                                   Bogdan Trawiński
                                                                      Jason J. Jung

Contents

Part II: Data Mining and Computational Intelligence

Part III: Service Composition and User-Centered Approach

Part IV: Intelligent Management and e-Business

Part I

Information Retrieval and Management

Algorithm to Determine Longest Common Subsequences of Two Finite Languages

Dang Quyet Thang

Namdinh University of Technology Education, Namdinh, Vietnam
thangdgqt@gmail.com

Abstract. In pattern matching problems, determining exact or approximative longest common subsequences of two languages appears in many practical problems. Applying weighted finite automata, as a modification of Mohri's method (2003) in determining Levenstein edit distance of two languages, in this article, we propose an effective method which allows us to compute the longest common subsequences of two finite languages accepted by two finite automata A_1 and A_2 with the time complexity $O(|A_1||A_2|)$, in that, $|A_i|$ is the number of states and edges of automata A_i, $i = 1,2$.

Keywords: weighted automaton, transducer, LCS, algorithm, finite language.

1 Introduction

The problem to determine a longest common subsequence (LCS) of two strings or its length plays an important part in theoretical research and applications. There has been a lot of research on this problem and many interesting results [6, 7, 8, 9, 10, 11, 12, 13]. The classic dynamic programming solution to LCS problem has running time $O(n^2)$ [10] in the worst case. Since then, several algorithms existing with complexities depending on other parameters can be found in the literature. An interesting and relevant parameter to this problem is R, which is the total number of ordered pairs of positions at which the two strings match. In [13] presented an algorithm to solve problem LCS in $O((R+n)\log n)$. In [9], an algorithm running in $O(R\log\log n+n)$ has been introduced. In [11], the LCS of two rooted, ordered, and labeled trees F and G were considered. Their first algorithm runs in time $O(r\text{height}(F)\ \text{height}(G)\ \lg\lg|G|)$, where r is the number of pairs $(v \in F,\ w \in G)$ such that v and w have the same label, and their second algorithm runs in time $O(Lr\lg r\lg\lg|G|)$, where L is the size of the LCS of F and G.

For the case of two sets of finite strings, many problems lead to the problem of determining LCS of two finite languages. For example: determining the distance between two sets of key words of the two texts, between a set of key words and a text file, or two sets of gene sequences, two sets of features of the considered objects... by LCS approach. As mention above, the problem to determine the LCS of two strings has the best time complexity about $O(n)$. However, applying directly these algorithms to two finite sets of strings L_1, L_2 can lead to multiplication with a factor of $|L_1||L_2|$. Then the problem to determine the LCS of two finite languages solved in a common way has a time complexity about $O(n^3)$. In this article, as a modification from Mohri's method in [2](2003), we

N.T. Nguyen et al. (Eds.): New Challenges for Intelligent Information, SCI 351, pp. 3–12.
springerlink.com   © Springer-Verlag Berlin Heidelberg 2011

propose an effective method which allows to determine the LCS of two finite languages with a time complexity of $O(|A_1||A_2|)$, approximative $O(n^2)$, where A_1 and A_2 are finite automata accepting finite languages L_1 and L_2, $|A_i|$ are the number of states and edges of A_i, $i = 1, 2$. In [2], Mohri introduced an algorithm to determine Levenstein edit distance of two languages. In this algorithm, an algorithm to determine the shortest paths in a finite graph composed by automata and transducers is used. The Levenstein edit operations are insertions, deletions, or substitutions. In our paper, edit operations are insertions, deletions, or coincidences and we apply an algorithm to determine the longest paths in an acyclic finite graph composed by these A_1, A_2 and a suitable weighted transducer.

Given two strings x and y on an alphabet Σ, we denote by $LCS(x, y)$ the set of all LCS of x and y, $L(x, y)$ by the length of a LCS of x and y.

We recall the classical definition of the length of LCS [10]. Let s be a string, $s=a_1a_2...a_k$ where a_i is a letter in an alphabet Σ, $i = 1...k$. We denote the length of s by $|s|$, by $s[i]$ the letter a_i, by s_i the prefix of s of length i, $s_i=a_1..a_i$. Suppose that x, y are two strings with lengths m, n respectively. Then $x=x_m$, $y =y_n$ and $L(x, y)$- the length of the LCS of x, y can be determined recursively from $L(x_i, y_j)$, $i=1..m$, $j=1..n$ as follows:

$$L(x_m, y_n) = \begin{cases} 0 & \text{if } m = 0 \text{ or } n = 0 \\ L(x_{m-1}, y_{n-1})+1 & \text{if } x[m] = y[n] \\ \max(L(x_{m-1}, y_n), L(x_m, y_{n-1})) & \text{otherwise} \end{cases} \quad (1)$$

Given a finite alphabet Σ and two finite languages $X, Y \subseteq \Sigma^*$. We formalize the problem of the LCS of the two finite languages as follows:

Problem 1: to determine

a) *The set of the longest common subsequences of X and Y*, denoted as $LCS(X, Y)$ which is determined by: $LCS(X, Y) = \cup_{a \in X, b \in Y : L(a, b) \geq L(x, y), \forall x \in X, y \in Y} LCS(a, b)$.

b) to find a concrete couple (a,b) satisfying $LCS(a, b) \subseteq LCS(X, Y)$, $a \in X$, $b \in Y$ together with some string $u \in LCS(a,b)$.

Problem 2: to determine *the length of the LCS of X and Y*, denoted as $L(X, Y)$.

In this article, we focus on solving the problem (1.b) and problem (2). The problem (1.a) is followed as a consequence.

In the section 2, we give the definition of the LCS of two finite languages, recall some concepts of semiring and transducer due to [2]. In the section 3 two algorithms on finite transducers are presented. In the section 4 we proposes an effective algorithm which allows us to determine LCS (problem 1.a) and the length of LCS of two finite languages (problem 2) and the end of the article is the conclusion.

2 Definitions and Concepts

2.1 The Longest Common Subsequences of Two Languages

In this part, we introduce definitions and notations which are needed in the next sections.

Let Σ be a finite alphabet, $\Gamma=\{(a, a)/a \in \Sigma\}$, consider the set $\Omega=\{(\Sigma \times \{\varepsilon\}) \cup (\{\varepsilon\} \times \Sigma) \cup \Gamma$. An element $\omega = (a_1, b_1)...(a_n, b_n) \in \Omega^*$ is considered as an element of $\Sigma^* \times \Sigma^*$ via the morphism $h: \Omega^* \to \Sigma^* \times \Sigma^*$ determined by $h(\omega) = (a_1...a_n, b_1...b_n)$.

Definition 1. *An alignment ω of two strings x and y over the alphabet Σ is an element of Ω^* such that $h(\omega)=(x, y)$.*

Example. $\omega =(a, \varepsilon)(\varepsilon, b)(a, a)(b, b)$ is an alignment of two strings $x = a\varepsilon ab$, $y = \varepsilon bab$, with $a, b \in \Sigma$.

Each element of Ω presents an edit operation including: an *insertion* (ε, a), a *deletion* (a, ε), a *coincidence* (a, a), $\forall a \in \Sigma$. A function c to compute the cost of each edit operation is determined as follows:

$$\forall a, b \in \Sigma, c((a, b)) = 1 \text{ if } a = b, \text{ and } c((a, b)) = 0 \text{ otherwise.} \tag{2}$$

The cost of the alignment $\omega = \omega_0\omega_1 \dots \omega_n \in \Omega^*$ is the total cost of its components:

$$c(\omega) = \sum_{i=0}^{n} c(\omega_i). \tag{3}$$

Proposition 1. Let two strings x and y on the alphabet Σ. For each $u \in \Omega^*$ satisfying $h(u) = (x, y)$, $L(x, y) = c(u)$ if and only if $c(u) = \max_{\omega \in \Omega^*}\{c(\omega) \mid h(\omega)=(x, y)\}$.

Proof. First, we prove the property: *For each string $x, y \in \Sigma^*$ with the length m, n respectively, there exists $\omega_0 \in \Omega^*$ such that $h(\omega_0) = (x, y)$, $L(x_m, y_n) = c(\omega_0)$ and $c(\omega_0) = \max_\omega\{c(\omega) \mid h(\omega) = (x, y)\}$*. Basing this property, the proof of Proposition 1 is directly implied.

Indeed, we prove this property by induction on both m, n.

In the case $(i = 0 \,\&\, j = 0, \dots, n)$ or $(i = 1,.., m \,\&\, j = 0)$: It is obviously true.

Suppose that the property is true with i, j $(i \le m, j < n)$ or $(i < m, j \le n)$.

We have to prove that the property is true with $i = m, j = n$.

* If $x[m] = y[n]$, according to inductive assumption, there exists $\omega_2 \in \Omega^*$ satisfying: $h(\omega_2)=(x_{m-1}, y_{n-1})$, $L(x_{m-1}, y_{n-1}) = c(\omega_2)$ and $c(\omega_2) = \max_{\omega \in \Omega^*}\{c(\omega) \mid h(\omega)=(x_{m-1}, y_{n-1})\}$.

Denote $\omega_0 = \omega_2 u$ with $u = (x[m], y[n]) \in \Omega \Rightarrow h(\omega_0) = (x_m, y_n)$, $c(\omega_2)+1 = c(\omega_0) = L(x_m, y_n)$ and $c(\omega_0) = \max_{\omega \in \Omega^*}\{c(\omega) \mid h(\omega) = (x_m, y_n)\}$.

* If $x[m] \ne y[n]$, according to inductive assumption, there exists $\omega_1, \omega_3 \in \Omega^*$ such that: $h(\omega_1) = (x_{m-1}, y_n)$, $L(x_{m-1}, y_n) = c(\omega_1) = \max_{\omega \in \Omega^*}\{c(\omega) \mid h(\omega)=(x_{m-1}, y_n)\}$

$h(\omega_3) = (x_m, y_{n-1})$, $L(x_m, y_{n-1}) = c(\omega_3) = \max_{\omega \in \Omega^*}\{c(\omega) \mid h(\omega)=(x_m, y_{n-1})\}$

Denote $u_1 = (x[m],\varepsilon) \in \Omega$, $\omega_1' = \omega_1 u_1$, $u_3 = (\varepsilon, y[n]) \in \Omega$, $\omega_3' = \omega_3 u_3$, $\Rightarrow$ $h(\omega_1') = h(\omega_3') = (x_m, y_n)$. Choose $\omega_0 \in \{\omega_1', \omega_3'\}$ such that: $c(\omega_0) = \max\{c(\omega_1'), c(\omega_3')\}$ then $c(\omega_0) = L(x_m, y_n)$ and $c(\omega_0) = \max_{\omega \in \Omega^*}\{c(\omega) \mid h(\omega) = (x_m, y_n)\}$. $\square$

Definition 2. *The length of the longest common subsequence $L(x, y)$ of two strings x and y on the alphabet Σ is determined as follows:*

$$L(x, y) = \max_{\omega \in \Omega^*} \{c(\omega) \mid h(\omega) = (x, y)\}. \tag{4}$$

Definition 3. *The length of the longest common subsequence of two finite languages $X, Y \subseteq \Sigma^*$, is determined by:*

$$L(X, Y) = \max\{L(x, y) / \forall x \in X, y \in Y\}. \tag{5}$$

Definition 4. *Let X and Y be two finite languages which are accepted by two finite automata A_1, A_2 respectively. The length of the longest common subsequence of A_1 and A_2, denoted as $L(A_1, A_2)$, is determined by:*

$$L(A_1, A_2) = L(X, Y). \tag{6}$$

2.2 Semiring and Transducer

In this part we recall the concepts of semiring, transducer and weighted automaton similar to definitions in [2] which are used in this article.

Definition 5. *A system* $(K, \oplus, \otimes, \overline{0}, \overline{1})$ *is a semiring if:* $(K, \oplus, \overline{0})$ *is a commutative monoid with identity element* $\overline{0}$; $(K, \otimes, \overline{1})$ *is a monoid with identity element* $\overline{1}$; $\otimes$ *distributes over* $\oplus$; $\overline{0}$ *is an annihilator for* $\otimes$: $\forall a \in K, a \otimes \overline{0} = \overline{0} \otimes a = \overline{0}$.

Example. The *tropical semiring* $(\mathbb{R} \cup \{-\infty, +\infty\}, \min, +, +\infty, 0)$ [1], [2] and [4].

Definition 6. *A weighted finite-state transducer T (transducer) over a semiring* K *is an 8-tuple* $T = (\Sigma, \Delta, Q, I, F, E, \lambda, \rho)$ *where* Σ *is the finite input alphabet of the transducer;* Δ *is the finite output alphabet;* Q *is a finite set of states;* $I \subseteq Q$ *the set of initial states;* $F \subseteq Q$ *the set of final states;* $E \subseteq Q \times (\Sigma \cup \{\varepsilon\}) \times (\Delta \cup \{\varepsilon\}) \times K \times Q$ *a finite set of edges;* $\lambda: I \to K$ *the initial weight function;* $\rho: F \to K$ *the final weight function mapping F to K.*

A weighted automaton over a semiring K is a 7-tuple $A = (\Sigma, Q, I, F, E, \lambda, \rho)$, where $\Sigma, Q, I, F, \lambda, \rho$ is defined as in Definition 6 and $E \subseteq Q \times (\Sigma \cup \{\varepsilon\}) \times K \times Q$.

An unweighted automaton is a 5-tuple $A = (\Sigma, Q, I, F, E)$, where Σ, Q, I, F is defined as in Definition 6 and $E \subseteq Q \times (\Sigma \cup \{\varepsilon\}) \times Q$.

Let an edge (a transition) $e = (q_1, a, b, w, q_2) \in E$. q_1, q_2 is in turn called *the initial (previous) state* and *the final (next) state* of the edge e, we say that e *leaves* q_1 and *comes to* q_2, the edge e is called *a loop* if $q_1 = q_2$. We denote: $i[e]$ - the input label of e; $p[e]$ - the initial or previous state of e; $n[e]$ - the final or next state of e; $w[e]$ - the weight of e (used in the case of transducer and weighted automaton); $o[e]$ - the output label of e (used in the case of transducer).

For each state $q \in Q$, we denote $E[q]$ as the set of edges leaving q, $A[q]$ is the set of states $q' \in Q$ having the edge leaving q and coming to q'.

The path $\pi = e_1...e_k$ is an element of E^* such that $n[e_{i-1}] = p[e_i]$, $i = 2,..., k$. π is called *the path from* q *to* q' if $p[e_1] = q$, $n[e_k] = q'$. We also extend n and p for any path π as follows: $n[\pi] = n[e_k]$ and $p[\pi] = p[e_1]$. A cycle π is a path such that $n[\pi] = p[\pi]$.

We denote $P(q, q')$ as the set of paths from q to q', $P(q, x, q')$ is the set of paths from q to q' with input label $x \in \Sigma^*$, $P(q, x, y, q')$ is the set of paths from q to q' with input label $x \in \Sigma^*$ and output label $y \in \Delta^*$ (in the case of transducer). These definitions can be extended to subsets $R, R' \subseteq Q$ by: $P(R, x, R') = \cup_{q \in R, q' \in R'} P(q, x, q')$, $P(R, x, y, R') = \cup_{q \in R, q' \in R'} P(q, x, y, q')$.

The input labeling function i, the output labeling function o and the weight function w can also be extended. Input labeling, output labeling of path π is the concatenation of the corresponding input labels and output labels of its constituent edges, the weight of a path as $\otimes$- product of the weight of its constituent edges: $i[\pi]=$ $i[e_1]...i[e_k]$, $o[\pi] = o[e_1]...o[e_k]$, $w[\pi] = w[e_1] \otimes... \otimes w[e_k]$. We also extend w to any finite set of paths Π by setting: $w[\Pi] = \oplus_{\pi \in \Pi} w[\pi]$.

A weighted automaton A is regulated if:

$$[\![A]\!](x) = \bigoplus_{\pi \in P(I,x,F)} \lambda(p[\pi]) \otimes w[\pi] \otimes \rho(n[\pi]) . \tag{7}$$

is well-defined and in K. $[\![A]\!](x)$ is defined to be $\bar{0}$ when $P(I, x, F) = \phi$.

A transducer T is regulated if:

$$[\![A]\!](x, y) = \bigoplus_{\pi \in P(I,x,y,F)} \lambda(p[\pi]) \otimes w[\pi] \otimes \rho(n[\pi]) . \tag{8}$$

is well-defined and in K. $[\![A]\!](x, y)$ is defined to be $\bar{0}$ when $P(I, x, y, F) = \phi$. In the following, we will assume that all the automata and transducers considered are regulated. We denote $/M/$ as the number of states together with edges of an automaton or transducer M. A *successful path* in an automaton or in a transducer M is a path from an initial state to a final state.

3 Algorithm on Transducers

In this section we present an algorithms of composition of transducers and an algorithm finding a source-single longest path.

3.1 Composition of Transducers

Composition is a fundamental operation on transducers that can be used in many applications to create complex transducers from simpler ones. Let K be a commutative semiring and let T_1, T_2 be two transducers defined over K, such that the input alphabet of T_2 coincides with the output alphabet of T_1. Then, the composition of T_1 and T_2 is a transducer $T_1 \circ T_2$ defined for all x, y by [3] and [4]:

$$[\![T_1 \circ T_2]\!](x, y) = \oplus_z T_1(x, z) \otimes T_2(z, y) . \tag{9}$$

There exists a general and efficient composition algorithm for two transducers T_1 and T_2 [3]. Each state in the composition $T_1 \circ T_2$ are a pair of a state in T_1 and a state in T_2. The following rule specifies how to compute an edge of $T_1 \circ T_2$ from appropriate edges of T_1 and T_2: (q_1, a, b, w_1, q_2), $(q'_1, b, c, w_2, q'_2) \Rightarrow ((q_1, q'_1) a, c, w_1 \otimes w_2, (q_2, q'_2))$.

In the worst case the time complexity of this algorithm is $O(|T_1||T_2|)$ [3].

To construct the composition of a weighted automaton or an unweighted automaton with a transducer, we consider this automaton as a particular case of transducers, where the input label, the output label of each edge are identical and the weight of these edges is 0 for the case of unweighted automaton. Let us note that, the edges with label ε of an automaton from a state to itself are also used in composition. Figure 1 illustrates the algorithm above.

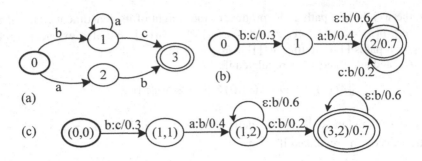

Fig. 1. Composition of the automaton in figure 1-(*a*) with the transducer in figure 1-(*b*) giving the composition transducer in figure 1-(*c*) On *Tropical* semiring.

3.2 The Algorithm Finding a Source-Single Longest Path

In the graph theory, we recall the following basic proposition:

Proposition 2. Let G be an acyclic directed graph, then the set of vertices of G can be numbered such that each edge of the graph goes from the vertex having small index to the one having bigger index.

In our problem, to find the source-single longest path, we consider the finite state transducer without a cycle. This allows us to obtain the algorithm to number states and computing source-single longest paths over the transducer basing on the features of acyclic directed graphs.

We consider a transducer T as an acyclic directed graph, each state in T is a vertex of the graph. The algorithm below allows us to number states of the transducer T satisfying the conditions of Proposition 2. The idea of the algorithm is follows:

First, we find states without an edge coming to them on T (the transducer T has no cycle, so there exists a state without an edge coming to it), then number them arbitrarily. Next, we remove from T numbered states and edges leaving them and receive T without a cycle.

The procedure is repeated until all the states on T are numbered.

Below is the algorithm to number states of the transducer.

Procedure Numbering
Input: A transducer $T = (\Sigma, \Delta, Q, I, F, E, \lambda, \rho)$ without a cycle;
OutPut: The transducer which has numbered states satisfying Proposition 2.
  1.   For $q \in Q$ do $inp[q] = 0$; //$inp[q]$ *contain the number of edges coming to q*
  2.   For $q \in Q$ do For $p \in A[q]$ do $inp[p] = inp[p]+1$
  3.   QUEUE $= \phi$; num $=0$;   //*counting variable*
  4.   For $q \in Q$ do if $inp[q] = 0$ then QUEUE $\Leftarrow q$;
  5.   While QUEUE $<> \phi$ do   //$Nr[q]$ *numbers the index of state q*
          $q \Leftarrow$ QUEUE; num $=$ num $+1$; $Nr[q] =$ num;
             For $p \in A[q]$ do $\{inp[p] = inp[p] -1$; if $inp[p] =0$ then QUEUE $\Leftarrow p;\}$

The algorithm browse all edges of the transducer, so the algorithm has the time complexity of $O(|E|)$.

Below we present the algorithm finding a source-single longest path on the transducer T, on which the states are numbered by the Numbering algorithm above.

Procedure LongestPath;
Input: A transducer $T = (\Sigma, \Delta, Q, I, F, E, \lambda, \rho)$ without a cycle which has been numbered satisfying Proposition 2;
OutPut: The longest distance from the state $q_1 \in I$ to the rest states of the transducer saved in the array $d[q_i]$ $i = 1, .., n\}$
1. $d[q_1] = 0$;
2.   For $i = 2$ to $|Q|$ do $d[q_i] = w[e]$; $//e = (q_1, a, b, w, q_i)$, *if there is no edge leaving*
                                                      $//q_1$ *and coming* to q_i *then put* $d[q_i] = -\infty$
3.   For $i = 2$ to $|Q|$ do
          For $p \in A[q_i]$ do $d[p] = \max(d[p], d[q_i] + w[e])$ // $e = (q_i, a, b, w, p)$

The algorithm browse all edges of the transducer, so the algorithm has the time complexity of $O(|E|)$.

Let us note that, to implement effectively the above algorithms, we simply transform T into a transducer with a beginning state and an end state.

4   Determining LCS of Two Languages

Let two finite languages X and Y over the alphabet Σ which are accepted by two finite automata A_1 and A_2 without cycles respectively. The problems (1.b) and (2) are solved owing to the composition of two transducers, the algorithm finding source-single longest paths on a transducer without a cycle. The main results are proposed in this part.

4.1   The Cost of an Alignment on a Semiring Tropical

Let Ψ be a formal power series defined on the alphabet Ω and the semiring *Tropical* by: $(\Psi, (a, b)) = c((a,b))$, $\forall (a,b) \in \Omega$. The below lemma is improved from lemma 7 in [2], each operation is an insertion, a deletion or a coincidence.

Lemma 1. *Let* $\omega = (a_0, b_0) ...(a_n, b_n) \in \Omega^*$ *be an alignment, then* (Ψ^*, ω) *is the cost of the alignment* ω.

Proof. According to the definition of + - operation of the power serial on the semiring *Tropical*, we have:

$$(\Psi^*, \omega) = \min_{u_0...u_k = \omega} (\Psi, u_0) + ... + (\Psi, u_k)$$
$$= (\Psi, (a_0, b_0)) + ... + (\Psi, (a_n, b_n)).$$
$$= \sum_{i=0}^{n} c((a_i, b_i)) = c(\omega)$$
(10)

The lemma is proved.   □

Ψ^* is a rational power serial. According to the theorem Schützenberger [5], there exists a weighted automaton A defined over the alphabet Ω and a semiring *Tropical* realizing Ψ*. The automaton A can be considered the transducer T with the same input and output alphabet Σ. Figure 2 shows that the simple transducer realizing Ψ* with the cost function c and $\Sigma = \{a, b\}$.

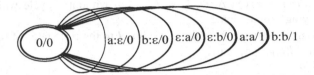

Fig. 2. Transducer T realizing Ψ*

4.2 Algorithm Determining LCS of Two Languages

Theorem 1. *Let the transducer* $U = A_1 \circ T \circ A_2$ *over the semiring Tropical, then:*

1. U has no cycle;

2. If π *is a longest path from the initial to the final state in U, then some longest common subsequence of two strings* $i[\pi]$ *and* $o[\pi]$ *is a longest common subsequence of* A_1 *and* A_2 *and* $L(A_1, A_2) = w[\pi]$.

Proof. 1. We prove that: a) U has no loop: Suppose the opposite, U has a loop at a state $((q_i, 0), p_j)$. In that q_i is a state in A_1, according to the above part the transducer T has only one state denoted as 0, p_j is a state in A_2, the label of this loop only has one of the three following types: $a{:}a$, $a{:}\varepsilon$, $\varepsilon{:}a$. We consider the case when the label of the loop is $a{:}a$. According to the definition of the composition of transducers, the automaton A_2 has a loop with the label a at the state p_j, which is a contradiction to the fact that A_2 has no loop. Similarly, we consider the cases when the label of the loop is $a{:}\varepsilon$, $\varepsilon{:}a$. So the assumption is false and U has no loop.

b) U has no cycle: Suppose the opposite, U has a cycle $\pi = e_1...e_k \Rightarrow p[e_1] = n[e_k]$. U has no loop, so $k > 1$ and $e_{i-1} \neq e_i$, with $i = 2,...,k$. Corresponding to the cycle π, we have a range of states in U: $((q_t, 0), p_s), ((q_{t+1}, 0), p_{s+1}), ..., ((q_{t+k}, 0), p_{s+k})$ and $n[e_i] = ((q_{t+i}, 0), p_{s+i}), p[e_i] = ((q_{t+i-1}, 0), p_{s+i-1}), i = 1,...,k$. According to the definition of the composition of two transducers, A_2 has a cycle $\pi' = e_1'...e_k'$ corresponding to a range of states $p_s, p_{s+1}, ..., p_{s+k}$ ($n[e_i'] = p_{s+i}, p[e_i'] = p_{s+i-1}, i = 1,.., k$), which is a contradition to the fact that A_2 has no cycle. So the assumption is false and U has no cycle.

2. The transducer T realizes Ψ*, according to the definition of the composition of transducers, a set of successful paths of the transducer U is $P = \cup_{x \in X, y \in Y} P(I, x, y, F)$. For each successful path in P, corresponding to an alignment $\tau \in \Omega^*$ is a range of notation edit operations, transforming the string $x \in X$ into the string $y \in Y$ (or $h(\tau)=(x, y)$). If we only consider each string $x \in X$ and $y \in Y$, then P contains all successful paths corresponding to alignments $\tau \in \Omega^*$ such that $h(\tau)=(x, y)$. On the other hand, π is a longest path from the initial to the final state in U, according to Proposition 1 and Lemma 1 the cost of the alignment ω corresponding to the path π is the length $L(x,y)$ of LCS of two strings x, y with $h(\omega)=(x, y)$, the cost of the alignment ω is also the length $L(X,Y)$ of the LCS of two finite languages X, Y and $L(X,Y) = L(A_1, A_2) = w[\pi]=c(\omega)$.   □

The above theorem is the base for the following algorithm.

Algorithm finding LCS_L(A_1, A_2, T) //resolve the problem 1.b and 2
Input: A_1, A_2 are two unweighted automata without cycles, T is a transducer realizing $\Psi*$.
Output: A string belonging to $LCS(A_1, A_2)$ and $L(A_1, A_2)$.

1. $U = A_1 \circ T \circ A_2$; Call Numbering; Call LongestPath;
2. According to π we determine a string u belonging to $LCS(A_1, A_2)$ and the couple (a, b) due to (2) Theorem 1;
3. According to π we determine $L(A_1, A_2)$; //also $L(X,Y)$, since $L(A_1, A_2)=L(X,Y)$

The time complexity of the algorithm
Similar to the evaluation in [2] applied to two automata A_1 and A_2, the algorithm to build U has the time complexity of $O(|\Sigma|^2|A_1||A_2|)= O(|A_1||A_2|)$, with $O(|A_1|) = O(|Q_1|+|E_1|)$, $O(|A_2|)=O(|Q_2|+|E_2|)$. The numbering algorithm to number states on U and the algorithm finding a source-single longest path (LongestPath) have the time complexity of $O(|E|)$, in which E is the set of edges in U. Thus, the time complexity of the algorithm to determine the LCS of two languages is approximately $O(|A_1||A_2|)$.

4.3 Example

Let $X = \{aba, ba\}$, $Y = \{aa\}$, with $\Sigma=\{a, b\}$, the cost function c. We compute some LCS of two languages X, Y and $L(X,Y)$. $A_1=(\Sigma, Q=\{q_0, q_1, q_2, q_3\}, I=\{q_0\}, F=\{q_3\}, E=\{(q_0, a, q_1), (q_1, b, q_2), (q_2, a, q_3), (q_0, b, q_2)\})$ accepting X, $A_2=(\Sigma, Q=\{p_0, p_1, p_2\}, I=\{p_0\}, F=\{p_2\}, E=\{(p_0, a, p_1), (p_1, a, p_2))$ accepting Y.

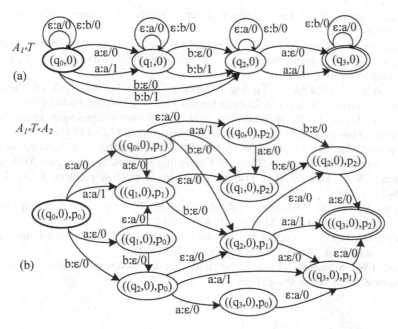

Fig. 3. Figure (a) is the composition result $A_1 \circ T$, figure (b) is the composition result $U= A_1 \circ T \circ A_2$, over the semiring *Tropical*.

In Figure 3-(b) it is realized that $\pi = e_1e_2e_3$ is the longest path from the initial to the final state. According to Theorem 1, the length of LCS of two languages is equal to 2, one LCS is "aa", in that: $e_1 = (((q_0,0),p_0), a, a, 1, ((q_1,0),p_1))$;

$$e_2 = (((q_1,0),p_1), b, \varepsilon, 0, ((q_2,0),p_1)); \; e_3 = (((q_2,0),p_1), a, a, 1, ((q_3,0),p_2)).$$

5 Conclusion

Studying advanced automata models and applications of them are paid much attention by scientists and technologists. This article proposes a method to determine LCS of two finite languages and its length owing to the composition of transducers, the algorithm finding source-single longest paths. This is a significant problem in terms of theory as well as applications.

References

1. Berstel, J., Reutenauer, C.: Rational Series and Their Languages. Springer, Berlin (1988)
2. Mohri, M.: Edit-Distance of Weighted Automata: General Definitions and Algorithms. International Journal of Foundations of Computer Science 14(6), 957–982 (2003)
3. Mohri, M., Pereira, F., Riley, M.: Speech Recognition with Weighted Finite-State Transducers. Springer Handbook of Speech Processing. Springer, Heidelberg (2007)
4. Salomaa, A., Soittola, M.: Automata-Theoretic Aspects of Formal Power Series. Springer, New York (1978)
5. Schützenberger, M.P.: On the definition of a family of automata. Information and Control 4 (1961)
6. Trung huy, P., Khang, N.Q.: Fuzy automaton and application to the problem of determining the longest common subsequence (in Vietnamese). In: The sixth National Conference of Maths, Vietnam (2002)
7. Aho, A.V.: Algorithms for Finding Patterns in String. In: Handbook of Theoretical Compter Science, ch. 5, vol. A. Elsevier Sciencs Publisher BV, Amsterdam (1990)
8. Jiang, T., Li, M.: On the approximation of shortest common supersequences and longest common subsequences. SIAM Journal of Computing 24(5), 1122–1139 (1995)
9. Iliopoulos, C.S., Sohel Rahman, M.: A New Efficient Algorithm for Computing the Longest Common Subsequence. In: Theory of Computing Systems. Springer, New York (2008)
10. Wagner, R.A., Fischer, M.J.: The string-to-string correction problem. J. ACM 21(1), 168–173 (1974)
11. Mozes, S., Tsur, D., Weimann, O., Ziv-Ukelson, M.: Fast algorithms for computing tree LCS. Theoretical Computer Science 410, 4303–4314 (2009)
12. Hirschberg, D.S.: An Information Theoretic Lower Bound for the Longest Common Subsequence Problem. Rice Technical Report No.7705 (1977)
13. Hunt, J.W., Szymanski, T.G.: A fast algorithm for computing longest subsequences. Commun. ACM 20(5), 350–353 (1977)

Detecting Changes in Stream Query Results

Majid Ghayoori, Khosro Salmani, and Mostafa S. Haghjoo

Department of Computer Engineering, Iran University of Science and Technology (IUST),
Narmak, Tehran, Iran, 1684613114
{ghayoori,haghjoom}@iust.ac.ir, Kh_salmani@comp.iust.ac.ir

Abstract. Nowadays, data stream processing is one of the basic requirements of many enterprises. In addition, many organizations can't develop data stream management systems internally and must outsource this service. Such organizations should assure about the accuracy and the honesty of the outsourced server. In this setting, the data stream is generated and sent to the server by the owner. The users have no access to the data stream and they should to verify the integrity of the results. In this paper, we present a probabilistic auditing model for the integrity control of an outsourced data stream server. In our architecture, the server is considered as a black box and the auditing process is fulfilled by contribution between the data stream owner and the users. We exploit existing Data Stream Management Systems (DSMS) without any modification. Our method has no significant cost on users' side and tolerates server load shedding accurately. Correctness and convergence of our probabilistic algorithm is proved. Our evaluation shows very good results.

Keywords: Data stream security, data stream integrity, Data stream management system, DSMS outsourcing.

1 Introduction

Many organizations and companies can't develop and run DSMSs internally and have to outsource their data stream processing services to a third party. Although some academic and commercial DSMS are developed [1],[2],[3],[4],[5], little work is done in outsourcing data stream services.

The most important challenge in outsourcing DSMS services is to ensure about accuracy and honesty of servers. An external server may send inaccurate or incomplete results for different reasons, including software bugs, increasing benefits, less resource allocation, misleading users, etc. Fig 1 shows the architecture of an outsourced DSMS. In this architecture, a Data Stream Owner (DSO) sends data streams to the server and the server applies requested queries on data stream. Here, the users should to control correctness, freshness and completeness of results[6],[7].

In this paper, we present a model to control the integrity of results by users. In our model, integrity control is accomplished via cooperation between the users and the DSO. The DSO merges a fake stream with real stream and sends merged stream to the server. The users, which can generate and identify fake stream, apply their query on fake data stream and compare the results with received fake records (FR).

N.T. Nguyen et al. (Eds.): New Challenges for Intelligent Information, SCI 351, pp. 13–24.

Fig. 1. System Architecture of an Outsourced DSMS

The main contributions of this paper are as follows:

- Presenting an architecture for the integrity control of outsourced DSMS server
- Formulating accuracy of auditing level based on received results
- Implementation of the system and evaluation of the results

This paper is organized as follows. Section 2 describes related works. Section 3 describes problem statement and preliminary assumptions. Sect. 4 presents our method, and Section 5 focuses on the integrity control algorithms. Section 6 experimentally evaluates our method, and finally, Section 7 concludes the paper.

2   Related Works

The proposed integrity control of stream query results can be categorized in two major categories, authenticated structure (AS) based methods and probabilistic methods.

Up to now, the most popular AS is Merkle Hash-Tree (MH-Tree)[8]. In [9] an extended model of MH-Tree that is based on B+-Tree is presented. In this model, record domains are divided to some ranges and records tree is built upon these ranges. This model is designed for executing continues range queries and supports one-time queries as well. This model can audit correctness and completeness of results. To eliminating tree refreshes, an algorithm for dividing data stream into some trees is presented in [7].

Based on our studies, there is one research based on probabilistic methods[10],[11]. This method is designed for analysis of network traffic data and named as Polynomial Identity Random Synopses (PIRS). This algorithm calculates synopsis of received data stream and compares it with the expected one. This method supports COUNT and SUM queries and has an acceptable computation and memory overhead. In PIRS, the users must access data stream for calculating synopsis.

3   Problem Statement and Preliminary Assumptions

In an outsourced DSMS, users must make sure about the integrity of results from external server as followings:

1. Completeness: None of the records is deleted from the results.
2. Correctness: There are no spurious records among results.
3. Freshness: The results are derived from the latest stream records.

The preliminary assumptions of this research are as bellow:

1. Executing queries by the DSMS is considered as a black box.
2. The server might be untrusted (compromised, malicious, running faulty software).
3. The records are transferred between the DSO, the server, and the users as encrypted.
4. The Users can register continues queries in the following form:

```
SELECT * FROM Sₚ WHERE condition
WINDOW SIZE n SLIDE EVERY t        (t>=n)
```

Based on the above, the main issue in this study is to design a method to control the integrity of results received from an untrusted outsourced DSMS server.

4   Our Method

In our method, the DSO sends some Fake Records (FRs) along with the real records. The users can also generate such records in parallel with the DSO. Records are sent to server in encrypted form, so the server or adversary is unable to distinguish FRs from real ones. Server executes registered queries on all records (real and the fake ones) and sends the results to the users. Because the users are aware of FRs, they are able to separate fake results from the real ones and judge the integrity of the results they receive.

4.1   Our System Architecture

Our System architecture is shown in Fig 2. According to the figure, there is a common part in the DSO and the users for the fake data stream (FDS) generation. The maximum Load Shedding Ratio (LSR) is agreed by the server and the users in the format of Service Level Agreement (SLA). The FR Ratio (FRR) is agreed by the DSO and the users.

To enable the users to check correctness of the results and separate FRs from the real ones, a header is added to all records stream as below:

$$\forall\, r \in RDS: H_r = Hash(a_1|a_2|...\,|a_n)$$

$$\forall\, r' \in FDS: H_{r'} = Hash(a_1|a_2|...\,|a_n) + 1 \tag{1}$$

To implement the above architecture, we have to overcome the following challenges:

1. Synchronization of the DSO and the users in FDS generation.
2. Constructing FRs so that there should be some FRs in each user's query.
3. Synchronizing the users with the server for applying queries on FDS.

In this section, we present our algorithms for first two challenges and next one is described in the next chapter.

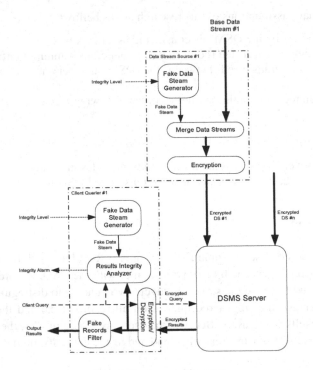

Fig. 2. System architecture

4.2 Fake Data Stream Generation

4.2.1 Fake Records Position

To achieve synchronization between DSO and users, the FR positions has to be determined by a deterministic function. In this research, we design an algorithm based on pseudo-random number generation for determining the FR positions. In our algorithm, the initial seed and the selected pseudo-random number generation function should be transferred as a symmetric keys between the DSO and the users via a secure channel. The main steps of determining the FR positions are:

1. The DSO adds a Record Sequence Number (RS#) to all records. This field is a small integer and has no significant effect on record size.
2. The DSO classifies the data stream in consecutive windows named General Window (GW). The length of GW (L_{GW}) is much bigger than the users query windows (QW) so that in each GWs, some result records are sent to users.
3. The DSO sets RS# of first record in all GWs to zero.

4.2.2 Fake Records Generation

The FRs must be generated by DSO and merged with real data stream (RDS). Moreover, the users should produce the same FDS synchronously with the DSO. Therefore to produce FDS, we must address the following problems:

1. Coordination of the DSO with the users in FDS generation.

    Method 1: A simple method to produce the *same* sequence of FDS by the DSO and the users is to pre-generate and save it. This method imposes a rather low computing overhead on the DSO and the users, but the system suffers a considerable storage, and more importantly, data transmission overhead via the limited secure channel between the DSO and the users.

    Method 2: The DSO and the users may use a deterministic function to generate FDS. The initial value of this function is transferred between the DSO and the users via the secure channel. This method has rather low storage and high computing overhead.

2. Maximum coverage of FDS in user query results.

    The more FRs in the query results, the more reliable integrity control. To generate FRs, we use records simulation methods that are based on a probability distribution function obtained from statistical analysis of real streams[12],[13],[14].

5 Integrity Control of Results

The users should determine FRs which must be appeared in results by applying their queries on the FDS. The server starts the user query execution with a delay that the user is not aware of it(Fig 3). Therefore, they must estimate the position of server QW.

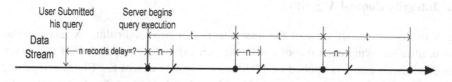

Fig. 3. Delay of server before starting query execution

In the rest of this section, we present our solution for above challenge and the complete algorithm for the integrity control of results.

5.1 Estimating Position of Server Query Window by Users

The users are not aware of the delay of query execution by the server, so they should estimate the position of server's QW using received records. We present a repetitive algorithm for estimating position of server's QW_0 by users. The positions of next QWs are computed based on QW_0. Here is the estimation algorithm:

```
void EstimateServerQWPosition(int RS#){
    int RS#=(RS# mod t)+t//Map RS# to first QW
    if (RS#<prevRS#) {   // GW is changed
            RS#es=RS#es-(Lgw mod t); RS#ee=RS#ee-(Lgw mod t);
            if (RS#es<t) {
            RS#es+=t; RS#ee+=t;
      }
    } else if ((RS#ee-RS#L>n | RS#L-RS#es>n) & !(RS#es==0 & RS#ee==0)){
            // some records are deleted or freshness error
        cer++;
        if (cer>threshold) {
           Alarm("Freshness Error");
           RS#ee=RS#L; RS#es=RS#L; cer=0;
        }
    } else {
            cer=0;
            if (RS#es==0 & RS#ee==0) {// First Received Record
                RS#es=RS#L; RS#ee=RS#L;
            } else {
                RS#es=Min(RS#L,RS#es); RS#ee=Max(RS#L,RS#ee);
            }
        }
    RS#cs=RS#es-((n-(RS#ee-RS#es))/2); RS#ce=RS#cs+n-1;
    prevRS#=RS#;
  } // End of EstimateServerQWPositions
```

5.2 Integrity Control Algorithm

Fig 4 shows an overall view of the user integrity control algorithm. As seen in the figure, after receiving each record result, the user estimates the position of the server QW. The user's query is applied on FDS and the results is sent to "Check Integrity". Simultaneously, the FRs in the results are selected and sent to "Check Integrity". To compare the records of the received FRs with the expected ones, users must execute an *EXCEPT* operation on the generated and received FRs. To remove this overhead, it is proved in [15] that the *COUNT* of the two sets can be compared instead of comparing them record by record. Based on requested integrity level (IL), a trigger is generated by "Generate Control Trigger" to start integrity evaluation of the next sliding check window.

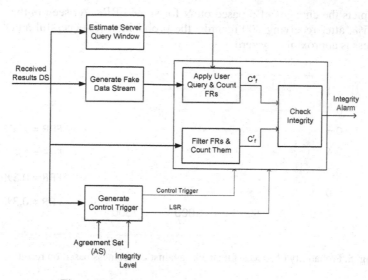

Fig. 4. User integrity control algorithm schema

5.3 Considering Load Shedding

The DSO, users and the server have an agreement about server's maximum load shedding schedule. For considering load shedding, let C^r_f be the number of received FRs and C^e_f be the number of expected FRs, then, if $C^e_f > (1 - LSR)*C^r_f$, some of the expected records are not received from the server.

5.4 Integrity Level

In our model, the integrity level is defined upon accuracy of system in detecting the integrity violation attacks. For this purpose, we first present a probabilistic method to estimate the accuracy of the integrity control algorithm, and then define an algorithm reach the requested accuracy.

5.4.1 Probabilistic Integrity Estimation
Let N be the average count of the result records in a predefined time range. If the FRs were distributed uniformly in the result records, the average count of real records is $N * (1 - FRR)$. The probability of deleting *only real records* from results is:

$$p = \prod_{i=0}^{N*(1-FRR)-1} \frac{N*(1-FRR)-i}{N-i} \tag{2}$$

In the above product, all terms are less than 1 and the largest term is equal to $\frac{N*(1-FRR)}{N}$. Therefore:

$$p < (1 - FRR)^{N*(1-FRR)} \tag{3}$$

Fig 5 depicts the changes of p based on N for some FRRs. As seen in the figure, for FRR=0.5%, after receiving 300 records, the probability of successful attacks against correctness is approximately zero.

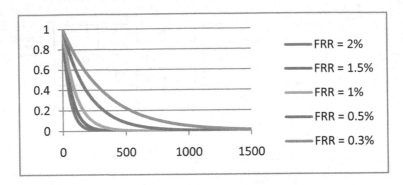

Fig. 5. Probability of successful attacks against correctness based on results count

Note 1: To considering load shedding, we must replace FRR with FRR-LSR:

$$p < (1 - (FRR - LSR))^{N*(1-(FRR-LSR))} \tag{4}$$

Note 2: The above results is based on the assumption of the uniform distribution of FRs in results.

5.4.2 Determination of Integrity Level
Let α be the expected accuracy. Therefore:

$$1 - a < (1 - FRR)^{N(1-FRR)} \tag{5}$$

As the FRR is agreed by the DSO and the users, to achieve accuracy α, we must select N so that the above formula satisfies. Therefore:

$$N > \frac{Ln(1-a)}{(1-FRR)*Ln(1-FRR)} \tag{6}$$

For example, to achieve accuracy of 0.99 with FRR=2%, system must check completeness for any 233 result records. "Generate Control Trigger" in Fig 4, computes minimum number of result records (N) for achieving expected accuracy based on FRR, LSR and expected integrity level (α), and fires a "Control Trigger".

6 Experimental Results

We use three Pentium IV 2.0GHz PCs with 2GB RAM and 200GB hard disk to simulate the DSO, the server and the client. The DSO and the server machines are connected with a local Ethernet network running at 100MBps. The Server and the client are connected in same setting. We use a generic query processor as kernel of DSMS server. Our algorithms are developed in JAVA with the JDK/JRE 1.6 develop kit and Runtime Library.

We generate a synthetic web click data stream to evaluate our probabilistic approach. The synthetic data stream is composed of 10,000,000 randomly generated records. The structure of records is shown in Fig 6.

IP Address	Request Date	Request Time	GMT	OS	Web Browser

Fig. 6. Structure of sample synthetic web click records

As there is no similar system in literature, it is impossible to compare results. Therefore, we evaluate the behavior and performance of our algorithm and show that our method has acceptable successes.

6.1 Query Window Location

Our experiments show that error of EstimateServerQWPosition approaches zero after receiving a finite number of result record. We measured this error as the number of records between the estimated QW and the real QW. Fig 7 depicts the effect of the FRR as well as deletion of results (completeness attack) on QW estimation error.

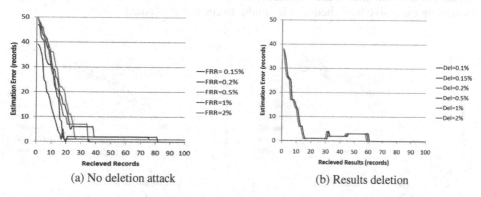

(a) No deletion attack   (b) Results deletion

Fig. 7. QW estimation error

The results shows that the two factors (result deletion and FRR) have no significant effect on the QW estimation error, and the estimation error approaches zero after receiving a small number of records (60-100 records).

6.2 Completeness Attacks Detection

Fig 8 shows the detection delay of completeness attacks versus deletion attack for different FRRs. As shown in the figure, our method detects the completeness attacks quickly. The maximum delay of detection is not high. As FRR increases, the initial

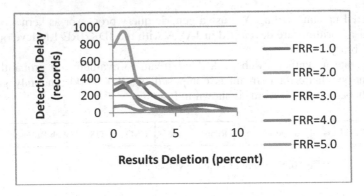

Fig. 8. Detection delay versus deletion attacks for different FRRs

points of the delay curves decreases, and negative slop of curves increases and approaches zero quickly. Also by increasing the deletion attack ratio, all curves approach zero.

Fig 9 shows the attack detection ratio based on deletion attack for different FRRs. As shown in the Fig 9.a all curves show a positive value for values 0-2 of deletion attack ratio. This means that all attacks are detected by the system even for small deletion attack ratio. Also, for all FRRs, the attack detection ratio approaches 100 that means the deletion attacks are detected quickly. Fig 9.b shows that the behavior of the system on excessive load shedding is similar to the deletion attack.

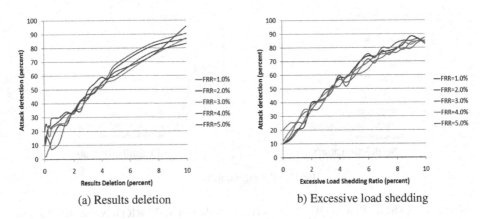

(a) Results deletion   b) Excessive load shedding

Fig. 9. Attack detection ratio versus deletion attack for different FRRs

Fig 10 shows the attack detection ratio based on user requested accuracy for different deletion attack ratios (FRR is constant). As shown in Fig 10.a,b, attack detection ratio for all requested accuracy is higher than 10%. By increasing requested accuracy, attack detection is increased, and the offset of attack detection curve is increased. In addition, Fig .b shows that higher FRRs have no significant effect on attack detection.

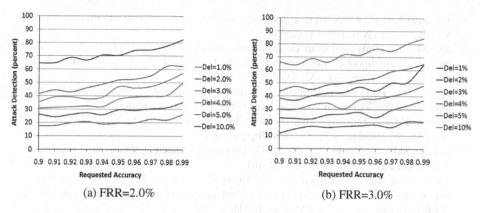

(a) FRR=2.0%   (b) FRR=3.0%

Fig. 10. Attack Detection ratio versus requested accuracy for different deletion attack

6.3 Freshness Attacks Detection

Our experiments on the freshness attack detection showed that the system detects all freshness attacks. It not an unexpected result because EstimateServerQWPosition algorithm uses RS# as a timestamp in GWs to check records sequence.

7 Conclusions

In this paper, we presented a solution for integrity control of query results from an outsourced DSMS server by adding FRs to data streams. First we presented an architecture based on this idea, and then we presented detailed algorithms for main part of the architecture. Finally we implemented our model and evaluated the results which showed very good results. The results shows that our method is a self tunning model for integrity auditing of stream query results and can tolerate practical exeptions.

This paper concentrates on applying queries on one stream. In future, we plan to consider multiple streams and investigate algorithms for integrity control of join queries as well.

References

1. Arasu, A., et al.: STREAM: The Stanford Stream Data Manager. In: International Conference on Management of Data, Proceedings of the 2003 ACM SIGMOD, San Diego, California (2003)
2. Chandrasekaran, S., et al.: TelegraphCQ: Continuous Dataflow Processing for an Uncertain World. In: CIDR (2003)
3. Schmidt, et al.: Robust real-time query processing with QStream. In: Proceedings of the 31st International Conference on Very Large Data Bases, VLDB, Trondheim, Norway (2005)
4. Abadi, D., et al.: The Design of the Borealis Stream Processing Engine. In: 2nd Biennial Conference on Innovative Data Systems Research, CIDR 2005 (2005)

5. Abadi, D.J., et al.: Aurora: a new model and architecture for data stream management. The VLDB Journal 12(2), 120–139 (2003)
6. Xie, M., et al.: Providing freshness guarantees for outsourced databases. In: Proceedings of the 11th International Conference on Extending Database Technology: Advances in Database Technology, pp. 323–332. ACM, Nantes (2008)
7. Li, F., et al.: Proof-Infused Streams: Enabling Authentication of Sliding Window Queries On Streams. In: VLDB Endowment, Very Large Data Bases, Vienna (2007)
8. Merkle, R.C.: A certified digital signature. In: Brassard, G. (ed.) CRYPTO 1989. LNCS, vol. 435, pp. 218–238. Springer, Heidelberg (1990)
9. Papadopoulos, S., Yang, Y., Papadias, D.: Continuous authentication on relational streams. The VLDB Journal 19(2), 161–180 (2010)
10. Yi, K., et al.: Randomized Synopses for Query Assurance on Data Streams. In: Proceedings of the 2008 IEEE 24th International Conference on Data Engineering, pp. 416–425. IEEE Computer Society, Los Alamitos (2008)
11. Yi, K., et al.: Small synopses for group-by query verification on outsourced data streams. ACM Trans. Database Systems 34(3), 1–42 (2009)
12. Fiorini, P.M.: Modeling telecommunication systems with self-similar data traffic, p. 56. Department of Computer Science, The University of Connecticut (1998)
13. Dill, S., et al.: Self-similarity in the web. ACM Trans. Internet Technology 2(3), 205–223 (2002)
14. Wang, X., Liu, H., Er, D.: HIDS: a multifunctional generator of hierarchical data streams. SIGMIS Database 40(2), 29–36 (2009)
15. Xie, M., et al.: Integrity Auditing of Outsourced Data. In: VLDB Endowment. Very Large Data Bases, Vienna (2007)

An Efficient Parallel Algorithm
for the Set Partition Problem

Hoang Chi Thanh[1] and Nguyen Quang Thanh[2]

[1] Department of Informatics, Hanoi University of Science, VNUH
334 - Nguyen Trai Rd., Thanh Xuan, Hanoi, Vietnam
thanhhc@vnu.vn
[2] Da Nang Department of Information and Communication
15 - Quang Trung str., Da Nang, Vietnam
thanhnq@dsp.vn

Abstract. In this paper we propose a new approach in organizing a parallel computing to find the sequence of all solutions to a problem. We split the sequence of desirable solutions into subsequences and then execute concurrently processes to find these subsequences. We propose a new simple algorithm for the set partition problem and apply the above technique to this problem.

Keywords: algorithm, complexity, parallel computing, set partition.

1 Introduction

When solving a problem on computer we have to construct a proper algorithm with the deterministic input and output. The algorithm is programmed. The input is put into a computer. Then the computer performs the corresponding program to hold all the solutions to the problem after some duration of time.

The scheme for computing a problem's solutions is as the following figure.

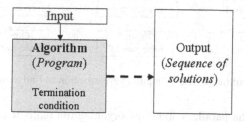

Fig. 1. The scheme for computing a problem's solutions

The practical meaning of a problem and an algorithm is better if the time duration of computing is shorter. One of good methods to reduce computing time is to organize a parallel computing where the computation environment allows. There are some methods of organizing a parallel computing to find quickly a problem's all solutions. For example, constructing a sequential algorithm first and then transforming it into

N.T. Nguyen et al. (Eds.): New Challenges for Intelligent Information, SCI 351, pp. 25–32.
springerlink.com   © Springer-Verlag Berlin Heidelberg 2011

concurrent one [6], splitting data into separate blocks and then computing concurrently on these blocks [8]... Such a parallel computing is an illustration of the top-down design.

For many problems, we can know quite well the number of all desirable solutions and their arrangement in some sequence. The set partition problem [2,3,5] is a typical example. This problem is broadly used in graph theory [1], in concurrent systems [4], in data mining [8] ...

So we can split the sequence of all desirable solutions into subsequences and use a common program (*algorithm*) in a parallel computing environment to find concurrently these subsequences. Therefore, the amount of time required for finding all the solutions will be drastically decreased by the number of subsequences. This computing organization is an association of the bottom-up design and the divide and conquer one.

2   Parallel Computing a Problem's Solutions by Splitting

To perform the above parallel computing we split the sequence of all desirable solutions to a problem into subsequences. The number of subsequences depends on the number of calculating processors. Let us split the sequence of all solutions of the problem into m subsequences ($m \geq 2$). The scheme of the parallel computing organization to find a problem's all solutions is illustrated as the following figure.

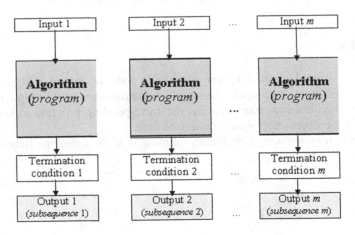

Fig. 2. The scheme of the parallel computing organization to find a problem's solutions

In order to make the parallel computing organization realistic and optimal, subsequences of solutions should be satisfied the two following criteria:

1)   It is easy to determine the input and the termination condition for the computing process of each subsequence.
2)   The difference of lengths of the subsequences is the less, the better.

Of course, the input 1 is indeed the input of the problem and the last termination condition is the termination condition of the algorithm.

The first criteria ensures that the splitting of solutions sequence is realistic. In many cases, we can use (a part of) the input of the computing process for the next subsequence as the termination condition of the computing process for the previous subsequence. The second criteria implements the balance of computing to processors. Then the parallel computing processes become optimal.

If the problem has many algorithms, we should choose the algorithm with the least number of variables. Just then, the input and the termination condition of computing processes of subsequences turn simple.

3 Application to the Set Partition Problem

We apply the above presented technique to the set partition problem. For simplicity of presentation, we split the sequence of all desirable solutions into two subsequences. If we want to split the sequence into many subsequences then we apply this technique to each subsequences.

3.1 Set Partition Problem

Let X be a set.

Definition 3.1: A *partition* of the set X is a family $\{A_1, A_2, ..., A_k\}$ of subsets of X, satisfying the following properties:

1) $A_i \neq \varnothing$, $1 \leq i \leq k$;
2) $A_i \cap A_j = \varnothing$, $1 \leq i < j \leq k$;
3) $A_1 \cup A_2 \cup ... \cup A_k = X$.

Problem: Given a set X. Find all partitions of the set X.

The set partition problem is broadly used in graph theory [1], in concurrent systems [4], in data mining [8] ...

The number of all partitions of an n-element set is denoted by Bell number B_n, calculated by the following recursive formula [3,5]:

$$B_n = \sum_{i=0}^{n-1} \binom{n-1}{i} B_i \text{ , where } B_0 = 1.$$

The number of all partitions of an n-element set grows up as quickly as the factorial function does. For example,

n	B_n
1	1
3	5
5	52
8	4.140
10	115.975
15	1.382.958.545
20	51.724.158.235.372

Given an n-element set X. Let identify the set $X = \{1, 2, 3, ..., n\}$.

Let $\pi = \{A_1, A_2, ..., A_k\}$ be a partition of the set X. Each subset A_i is called *a block* of the partition π.

To ensure the uniqueness of representation, blocks in a partition are sorted in the ascending of the least element in block. In a partition, the block A_i ($i = 1, 2, 3, ...$) has the index i and element 1 always belongs to the block A_1. Each element $j \in X$, belonging to some block A_i has also the index i. It means, every element of X can be represented by the index of a block that includes the element.

Of course, the index of element j is not greater than j. Each partition can be represented by a sequence of n indexes. The sequence can be considered as a word with the length of n on the alphabet X. So we can sort these words in the ascending order. Then,

- The smallest word is 1 1 1 ... 1. It corresponds to the partition $\{\{1,2,3, ... , n\}\}$. This partition consists of one block only.
- The greatest word is 1 2 3 ... n. It corresponds to the partition $\{\{1\}, \{2\}, \{3\}, ... , \{n\}\}$. This partition consists of n blocks, each block has only one element. This is an unique partition that has a block with the index n.

Theorem 3.1: For $n \geq 0$,

$$B_n \leq n!.$$

It means, the number of an n element set's all partitions is not greater than the number of all permutations on the set.

Proof: Follow from the index sequence representation of partitions.   □

We use an integer array $AI[1..n]$ to represent a partition, where $AI[i]$ stores the index of the block that includes element i. Element 1 always belongs to the first block, element 2 may belong to the first or the second block. If element 2 belongs to the first block then element 3 may belong to the first or the second block only. And if the element 2 belongs to the second block then element 3 may belong to the first, the second or the third block.

Hence, the element i may only belong to the following blocks:

$$1, 2, 3, ..., \max (AI[1], AI[2], ..., AI[i\text{-}1]) + 1.$$

It means, for every partition:

$$1 \leq AI[i] \leq \max (AI[1], AI[2], ..., AI[i\text{-}1]) + 1 \leq i \text{ , where } i = 2, 3, ..., n.$$

This is an invariant for all partitions of a set. We use it to find partitions.

Example 3.2: The partitions of a three-element set and the sequence of their index representations.

No	Partitions	$AI[1..3]$
1	$\{\{1, 2, 3\}\}$	1 1 1
2	$\{\{1, 2\}, \{3\}\}$	1 1 2
3	$\{\{1, 3\}, \{2\}\}$	1 2 1
4	$\{\{1\}, \{2, 3\}\}$	1 2 2
5	$\{\{1\}, \{2\}, \{3\}\}$	1 2 3

3.2 A New Algorithm for Partition Generation

It is easy to determine a partition from its index array representation. So, instead of generating all partitions of the set X we find all index arrays $AI[1..n]$, each of them can represent a partition of X. These index arrays will be sorted in the ascending order.

The first index array is 1 1 1 ... 1 1 and the last index array is 1 2 3 ... n-1 n. So the termination condition of the algorithm is:

$$AI[n] = n$$

Let $AI[1..n]$ be an index array representing some partition of X and let $AI'[1..n]$ denote the index array next to AI in the ascending order.

To find the index array AI' we use an integer array $Max[1..n]$, where $Max[i]$ stores $\max(AI[1], AI[2], ..., AI[i\text{-}1])$. The array Max gives us possibilities to increase indexes of the array AI. Of course,

$$Max[1] = 0 \text{ and } Max[i] = \max\,(Max[i\text{-}1], AI[i\text{-}1]), i = 2, 3, ..., n.$$

Then,

$$AI'[i] = AI[i], i = 1, 2, ..., p\text{-}1, \text{ where } p = \max\,\{\ j\ |AI[j] \leq Max[j]\ \};$$
$$AI'[p] = AI[p] + 1 \text{ and } AI'[j] = 1,\ j = p+1, p+2, ..., n.$$

Basing on the above properties of the index arrays and the inheritance principle presented in [7], we construct the following new algorithm for generating all partitions of a set.

Algorithm 3.2 (*Generation of a set's all partitions*)

Input: A positive integer n.
Output: A sequence of an n-element set's all partitions, whose index representations are sorted by ascending.

Computation:

```
1  Begin
2    for i ← 1 to n-1 do AI[i] ← 1 ;
3    AI[n] ← Max[1] ← 0 ;
4    repeat
5      for i ← 2 to n do
6        if  Max[i-1] < AI[i-1]  then  Max[i] ← AI[i-1]
                                  else  Max[i] ← Max[i-1] ;
```

```
7        p ← n ;
8        while  AI[p] = Max[p]+1  do
9               { AI[p] ← 1 ; p ← p-1 } ;
10        AI[p] ← AI[p]+1 ;
11        Print the corresponding partition ;
12    until AI[n] = n ;
13 End.
```

The algorithm's complexity:

The algorithm finds an index array and prints the corresponding partition with the complexity of $O(n)$.

Therefore, the total complexity of the algorithm is $O(B_n.n)$. It approximates to $O(n!.n)$.

The algorithm 3.2 is much simpler and faster than the pointer-based algorithm 1.19 presented in [3].

3.3 Parallel Generation of Partitions

In order to paralellize the sequential algorithm above presented, we split the sequence of all desirable partitions of the set X into two subsequences. The pivot is chosen as a partition represented by the following index array:

$$1\ 2\ 3\ ...\ [n/2]-1\ [n/2]\ 1\ 1\ ...\ 1\ 1\ 2$$

So, the last partition of the first subsequence corresponds to the index array:

$$1\ 2\ 3\ ...\ [n/2]-1\ [n/2]\ 1\ 1\ ...\ 1\ 1\ 1$$

The chosen pivot and the last index array of the first subsequence are illustrated in the following figure.

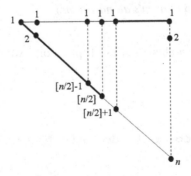

Fig. 3. The pivot and the last index array of the first subsequence

We have to determine a termination condition for the first computing process and an input of the second one.

The termination condition for the first computing process in the instruction 12 is replaced by:

$$AI[i] = i \ , \ i = 2, 3, \ \ldots, \ [n/2]-1, \ [n/2].$$

The input of the second computing process in the instruction 2 and 3 will be:

$$AI[i] \ \leftarrow \ i \ , \ i = 2, 3, \ \ldots \ , \ [n/2]-1, \ [n/2] \ ;$$
$$AI[j] \ \leftarrow \ 1 \ , \ j = [n/2]+1, \ [n/2]+2, \ \ldots \ , \ n-1 \ ;$$
$$AI[n] \ \leftarrow \ 2 \ ;$$

Experimental results point out that the pivot is nearly in the middle of the sequence of all index arrays. So the above splitting is appropriate.

4 Conclusion

In this paper we propose a new approach to organize a parallel computing for finding all solutions of a problem, whose sequential algorithm takes too long finding all solutions. The parallel computing organization above presented is an combination of the bottom-up design and the divide and conquer design. We also propose a new efficient and simple algorithm for the set partition problem and paralellize the algorithm.

In future, we will apply this technique to other problems, namely time-series processing, time-series matching, scheduling problem and system control...

Acknowledgment

This paper was supported by Vietnam National University, Hanoi (Project No. QG-09-01).

References

[1] Cameron, K., Eschen, E.M., Hoang, C.T., Sritharan, R.: The list partition problem for graphs. In: Proceedings of the Fifteenth Annual ACM-SIAM Symposium on Discrete Algorithms, New Orleans, pp. 391–399 (2004)
[2] Cornen, T.H., Leiserson, C.E., Rivest, R.L., Stein, C.: Introduction to Algorithms. The MIT Press, Cambridge (2001)
[3] Lipski, W.: Kombinatoryka dla programistów. WNT, Warszawa (1982)
[4] Rajasekaran, S., Lee, I.: Parallel algorithms for relational coarsest partition problems. IEEE Trans. Parallel Distributed Systems 9(7), 687–699 (1998)
[5] Thanh, H.C.: Combinatorics. VNUH Press (1999) (in Vietnamese)

[6] Thanh, H.C.: Transforming Sequential Processes of a Net System into Concurrent Ones. International Journal of Knowledge-based and Intelligent Engineering Systems 11(6), 391–397 (2007)
[7] Thanh, H.C.: Bounded sequence problem and some its applications. In: Proceedings of Japan-Vietnam Workshop on Software Engineering, Hanoi 2010, pp. 74–83 (2010)
[8] Thanh, H.C., Thanh, N.Q.: A Parallel Dimensionality Reduction for Time-Series Data and Some Its Applications. International Journal of Intelligent Information and Database Systems, InderScience 5(1), 39–48 (2011)

Fuzzy Ontology Integration Using Consensus to Solve Conflicts on Concept Level

Trong Hai Duong[1], Ngoc Thanh Nguyen[2],
Adrianna Kozierkiewicz-Hetmańska[2], and Geun Sik Jo[1]

[1] School of Computer and Information Engineering,
Inha University, Korea
haiduongtrong@gmail.com, gsjo@inha.ac.kr
[2] Institute of Informatics,
Wroclaw University of Technology, Poland
{thanh,adrianna.kozierkiewicz}@pwr.wroc.pl

Abstract. Nowadays, ontology has been backbone of Semantic web. However, the current ontologies are based on traditional logic such as first-order logic and description logic. The conceptual formalism of the ontologies cannot be fully representative for imprecise and vague information (e.g. "rainfall is very heavy") in many application domains. In this paper, a domain fuzzy ontology is defined clearly, and its components such as fuzzy relation, concrete fuzzy concept, and fuzzy domain concept as well as similarity measures between the components are addressed. Fuzzy ontology integration on concept level using consensus method to solve conflicts among the ontologies is proposed. In particular, the postulates for integration are specified and algorithms for reconciling conflicts among fuzzy concepts in ontology integration are proposed.

Keywords: Fuzzy Ontology, Conflict, Consensus, Ontology Matching, Ontology Merge, Ontology Integration.

1 Introduction

World Wide Web (WWW) was contained over 15 billion pages in 2006. However, most of which are to be represented in the form of human understandable only. Mid-90s, Extensible Markup Language (XML) was developed from GSML (Standard generalized Markup Language (SGML ISO 8879:1986)) to provide a set of metadata tags expecting representing semantic web data, but XML does not define the meaning of the tags. Thus, available information on the Web can only be accessed with syntactic interoperability. Software agents can not understand and process information effectively. In 2001, Tim Berners-Lee and colleagues have established a vision of the Semantic Web [3] - an extended Web of machine-readable information and automated services that extend far beyond current capabilities. The explicit representation of the semantic underlying data, programs, papers, and other Web resources will become a knowledge-based Web that improves in their capacity to assist human and machine to be more well-collaborative works. Recently, Ontologies have been developed to provide a machine-processable semantics of information sources that can be communicated between different agents

N.T. Nguyen et al. (Eds.): New Challenges for Intelligent Information, SCI 351, pp. 33–42.
springerlink.com   © Springer-Verlag Berlin Heidelberg 2011

(solfware and humans). The idea of ontologies is to allow users to organize information on the taxonomies of concepts, with their own attributes, to describe relationships between the concepts. When data is represented by using the ontologies, software agents can better understand the content of the data and messages, smarter integrate data for multiple jobs than the conventional web used. In this way, the sites not only performs the function of receiving and displaying information but also the ability to automatically extract information, query, argue the knowledge base to give accurate information automatically. However, those ontologies are based on traditional logic such as description logic [1], so it can not afford to provide well-defined means. Conceptual formalism of the ontologies can not be fully representative for imprecise and vague information (e.g. "very expensive book") in many application domains. Therefore, In 2006, Umberto Straccia [11] integrated the foundation of description logic and fuzzy logic of Zadeh [12] to obtain the another fuzzy logic (called Fuzzy Description Logic) in order for the processing of uncertain knowledge on the semantic web. In recent years, several studies on the methods described fuzzy logic integrated into the ontology to extend the traditional ontology more suitable for resolving problems of uncertain inferences, for example, [4] first steps toward the realization of a theoretical model and a complete structure-based ontologies can consider the nuances of natural language by integrating fuzzy logic to ontology. In which the authors clearly defined the fuzzy description logic, fuzzy ontology and fuzzy OWL. Lu and co-authors [6] by integrating -connection to describe fuzzy logic, the authors propose a new approach to combine both features fuzzy logic and distribution within the description. Their main contribution is to propose an algorithm to achieve discrete tableau inference fuzzy logical ontology system. This research aims at investigating an effective methodology for fuzzy ontology integration in which our main contributions are follows:

- A fuzzy domain ontology is defined clearly, and its components such as *fuzzy relation, concrete fuzzy concept, and fuzzy domain concept* as well as similarity measures between the components are addressed. It provides a solid basis for studies about fuzzy ontology integration.
- Fuzzy ongtology integration using consensus method is proposed in which postulates for integration are specified and algorithms for reconciling conflicts among fuzzy concepts in ontology integration are proposed.

2   Consensus Methods

Consensus methods were known in ancient Greece and were applied mainly in determining results of voting (Day 1987). Along with the development of software methods consensus has found many fields of applications, especially in solving conflicts and reconciling inconsistent data. One well-known conflict situation is called a conflict profile, and a consensus method is an effective approach that can be used to solve this conflict. In a conflict profile, there is a set of different versions of knowledge that explains the same goal or elements in the real world. The consensus aims at determining a reconciled version of knowledge which best represents the given versions. For example, many participants share their own knowledge to solve the same problem and there may be conflicts between their solutions. In this situation, consensus is used to find the best solution in a compromise of all the conflicted participants' viewpoints. Barthelemy et al [2] described two classes of problems that are considered in the consensus theory as follows:

- Problems in which a certain and hidden structure is searched for.
- Problems in which inconsistent data related to the same subject are unified.

The first class consists of problems with searching the structure of a complex or internally organized object. This object can be a set of elements and the searched structure to be determined can be a distance function between these elements. Data that is used to uncover this structure is usually based on empirical observations that reflect this structure, but not necessarily in a precise and correct manner. The second class consists of problems that arise when the same subject is represented (or voted on) in a different manner by experts (or agents or sites of a distributed system). In such a case, a particular method is desired that makes it possible to deduce one alternative from the set of given alternatives. In short the scheme of using consensus methods in a process of solving conflict or data inconsistency may be presented as follows Nguyen [8]:

1. Defining the set of potential versions of data
2. Defining the distance function between these versions
3. Selecting a consensus choice function
4. Working out an algorithm for consensus choice

Some comments should be done to these steps. Referring to first point notice that up to now the following structures of data versions have been investigated: rankings, partitions, ordered partitions and coverings, sets, n-trees, semilattices. A wide overview of these structures has been done in work [10, 11]. The logical structure presented in this chapter is a new one and has not been considered before. The definition of distance functions is dependent on the specific structure. It seems to be impossible to define a universal distance function for all structures. Referring to selecting a consensus choice function there are known 2 consensus choice functions. The first of them is the median function defined by Kemeny [7], which minimizes the sum of distances between the consensus and given inconsistent versions of data. The second function minimizes the sum of these distances squared [8]. As the analysis has shown, the first function in the best way represents the conflict versions while the second should be a good compromise of them [8]. We denote the consensuses chosen by these functions by O1-consensus and O2-consensus, respectively. The choice of a consensus function should be dependent on the conflict situation. With assumption that the final version which is to be determined represents a unknown solution of some problem then there two cases:

- In the first case the solution is independent on the opinions of conflict participants. Thus the consensus should at best represent the conflict versions of data. For this case the criterion for minimizing the sum of distances between the consensus and the conflict versions should be used, thus O1-consensus should be determined.
- In the second case the solution is dependent on the opinions of conflict participants. Then the consensus should be a compromise acceptable by the conflict participants. For this case a O2-consensus should be determined. However, it is worth to notice that determining an O2-consensus is often a complex problem and requires working out heuristic algorithms.

3 Fuzzy Ontology Model

3.1 Fuzzy Ontology Notion

We assume a real world (A, V) where A is a finite set of attributes and V is the domain of A. Also, V can be expressed as a set of attribute values, and $V = \bigcup_{a \in A} V_a$ where V_a is the domain of attribute a.

Definition 1 (Fuzzy Ontology). *An fuzzy ontology is defined as a 6-tuple:*

$$O = (C, \bar{\bar{\wedge}}, A, V, R, Z, F)$$

where,

- C *is set of concepts (classes). A concept is a fuzzy concept, if all its properties belong to the concept with a degree or the concept is set of fuzzy set (called concrete fuzzy concept);*
- A *is set of attributes belonging to the concepts in* C *(called data properties from the concepts in* C *to the values defined in a standard or user-defined data type). A attribute is a fuzzy property, if whose value is a fuzzy set;*
- V *is the domain of* A*. Also,* V *can be expressed as a set of attribute values, and* $V = \bigcup_{a \in A} V_a$ *where* V_a *is the domain of the the attribute a;*
- F*: is set of membership functions. Each components of the fuzzy ontology a has a membership function* $f_a \in F$*, and its range is a concrete fuzzy concept,* $\forall v \in V_a$*,* $f_a(v) \in V'_a \subset (0..1]$*.*
- $< C, \bar{\bar{\wedge}} >$ *is the taxonomic structure of the fuzzy concepts from* C *where* $\bar{\bar{\wedge}}$ *is the collection of subsumption relationship ($\sqsubseteq$) between any two concepts in* C*. For two concepts* c_1 *and* c_2 $\in C, c_2 \sqsubseteq c_1$ *if and only if any instances that are members of concept* c_2 *are also members of concept* c_1*, and the converse is not true. For each subsumption relation* $S \in \bar{\bar{\wedge}}$*, we have*

$$S \subset C \times C \times (0,1] \qquad (1)$$

- R *is a set of fuzzy relations between concepts,* $R = \{R_1, R_2, \ldots, R_m\}$ *where*

$$R_i \subset C \times C \times (0,1] \qquad (2)$$

for $i = 1, 2, \ldots, m$*. A relation is then a set of pairs of concepts with a weight representing the degree to which the relationship should be.*
- Z *is set of axioms, which can be interpreted as integrity constraints or relationships between instances and concepts. This means that* Z *is a set of restrictions or conditions (necessary & sufficient) to define the concepts in* C*;*

Definition 2 (Fuzzy Concept). *A fuzzy concept is defined as a triple:*

$$concept = (c, A^c, V^c, f_c) \qquad (3)$$

where c is the unique name of the concept, $A^c \subset A$ *is a set of attributes describing the concept and* $V^c \subset V$ *is the attributes domain:* $V^c = \bigcup_{a \in A^c} V_a$ *where* V_a *is the domain of the the attribute a and* f_c *is a fuzzy function:*

$$f_c : A^c \to [0,1] \qquad (4)$$

representing the degrees to which concept c is described by attributes. Triple (A^c, V^c, f_c) is called the fuzzy structure of concept c.

Definition 3 (Concrete Fuzzy Concept). *The concrete fuzzy concept $cf \subseteq C$ is defined as a following 4-tuple:*

$$cf = (V_{cf}, V'_{cf}, L_{cf}, f_{cf}) \tag{5}$$

where, cf is the unique identifier for the concept. The $V_{cf} \subseteq V$ is a concrete set, which called domain of the concept. $V'_{cf} \subseteq (0, 1]$ presents fuzzy values of the concrete set V_{cf}. The $f_{cf} \in F$ is a concrete fuzzy predicate on V_{cf} (considered as a membership function), $\forall v \in V_{cf}, f_{cf}(v) \in V'_{cf}. L_{cf} \subseteq V$ models linguistic qualifiers, which is determined by the strength of the property value in V_{cf}.

Example 1. We use concrete fuzzy concepts such as *Nil, Trace, Light, Moderate,* and *Heavy* to talk about the depth of rainfall. The two concepts *Moderate* and *Heavy* are defined as follows:

Table 1. Concrete Fuzzy Concepts for Rainfall

Fuzzy Concepts	Moderate	Heavy
V_{cf} (mm)	$4.9 \leq d < 25.0$	$25 < d \leq 100$
V'_{cf}	$[0, 1]$	$[0, 1]$
L_{cf}	little... very	little... very
f_{cf}	$\frac{x-4.9}{20.1}$	$\frac{x-25}{75.0}$

3.2 Similarity between Components of Fuzzy Ontology

Concrete Fuzzy Concept/ Property According to aforementioned definitions for concrete fuzzy concept/property, the degree of similarity between the concepts/properties depends on their fuzzy sets.

We denote 3-tuples $\{A, V, f_A\}$ and $\{A', V', f_{A'}\}$ belonging to the concrete fuzzy concepts/properties a and a' respectively. Where V and V' are the concrete sets, and A and A' are the fuzzy sets of the sets V and V' respectively. Denotes $U = V \cup V'$, defining the following operation between fuzzy subsets.

- $\forall x \in U, f_{A \cap A'}(x) = min[f_A(x), f_{A'}(x)]$
- $\forall x \in U, f_{A \cup A'}(x) = max[f_A(x), f_{A'}(x)]$
- $\forall x \in U, f_{A/A'}(x) = max[min(f_A(x), 1 - f_{A'}(x)), min(1 - f_A(x), f'_A(x))]$

A/A' is the fuzzy subset of elements that approximately belong to A and not to A' or conversely.

The similarity measures between the concepts/properties as follows:

$$Sim(a, a') = 1 - |A/A'| \tag{6}$$

$$Sim(a, a') = 1 - sup_{x \in U} f_{A/A'}(x) \tag{7}$$

$$Sim(a, a') = sup_{x \in U} f_{A \cap A'}(x) \tag{8}$$

Domain Fuzzy Concept. Similarity measure between the domain fuzzy concepts should be considered in multiple scenes of ontology such as semantic lable, properties's domains... Here, we only present similarity measure of concepts based on their properties as follows:

$$Sim(c, c') = \frac{\sum_{i=1}^{n} Sim(a_i^c, a_i^{c'})}{max(|c|, |c'|)} \tag{9}$$

where

- $Sim(a_i^c, a_i^{c'})$ is as defined in equations (6), (7) and (8) where $a_i^c, a_i^{c'}$ are properties belonging to two concepts c and c', respectively.
- $|c|$ and $|c'|$ are the number of the two concepts, respectively.
- n is the number of pairs of properties that their similarty degree is greater than a specific threshold.

4   Conflict Conciliation on Fuzzy Ontology Integration

We consider a real world $(\mathbf{A}, \mathbf{V})$ where $\mathbf{A}$ is a finite set of the fuzzy properties or the concrete fuzzy concepts (Notice that a concrete concept can be considered as a property). By $\mathbf{V}$ we denote the fuzzy universe which is defined as:

$$\mathbf{V} = \mathbf{A} \times [0, 1]. \tag{10}$$

An element of set $\mathbf{V}$ is called a fuzzy property value of the fuzzy universe. By $\prod(\mathbf{V})$ we denote the set of all finite and nonempty subsets with repetitions of set $\mathbf{V}$. An element of set $\prod(\mathbf{V})$ is called a conflict fuzzy profile of a concept.

Example 2. We consider a stuation of a concrete fuzzy concept *Expensive* in different ontologies for a concept *Car* (Denotes E_i be the concept *Expensive* in ontology i). A conflict fuzzy profile for the concept *Expensive* among the ontologies is as follow: $\{(E_1, 0.8), (E_1, 0.9), (E_1, 1), (E_2, 0.6), (E_2, 0.7), (E_2, 0.8), (E_3, 0.8), (E_3, 0.9)\}$

In a fuzzy profile, we can perform a partition of its elements referring to their first components. That is a profile $X \in \prod(\mathbf{V}$ can be divided into such sets (with repetitions) $X(x)$ where $x \in \mathbf{A}$, defined as follows:
Let $S_x = \{x \in A : x \text{ occurs in } X\}$ be a set without repetitions, then
$X(x) = (x, v) : (x, v) \in X$

Example 3. With the concrete fuzzy concept *Expensive* aforementioned, we have $X(E_1)$ = { $(E_1, 0.8), (E_1, 0.9), (E_1, 1)$ }, $X(E_2)$ = { $(E_2, 0.6), (E_2, 0.7), (E_2, 0.8)$ } and $X(E_3)$ = { $(E_3, 0.8), (E_3, 0.9)$} belonging to ontology 1, 2 and 3, respectively.

By an integration function we mean the following function:

$$C : \prod(\mathbf{V}) \to 2^{\mathbf{V}} \tag{11}$$

That is we assume that the result of the integration process for a concrete fuzzy concept is a fuzzy set (without repetitions). For a concrete fuzzy concept X, a set C(X) is called the integration of X. Notice that the integration $C(X)$ of a concrete fuzzy concept X is a set without repetitions. By $C(\mathbf{V})$ we denote the set of all integration functions for fuzzy universe $\mathbf{V}$.

Example 4. The result of the integration for the fuzzy concept *Expensive* above-mentioned are probably $\{(E, 0.8), (E, 0.9)\}$.

4.1 Postulates for Consensus

Now we present the postulates for integration functions for fuzzy profiles. Postulates on positions 1-10 are needed for inconsistency resolutions. They have been defined for consensus choice [8, 10]. Postulates on positions 9-11 are specific for integration process [9].

Definition 4. *An integration choice function $C \in C(\mathbf{V})$ safisfies the postulates of:*

1. *Unanimity (Un), iff $C(\{n * x\}) = \{x\}$*
   for each $n \in N$ and $x \in \mathbf{V}$
2. *Quasi-unanimity (Qu) iff*
   $(x \in C(X)) \Rightarrow (\exists n \in N : x \in C(X \cup (n * x)))$
   for each $x \in \mathbf{V}$.
3. *Consistency (Co) iff*
   $(x \in C(X)) \Rightarrow (x \in C(X \cup \{x\}))$
   for each $x \in \mathbf{V}$
4. *Reliability (Re) iff $C(X) \neq \varnothing$*
5. *Condorcet consistency (Cc), iff*
   $(C(X_1) \cap C(X_2) \neq \varnothing) \Rightarrow (C(X_1 \cup X_2) = C(X_1) \cap C(X_2)$
   for each $X_1, X_2 \in \prod(\mathbf{V})$.
6. *General consistency (Gc) iff*
   $C(X_1) \cap C(X_2) \subset (C(X_1 \cup X_2) \subset C(X_1) \cup C(X_2)$
   for any $X_1, X_2 \in \prod(\mathbf{V})$.
7. *Simplification (Si) iff*
   (Profile X is a multiple of profile Y) $\Rightarrow C(X) = C(X)$.
8. *Proportion (Pr) iff*
   $(X_1 \subset X_2 \wedge x \in C(X_1) \wedge y \in C(X_2)) \Rightarrow (d_F(x, X_1) \leq d_F(y, X_2))$for any $X_1, X_2 \in \prod(\mathbf{V})$.
9. *1-Optimality (O_1) iff for any $X \in \prod(\mathbf{V})$*
   $(x \in C(X)) \Rightarrow (d_F(x, X) = min_{y \in U_F} d_F(y, X))$
   where $d_F(y, X) = \sum_{y \in X} d_F(z, y)$ for $z \in \mathbf{V}$.
10. *2-Optimality (O_2) iff for any $X \in \prod(\mathbf{V})$*
    $(x \in C(X)) \Rightarrow (d_F^2(x, X) = min_{y \in V} d_F^2(y, X))$
    where $d_F^2(y, X) = \sum_{y \in X}(d_F(z, y))^2$ for $z \in \mathbf{V}$.
11. *Closure (Cl) iff $S_{C(X)} = S_X$*

Postulate 1-Optimality plays a very important role because it requires the consensus to be as near as possible to elements of the profile. Postulate *Cl* requires that in an integration for a fuzzy profile X all (and only those) elements from set S_X must appear. This means that the integration must refer to all and only those instances or properties for fuzzy ontologies

4.2   Conflict Conciliation on Concrete Fuzzy Concepts

Here, we present an algorithm for determining an integration for a fuzzy profile X satisfying postulate O_1 and postulate CL.

The problem: *For given, a profile $X = \{(c^i, v) : (v \in V^i)$ is the fuzzy set of the concrete fuzzy concept c belonging to the ontology O_i for $i = 1, \ldots, n\}$, we need to determine c^* for the profile X.*

This algorithm is presented as follows:

---

input : For given, a profile $X = \{(c^i, v) : (v \in V^i)$ is the fuzzy set of the concrete fuzzy
          concept c belonging to the ontology O_i for $i = 1, \ldots, n\}$
output: *An integration c^* for the profile X satisfying O_1 and CL.*

1 Create set (without repetitions)
2 $S_X = \{c \in \mathbf{A} : c$ occurs in $X\}$;
  ```
  /* notice that although the same concept or property but
  if they belong to different ontologies, they are
  considered as different elements in S_c */
  ```
3 **foreach** $c \in S_X$ **do**
4   Create $X(c) = \{(c, v) : (c, v) \in X\}$; and increasing sequence $V(c)$ consisting of
      numbers which appear in $X(c)$;
5   Let $n_c = card(X(c))$; As the consensus (c, v_c) for sub-profile $X(c)$ accept value v_c
      equal to the element of sequence $V(c)$ with index k satisfying inequality
      $\lfloor n_c/2 \rfloor \leq k \leq \lfloor (n_c + 1)/2 \rfloor$;
6 **end**
7 Create the integration c^* for X as follows:
8 $c^* = \{(c, v_c) : c \in S_c\}$
9 Return(c^*);

Algorithm 1. Conflict Conciliation on Concrete Fuzzy Concepts

4.3   Conflict Conciliation on Domain Fuzzy Concepts

We assume that 2 ontologies differ from each other in the structure of the same concept. That means they contain the same concept but its structure is different in each ontology.

Definition 5. *Let O_1 and O_2 be (A,V)-based ontologies. Assume that the concept c_1 belongs to O_1 and the concept c_2 belongs to O_2. We say that a conflict takes place in domain fuzzy concept level if $c_1 = c_2$ but $A^1 \neq A^2, V^1 \neq V^2$, or $f^1 \neq f^2$.*

Definition 5 specifies such situations in which two ontologies define the same concept in different ways. For example concept Weather in one system may be defined by attributes: hasRain(0.8), hasSnow(0.7), hasTemp(0.9), hasSuny(0.6), hasCloudy(0.6), while in other system it is defined by attributes: hasRain(0.7), hasSnow(0.5), hasTemp(0.9), hasSuny(0.8), hasCloudy(0.8),. Note that the same attributes can occur in both ontologies for the same concept, but the weights may be different.

Thus on concept level the problem of fuzzy ontology integration is the following:

The problem: *For a given set of pairs* $X = \{(A^i, V^i, f^i) : (A^i, V^i, f^i)$ *is the structure of concept c belonging to the ontology* O_i *for* $i = 1, \ldots, n\}$, *we need to determine the pair* (A^*, V^*, f^*) *which best represents those given.* A solution of this problem is shown as *Algorithm 2*.

---

input : $C^* = \bigcup c^i, i = 1 \ldots n$ is the set of concepts that are recognized as the same
   concept, but the associated structures $\{(A^i, V^i, f^i), i = 1 \ldots n\}$ are different.
output: triplet (A^*, V^*) which is the integration of the given pairs $\{(A^i, V^i), i = 1 \ldots n\}$.

1 $A^* = \bigcup A^i, i = 1 \ldots n$ where A^i is the set of attributes of the concept $c^i \in C$;
2 **foreach** *pair* $a_1, a_2 \in A^*$ **do**
3   **if** $R(a_1, \Leftrightarrow, a_2)$ **then** $A^* \setminus \{a_2\}$ and $X_a = X_a \cup \{f^i(a_2)\}$;   /* eg.,
      job $\Leftrightarrow$ *occupation* */
4   **if** $R(a_1, \sqsubseteq, a_2)$ **then** $A^* \setminus \{a_1\}$ and $X_a = X_a \cup \{f^i(a_1)\}$;   /* eg., *age* $\sqsubseteq$ *birthday*
      */
5   **if** $R(a_1, \sqsupseteq, a_2)$ **then** $A^* \setminus \{a_1\}$ and $X_a = X_a \cup \{f^i(a_1)\}$; /* eg., *sex* $\sqsupseteq$ *female* */
6   **if** $R(a_1, \perp, a_2)$ **then** $A^* \setminus \{a_1\}$ and $X_a = X_a \cup \{f^i(a_2)\}$; /* eg., *single* $\perp$ *married*
      */
7 **end**
8 **foreach** *attribute a from set* A^* **do**
9   **if** *the number of occurrences of a in pairs* (A^i, V^i, f^i) *is smaller than* $n/2$ **then** set
      $A^* := A^* \setminus \{a\}$;
10 **end**
11 **foreach** *attribute a from set* A^* **do**
12   Calculate $f(a) = \frac{1}{card(X_a)} \sum_{v \in X_a} v$;
13 **end**
14 Return($c^* = (A^*, V^*, f^*)$);

Algorithm 2. Conflict Conciliation on Domain fuzzy Concepts

5   Conclusions and Remarks

This research, we proposed a model of fuzzy ontology with their features which provide a solid basis for studies about fuzzy ontology integration. Consensus method is applied to reconcile conflicts among fuzzy concepts in ontology integration problem. The idea of the *Algorithm* 1 is based on determining for each sub-profile $X(c)$ where $c \in S_X$ a consensus satisfying postulate O_1, that is it minimizes the sum of distances to all elements of $X(c)$. The integration then consists of all such consensuses. Postulate Cl is satisfied since we have $S_{C(X)} = S_X$. In the *Algorithm* 2, words "at best" mean one or more postulates for satisfying by triplet. In general, we would like to determine such A^* of final attributes that all attributes which appear in sets A^i (i=1,...,n) are taken into account in this set. However, we cannot simply make the sum of A^i.

Acknowledgements

This paper was partially supported by Polish Ministry of Science and Higher Education under grant no. N N519 407437.

References

1. Baader, F., Nutt, W.: Basic Description Logics. In: Baader, F., Calvanese, D., McGuinness, D., Nardi, D., Patel-Schneider, P.F. (eds.) The Description Logic Handbook: Theory, Implementation, and Applications, pp. 43–95. Cambridge University Press, Cambridge (2003)
2. Barthelemy, J.P., Janowitz, M.F.: A Formal Theory of Consensus. SIAM J. Discrete Math. 4(4), 305–322 (1991)
3. Berners-Lee, T., Hendler, J., Lassila, O.: The Semantic Web. Scientific American 284(5), 35–43 (2001)
4. Calegari, S., Ciucci, D.: Integrating Fuzzy Logic in Ontologies. In: Manolopoulos, Y., Filipe, J., Constantopoulos, P., Cordeiro, J. (eds.) ICEIS, pp. 66–73. INSTICC Press (2006)
5. Duong, T.H., Jo, G.S., Jung, J.J., Nguyen, N.T.: Complexity Analysis of Ontology Integration Methodologies: A Comparative Study. Journal of Universal Computer Science 15(4), 877–897 (2009)
6. Lu, J., Li, Y., Zhou, B., Kang, D., Zhang, Y.: Distributed reasoning with fuzzy description logics. In: Shi, Y., van Albada, G.D., Dongarra, J., Sloot, P.M.A. (eds.) ICCS 2007. LNCS, vol. 4487, pp. 196–203. Springer, Heidelberg (2007)
7. Kemeny, J.G.: Mathematics without numbers. Daedalus 88, 577–591 (1959)
8. Nguyen, N.T.: Using Distance Functions to Solve Representation Choice Problems. Fundamenta Informaticae 48, 295–314 (2001)
9. Nguyen, N.T.: A Method for Integration of Knowledge Using Fuzzy Structure. In: IEEE/ACM/WI/IAT 2007 Workshops Proceedings, pp. 11–14. IEEE Computer Society, Los Alamitos (2007)
10. Nguyen, N.T.: Advanced Methods for Inconsistent Knowledge Management. Springer, London (2008)
11. Straccia, U.: A Fuzzy Description Logic for the semantic Web. In: Sanchez, E. (ed.) Proc. in Capturing Intelligence: Fuzzy Logic and The Semantic Web, pp. 167–181. Elsevier, Amsterdam (2006)
12. Zadeh, L.A.: Fuzzy sets. Information and Control 8, 338–353 (1965)

Improving Availability in Distributed Component-Based Systems via Replication

Samih Al-Areqi, Amjad Hudaib, and Nadim Obeid

Department of Computer Information Systems,
King Abdullah II School for Information Technology,
The University of Jordan
obein@ju.edu.jo

Abstract. One of the important criteria that affect the usefulness and efficiency of a distributed system is availability, which mainly depends on how the components of the system are deployed on the available hosts. If the components that need lots of interaction are on the same host, the availability will definitely be higher given that all the components are working properly. In this paper, we extend the work of the Avala model in two ways: (1) we provide a dependency relation between the components as an additional factor and (2) we extend the Avala algorithm by implementing a replication mechanism.

Keywords: Distributed Systems, Avala, E-Avala, Availability, Redeployment, Replication.

1 Introduction

A Distributed System (DS) can be viewed as several interrelated software components distributed on various hardware hosts in different physical locations, but nonetheless work together as a whole unit to meet some requirements. One important characteristic of DS is dynamicity which allows us to continuously add, retract, and/or change the locations of components to meet changing requirements. This results in some problems such as availability, dependency management, and dynamic configuration. Availability is defined as the ratio of the number of successfully completed inter-component interactions in the system to the total number of attempted interactions over a period of time [9]. It depends on how the components are deployed on the available hosts. If the components that need lots of interaction are on the same host, the availability will definitely be higher given that all the components are working properly. Dependency refers to the case when one component depends on other components [6, 5]. Dependency management becomes an issue when the number of components grows due to the dynamic nature of distributed systems and/or components redeployment which occurs over time. It is hard to create a robust and efficient system if the dependencies among components are not well understood.

It is difficult for the system designer to predict, at design time, the applications/tasks with which the system will be involved. Dynamic reconfiguration is how to dynamically redeploy (some) components in an existing configuration in such a way that consistency of the system is preserved [3]. A replication mechanism [9, 8]

N.T. Nguyen et al. (Eds.): New Challenges for Intelligent Information, SCI 351, pp. 43–52.

allows us to make a replica of a component in the target host. Replication and redeployment can be employed when it is realized that the availability is not very high. It is important to know how they could have effects on each others and change with time. For instance, redeployment changes the configuration (e.g., location, topology and host) of components. This may seem to be useful at a particular time in a certain situation. However, at a later time it may prove to be a disadvantage or a problem. Replication, as a mechanism, is very useful, or even necessary, in some situations.

Following [2, 13], we distinguish between two notions of consistency: (1) Replica consistency that defines the correctness of replicas (i.e., a measure of the degree to which replicas of the same logical entity differ from each other), (2) Concurrency consistency which is a correctness criterion for concurrent access to a particular data item (local isolation), usually employed in the context of transactions and (3) constraint consistency which defines the correctness of the system with respect to a set of data integrity rules (application of defined predicates).

In this paper we extend the work presented in [9] in that: (1) our proposal considers more factors and requirements such as dependency (e.g., relations of each component to other applications, positive dependency, and negative dependency) and (2) it presents an improvement of the actual algorithm by implementing a replication mechanism where we make use of negative dependency. In section 2 we give a brief presentation of replication techniques. In section 3 we present the Avala model. Sections 4, 5 and 6 are dedicated to the E-Avala model, its experiments, its performance and its cost model.

2 Replication Techniques

Network-dependent systems have to face the problem of disconnected operations where the system must continue functioning in a near absence of the network. This presents a major challenge because each local subsystem is usually dependent on the availability of non-local resources. Lack of access to a remote resource can halt a particular subsystem or even make the entire system unusable [1, 12, 4].

Replication of objects increases system reliability and availability as required by many applications [1]. Replicating data over a cluster of workstations is a powerful tool to increase performance. It provide fault-tolerance for demanding database applications [11]. Clearly, if many identical copies of an object reside on several computers with independent failure modes and rates, then the system would be more reliable. However, one of the disadvantages is that the read and write operations performed on a replicated object by various concurrent transactions must be synchronized.

Furthermore, the Grid environment introduces new and different challenges such as high latencies and dynamic resource availability. Given the size of the data sets, it is impossible for a human to make decisions about where the data should be placed. Clearly, a good replication strategy is needed to anticipate and/or analyze user's data requests and to place subsets of the data (replicas) accordingly. Current technologies and initiatives use only static or user-driven replication services to manually replicate data files. The Grid environment, however, is highly dynamic because resources availability and the network performance change constantly and data access requests vary with each application and each user. Two key challenges that need to be addressed are scalability and adaptability.

3 The Avala Model

The Avala model is part of an integrated solution for increasing the availability of a DS during disconnection among hosts. Availability is defined as the ratio of the number of successfully completed interactions in the system to the total number of attempted interactions. Avala considers a subset of all possible constraints for components deployment. These constraints are as follows:

(1) Software component properties, which consist of memory requirements, frequency of communication between components and size of the exchanged data.
(2) Hardware host properties, which consist of memory capacity, network reliability (of the links between hosts) and network bandwidth.
(3) Location constraints, which consist of two relations that restrict the location of software components. These relations are: (i) *Loc:* which determines whether or not a component can be deployed in a host and (ii) *Colloc:* which determines whether or not some components can be collected on the same host.

Let $h_1, h_2, \ldots, h_k$ $(1 \le k)$ stands for hosts, $MEM(h_i)$ be the memory of h_i. $C_1, \ldots, C_n$ $(1 \le n)$ stands for components, $MEM(C_i)$ be the memory of Ci and $FREQ(Ci, Cj)$ be the frequency between components C_i and C_j. The Avala algorithm [9] starts by ranking all hardware nodes and software components. The initial ranking of hardware nodes is performed by calculating, for each hardware node i, the Initial Host Rank $(IHR_i,)$ as follows:

$$IHR_i = \sum_{j=1}^{k} REL(h_i,h_j)+MEM(h_i))$$ (1)

The ranking of software components is performed by calculating for each i, the Initial Component Rank (ICR_i) as follows:

$$ICRi = d * \sum_{j=1}^{n} FREQ(C_i,C_j) + \frac{E}{MEM(C_i)}$$ (2)

Where d denotes the respective contributions of host memory and E denotes the respective contributions of event size of interactions between C_i and C_j.

After the initial rankings are performed, the host with the highest value of *IHR* is selected as the current host h. A component with the highest value of *ICR* that satisfies the memory and *Loc* constraints is selected and assigned to h.

The next software component to be assigned to h, is the one with the smallest memory requirement and which would contribute maximally to the availability function if placed on h, i.e., the components that may have the highest volumes of interaction with the component(s) already assigned (mapped) to h. The selection is performed by calculating the value of Component Rank (*CR*) for each unassigned component as follows:

$$CR(Ci, h) = D_1(Ci, h,n) + D_2(Ci, h)$$ (3)

where $D_1(C_i,h,n) = d * \sum_{j=1}^{n} FREQ(C_i,MC_j)* REL(h,f(MC_j)$

and $\quad D_2(C_i,h,n) = \dfrac{E}{MEM(C_i)}$

MC_j is a shorthand for mapped Components j, $f(MC_j)$ is a function that determines the hosts of mapped components, $REL(h, f(MC_j))$ is a function that determines the reliability between selected host h and hosts of mapped components.

The selected component is the one with the highest value of CR and that satisfies *Loc*, and *Colloc* constraints with respect to the current host h and components already assigned to every host j such that $REL(h, f(MC_j))$ is maximum. This process is repeated until there is no component with a memory size small enough to fit on h. The next host to be selected is the one with the highest memory capacity and highest link quality (i.e., highest value of reliability) with the host(s) already selected. Host Rank *(HR)* is calculated as follows:

$$HR(h_i) = \sum_{j=1}^{m} REL(h_i, MH(h_j)) + MEM(h_i) \tag{4}$$

where m is number of hosts that are already selected.

4   The E-Avala-Model

The Avala model does not take dependency relation into account. We consider this to be a very important issue in DS because in some situations it is difficult to deploy certain components in certain hosts if there are one or more components in other host that depends on this component or this component depends on other components.
In the E-Avala, we the employ the notion Depend(C_i, C_j) which is defined as follows:

$$Depend(C_i, C_j) = \begin{cases} 1 & \text{if } Ci \text{ depends on } Cj \\ -1 & \text{if } Cj \text{ depends on } Ci \end{cases} \tag{5}$$

Furthermore, E-Avala takes into considration whether or not is a need for data consistancy check regarding a component C_i as shown below in (6):

$$Consis(C_i) = \begin{cases} 1 & \text{if } Ci \text{ does requires data consistency} \\ 0 & \text{Otherwise} \end{cases} \tag{6}$$

Let h be the selected host, l is the level of dependency for system configuration which is determined by the user or the system designer and nm denotes be number of mapped components such as c_j (i.e., the number of all components which have already been assigned to all selected hosts so far), the E-Avala algorithm uses the same equations ((1), (2) and (3)) of Avala to calculate the *IHR*, *ICR* and *HR*. It improves Avala by employing two additional functions: *RCR* (resp. *Consis-RCR*) that compute replicate component rank without (resp. with) consideration for data consistency.

$$RCR(C_i, h, n, nm) = D_3(C_i, h, n) + D_1((C_i, h, nm) \tag{7}$$

where $D_3(C_i, h, n) = \sum_{p=1}^{n} Depend(C_p, C_i) + \dfrac{l + 2E}{MEM(C_i)}$

$$Consis\text{-}RCR(C_i, h, n, nm) = D_3(C_i, h, n) * (1 - Consis(C_i) + D_1(C_i, h, nm) \tag{8}$$

The E-Avala algorithm is presented in Figure1.

```
E-Avala_algorithm (hosts, comps)
{ NumOfMappedComps= 0
 UnmappedComps = comps
 h = host with max (initHostRank)
 Unmapped Hosts = hosts – h
 NumOfMappedHosts = 1
 RC= component with max (initReplicateCompRank)
 DC = component with max (initDeployeCompRank)
while (numOfMappedComps < numOfComps and
       numOfMappedHosts < numOfHosts and h<>-1) and
       (h.memory>c.memory and
       numOfMappedComps < numOfComps and RC<>-1 and DC<>-1 )
 If RC >= DC
   unmappedComps = unmappedComps –RC
   numOfMappedComps = numOfMappedComps + 1
   h.memory = h.memory – RC.memory
   Replication = replicate (c to h)
   RC=nextReplicateComp (comps, unmappedComps,h,)
 Else
   unmappedComps = unmappedComps –DC
   numOfMappedComps = numOfMappedComps + 1
   h.memory = h.memory – RC.memory
   Deployment = Deployment (c to h)
   DC=nextDeplyeComp(comps, unmappedComps,h,)
   h = next_host (unmappedHosts)
   UnmappedHosts = hosts – h
   NumOfMappedHosts = numOfMappedHosts + 1
 if numOfMappedComps= numOfComps
 return Success
 else
 No Deployment And Replication  Was Found }
```

Fig. 1. E-Avala Algorithm

The function "next-replicated-comp" which computes the next component to be replicated is presented in Figure 2.

```
NextReplicateComp (comps, unmappedComps, currentHost)
{ bestReplicatCompRank = 0
 betsReplicatCompIndex =  -1
mappedComps = comps – unmappedComps
for idx to unmappedComps.length
if (unmappedComps[idx].memory <= currentHost.memory
and unmappedComps[idx] sastisfies loc ,component consistency and colloc
       constraints with mappedComps)
thisCompRank = compRank(unmappedComps[idx],
currentHost)
if  bestReplicatCompRank < thisCompRank
betsReplicatCompIndex = idx
bestReplicatCompRank = thisCompRank
if  betsReplicatCompIndex =-1
 return NULL
else
 return unmappedComps[bestCompIndex]}
```

Fig. 2. Next Replicate Component Algorithm

The E-Avala makes a comparison between the selected components for redeployment determined by *CR* (cf. (3)) and the components to be replicated determined by *RCR* (cf. (5)). The selected component will be the one with the highest value of *CR* and *RCR* and that satisfies the constraints *Loc* and *Colloc* with respect to the current host *h* and components which are already assigned. This process will be repeated until *h* is saturated (i.e., there is no component small enough that can be assigned to *h*).

5   The E-Avala Performance

The Avala algorithm significantly reduces the time complexity of the exact algorithm [7] which is exponential (e.g., $O(k^n)$), where *k* is the number of hardware hosts, and *n* the number of software components). By fixing a subset of $m \leq n$ components to selected hosts, the complexity of the exact algorithm reduces to $O(k^{n-m})$. Even with this reduction, this algorithm is computationally too expensive unless the number of hardware nodes and unfixed software components is very small. For example, even for a relatively small deployment architecture (where n=15 components, and k= 4 hosts), a Java JDK 1.4 implementation of the exact algorithm runs for more than eight hours on a mid-range PC (processor 1.8 GHz, and 1 GB). The complexity of the Avala and E-Avala algorithms in the most general are $O(n^3)$.

We have employed DeSi simulator [1] which is a visual deployment environment that supports specification, manipulation, visualization, and (re)estimation of deployment architectures for large-scale, highly distributed systems. We had to leverage DeSi to add additional input values (dependency values between components and data consistency value for each component) and to support replication.

We run 5 tests to assess the performance of the E-Avala. In each test, we have taken the average results for 30 different randomly generated architecture configurations by using the parameters presented below in Table 1.

Table 1. System Input Parameters

Input Parameter	Value		Input Parameter	Value
Number of Components	100		Min host reliability	0
Number of Hosts	10		Max host reliability	1
Min component memory	2		Min component event size	1
Max component memory	8		Max component event size	10
Min Host memory (in KB)	50		Min host bandwidth	30
Max Host memory (in KB)	100		Max host bandwidth	1000
Min component frequency	0		Level of dependency	3
Max component frequency	10			

Test (1):   In this scenario, all the input parameters are fixed except the host memory (HM). Figure 3 shows the availability for values that range from 100 to 600.

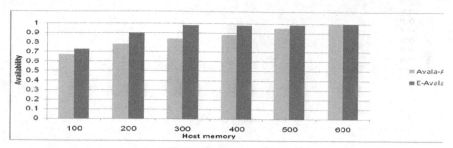

Fig. 3. Host memory Test

The average improvement of availability was 15%. However, when the host memory was high, the availability remained the same. Note that the distribution in Avala is not even. If the total available memory on hosts is significantly above the total memory required by the components, then some of the hosts will be filled to their full capacity, while others may contain only few components or even be empty. The uneven distribution of components among hosts results in higher overall availability of the system since it utilizes the maximum reliability for interactions between components residing on the same host. However, it may also result in some undesirable effects on the system, such as overloading the CPUs on hosts with large numbers of components, or overloading the used subset of network links. E-Avala addresses this concern via replication and using *Loc,* and *Colloc* constraints.

Test (2): in this scenario, all the input parameters are fixed except the event size E. Figure 4 shows the availability for values that range from 100 to 600.

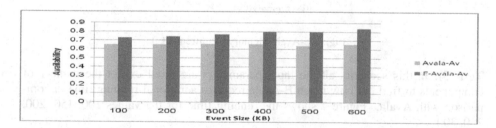

Fig. 4. Event Size Test

In this test the average improvement is over 18%, because E-Avala gave event-size more attention as reflected in (5) above.

Test (3): In this scenario, all the input parameters are fixed except the dependency level d. Figure 5 shows the availability for the values 2, 4, 6, 8 and 10.

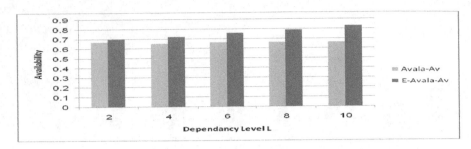

Fig. 5. Dependancy Level Test

The availability using the E-Avala algorithm is better than that when using Avala when the dependency level between the components is high. This is due to the fact that E-Avala employs replication.

Test (4): In this scenario, all the input parameters are fixed except the host memory to find out how much E-Avala requires additional memory in comparison with Avala. Figure 6 shows the results for values that range from 100 to 600.

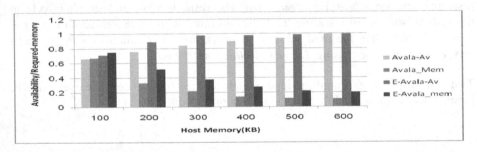

Fig. 6. E-Avala Memory Requirement

Test (5): In this scenario, all the input parameters are fixed except the number of components to find out how much E-Avala requires additional running time in comparison with Avala. Figure 7 shows the running time for the values 100, 150, 200, 250, 300.

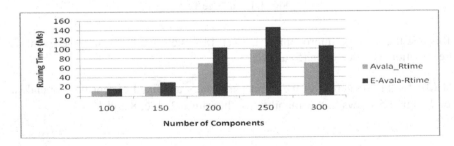

Fig. 7. E-Avala Time Cost

6 E-Avala Cost Model

In this section, we present a mathematical cost model [14] to quantify the performance of the E-Avala algorithm. There are three types of costs associated with the operations in servicing the requests between distributed components. The first one, $C_{I/O}$., is the cost of fetching an object from the local memory to the processor or saving an object from a processor to its local memory. The other two types of costs are control-messages transferring cost, $Cost_c$, and data-messages transferring cost $Cost_d$. Let p_i be a processor in host i, A_0 be the allocation schema of component C on the request and f_n is the number of request frequency between servers.

$$Cost_{E-AVala} = \begin{cases} 1 & if\ p_i \in A_0 \\ 1 + (\ Cost_c + Cost_d\ ) * f_n & otherwise \end{cases} \tag{9}$$

If $p_i \notin A_0$ then the cost is $1 + (Cost_c + Cost_d) * f_n$ as the request has to be retrieved fro the target server.

Formally an algorithm α is said to be c-competitive, if there exist constants c≥1 and d≥0 such that for any request sequence φ, $Cost(\alpha(\varphi)) \leq c * Cost_{opt}(\varphi) + d$ where $Cost_{opt}(\varphi)$ is the cost on φ for an optimal off-line algorithm which knows the entire request sequence in advance and can serve the requests with a minimum cost.

Theorem 1: The E-Avala algorithm is $\dfrac{1 + (\ Cost_c + Cost_d\ ) * f_n}{MEM_{comp}} * k$ -Competitive

where $k \geq 1$ is the number of distributed components and MEM_{comp} is the memory size of the components.

The proof is not presented here due to the lack of space.

7 Concluding Remark

Avala [9] has been proposed as an efficient algorithm for improving components' availability in a DS via redeployment. However, it has ignored the dependency relationships among components. In this paper, we have presented E-Avala as an extension on Avala by managing the dependency relationships among components. Negative dependency is considered using component replication and the positive dependency is considered as a constraint. We have also presented a component replication mechanism which is useful when the negative dependency is high. However, issues such as dealing with functional consistency of components and including additional system parameters such components structure (e.g. hierarchical representations of the components must be properly addressed.

References

1. Adly, N., Kumar, A.: A Hierarchical Propagation Protocol for Large Scale Replication in Wide Area Networks, Technical Report TR-331, University of Cambridge (1994)
2. Bernstein, P., Goodman, N.: An Algorithm for Concurrency Control and Recovery in Replicated Distributed Databases. ACM Transactions on Database Systems 9(4), 596–615 (1984)
3. Chen, X.: Dependency Management For Dynamic reconfiguration of component-based distributed System. In: Acceding of the 17th IEEE International Conference on Automated Software Engineering, ASE 2002 (2002)
4. Dahlin, M., Chandra, B., Gao, L., Khoja, A.A., Nayate, A., Razzaq, A., Sewani, A.: Using Mobile Extensions to Support Disconnected Services, University of Texas Department of Computer Sciences Tech. Report TR-2000-20 (2000)
5. Heydarnoori, A., Mavaddat, F.: Reliable Deployment of Component-Based Applications into Distributed Environments. In: Proceedings of the 3rd IEEE International Conference on Information Technology: New Generations, ITNG 2006 (2006)
6. Kon, F., Campbell, R.: Dependence Management in Component-Based Distributed Systems. IEEE Concurrency 8(1), 26–36 (2000)
7. Malek, S., Mikic-Rakic, M., Medvidovic, N.: An Extensible Framework for Autonomic Analysis and Improvement of Distributed Deployment Architectures. In: Proceedings of the 1st ACM SIGSOFT Workshop on Self-Managed Systems, pp. 95–99. ACM, New York (2004)
8. Malek, S., Mikic-Rakic, M., Medvidovic N.: A Decentralized Redeployment Algorithm for Improving the Availability of Distributed Systems. In: 3rd International Conference on Component Deployment, France (2005)
9. Mikic-Rakic, M., Malek, S., Medvidovic, N.: Improving availability in large, distributed component-based systems via redeployment. In: Dearle, A., Savani, R. (eds.) CD 2005. LNCS, vol. 3798, pp. 83–98. Springer, Heidelberg (2005)
10. Mikic-Rakic, M., Medvidovic, N.: A Classification of Disconnected Operation Techniques. In: Proceedings of the 32nd EUROMICRO Conference on Software Engineering and Advanced Applications (2006)
11. Tierney, B.: File and Object Replication in Data Grids. In: Proceedings of the Tenth International Symposium on High Performance Distributed Computing (HPDC-10). IEEE Press, Los Alamitos (2001)
12. Nelson, V.P.: Fault-Tolerant Computing Fundamental Concepts. IEEE Computer 23(7), 19–25 (2002)
13. Osrael, J., Froihofer, L., Goeschka, K.M.: What Service Replication Middleware Can Learn from Object Replication Middleware. In: Proceedings of the 1st Workshop on Middleware for Service Oriented Computing (MW4SOC), pp. 18–23. ACM, New York (2006)
14. Sleator, D., Tarjan, R.E.: Amortized Efficiency of List Uupdate and Paging Rules. Communications of the ACM 28(2), 202–208 (1985)

OPLSW: A New System for Ontology Parable Learning in Semantic Web

Navid Kabiri, Nasser Modiri, and NoorAldeen Mirzaee

Department of Computer Engineering,
Islamic Azad University of Zanjan, Iran
{Navid_kay2000,Nassermodiri,Mirza683}@Yahoo.com

Abstract. The Semantic Web is expected to extend the current Web by providing structured content via the addition of annotations. Because of the large
amount of pages in the Web, the manual annotation is very time consuming.
Finding an automatic or semiautomatic method to change the current Web to
the Semantic Web is very helpful. In a specific domain, Web pages are the parables of that domain ontology. So we need semiautomatic tools to find these
parables and fill their attributes. In this article, we propose a new system named
OPLSW for parable learning of an ontology from Web pages of Websites in a
common domain. This system is the first comprehensive system for automatically populating the ontology for websites. By using this system, any Website
in a certain domain can be automatically annotated.

Keywords: Semantic Web, Ontology, Parable Learning, Classification.

1 Introduction

Nowadays, the Web is rapidly growing and becoming a huge repository of information, with several billion pages. Indeed, it is considered as one of the most significant
means for gathering, sharing, and distributing information and services. At the same
time this information volume causes many problems that relate to the increasing difficulty of finding, organizing, accessing and maintaining the required information by
users. Recently, the area of the Semantic Web is coming to add a layer of intelligence
to the applications do these processes.

As defined in [1], "the Semantic Web is an extension of the current Web in which
information is given well defined meaning, better enabling computers and people to
work in cooperation." W3C has a more formal definition that "The Semantic Web is
the representation of data on the World Wide Web. It is a collaborative effort led by
W3C with participation from a large number of researchers and industrial partners. It
is based on the Resource Description Framework (RDF), which integrates a variety of
applications using XML for syntax and URIs for naming" (*W3C*).

Semantic Web provides a common framework that allows data to be shared and
reused in application, enterprise and community boundaries. A prerequisite for the
Semantic Web is the availability of structured knowledge, so methods and tools need
to be employed to generate it from existing unstructured content. Because the size of
the Web is too huge and the manual annotation of Web pages is too time and resource

N.T. Nguyen et al. (Eds.): New Challenges for Intelligent Information, SCI 351, pp. 53–61.
springerlink.com   © Springer-Verlag Berlin Heidelberg 2011

consuming, an automatic or semiautomatic tool or method to transform the current Web content to the Semantic Web content is very useful in the process of developing Semantic Websites.

Ontologies are important components of Semantic Web, because they allow the description of semantics of Web content. The work of changing the Web to Semantic Web could be done by methods and tools of ontology learning and automatic annotation tools.

This paper has organized as follows. In section 2, we will define and describe ontology and ontology learning and their usage. In section 3, we will introduce our parable learning system, OPLSW (Ontology Parable Learner for Semantic Web). Our experience on different classification algorithms on Web pages is discussed in section 4. In section 5, we will briefly review the related works in parable population and at last our conclusion is given in section 6.

2 Ontology Learning

Ontology is a philosophical discipline, a branch of philosophy that deals with the nature and the organization of being [11]. Ontologies are used for organizing knowledge in a structured way in many areas. We usually refer to an ontology as a graph/ network structure consisting from concepts, relations and parables. In a formal way, ontology can be defined as:

Definition. An ontology is defined by a tuple $O=(C,T,\leq C,\leq T,R,A,\sigma_R,\sigma_A,\leq R,\leq A)$ which consists:

C as a set of concepts aligned in a hierarchy $\leq C$, R as a set of relations R with $\leq R$, the signature $\sigma_R : R \rightarrow C^2$, a set of data type T with $\leq T$, a set of attributes A with $\leq A$, and signature $\sigma_A : A \rightarrow C \times T$. For a relation $r \in R$ the domain and range of ontology define as [8], :

$$dom(r) := \prod{}_1(\sigma_R(r)) \quad and \quad range(r) := \prod{}_2(\sigma_R(r)) \tag{1}$$

Today a completely automatic construction of good quality ontologies is in general impossible for theoretical, as well as practical reasons [7].

Depending on the different assumptions regarding the provided input data, ontology learning can be addressed via different tasks: learning just the ontology concepts, learning just the ontology relationship between the existing concepts, learning both concepts and relations at the same time, populating an existing ontology and other tasks. In [7], the ontology learning tasks defined in terms of mapping between ontology components where some of the components are given and some of them are missing, and with these tasks the missing ones induced. Some typical scenarios for these tasks are:

- Inducing concepts/clustering of parables (given parables).
- Inducing relations (giving concepts and the associated parables).
- Ontology population (giving an ontology and relevant, but no associated parables).

- Ontology generation (given parables and any other background information).
- Ontology updating/extending (given an ontology and background information, such as new parables or the ontology usage patterns).

Creating Semantic web from the current Web can be done in two ways: annotating the Web pages according to the domain ontology of that pages and constructing the web pages according to the domain ontology. In constructing Semantic Web pages, we can have a specified structure for each page and pages constructed according to that structure and all definitions in domain ontology.

Nowadays we have a large amount of Web pages, and if we want to annotate them manually or construct them again, it takes a lots of time and many mistakes may occurs. One of the efficient ways to do that is to annotate these pages automatically according to the domain ontology of those Web pages. This work is the scenario 3 which we have the domain ontology (concepts and relations) and we want to populate the ontology. For this work we should do some steps which we will define in the next part.

3   OPLSW: Our Proposed Ontology Parable Learner for Semantic Web

As we mentioned before, ontology parable learning or parable population is one of the important issues in ontology learning. Most of the works on ontology population have been done on information extraction and finding relation between pages in Web. But another problem in Web and in changing the current Web to the Semantic Web is for Website designers. It is hard and time consuming and also with lots of mistakes for designers to manually annotate their Websites. The number of pages in medium and large Websites occasionally is more than 1000 pages and this make reconstruction of these Websites difficult.

We propose a new system, named OPLSW (Ontology Parable Learner for Semantic Web) to do this task automatically in Websites with a common ontology (Figure 1). This system is a general system which different technologies can used in it's component.

As shown in Figure 1, as an input of such a system, we have an ontology belongs to the domain and Web pages of that domain (Web pages of Website) which we want to make them as parables of that domain.

The system work on ontologies which have an organization view. This means that this system work on websites that have this organizational structure, such as university web sites or companies websites. According to these ontologies and websites, each page in the website is an parable of the ontology class.

This system operates in two stages: training and operation. In training the system learns the rules and models and in operation the system uses these rules and models for annotating the Web pages. Training stage of system operation consists of the following steps:

- Learning the classifier: Using classification algorithms.
- Learning rules of information extraction of that domain: using information extraction algorithms.
- Learning rules of relation extraction of that domain: using methods of extraction relations.

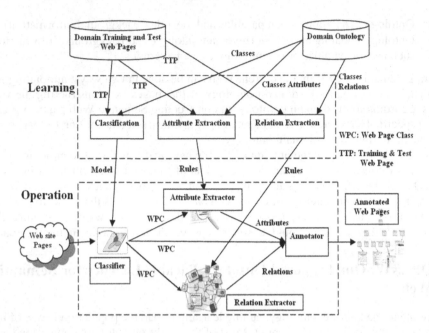

Fig. 1. Architecture of OPLSW

For learning the classifier we have the ontology classes as classes of classifier and Web pages of that domain for training and testing the classifier. We implement this part as the most important part of the system and will describe in chapter 4. For information extraction attributes of ontology and Web pages of domain are inputs and for extracting the relations the relations in ontology and Web pages are the inputs.

For information extraction component, we can use traditional IE or tools like Amilcare [5], and MnM [13], that extract the rules of information extraction according to the ontology with some revises. In rule extraction also we can use the link and hyperlink properties or traditional IE to extract the rules which the rule extractor in the system can use for finding the relation between the parables.

The operation stage has 4 main tasks: classification, information extraction, relation extraction and annotation. Steps of operation are:

- Relating each page to a class of the ontology (Classifier).
- Finding the attributes of each parable (Attribute extractor).
- Detect the relations between the parables that are the relations between the classes in ontology (Relation extractor).
- Annotating the Web pages (Annotator).

Inputs and outputs of each step are as follow: Inputs of classifier are ontology classes, Web pages and classifier model and the output is the pages classified in ontology classes.

The first step in operation use classification models we have learned in the training. In attribute extractor we can use the information extraction rules we have learned in training to extract the proper values for attributes. The input of this step is information extraction rules, and the classified pages and the output is the attributes. For relation extractor we have learned rules for relation extraction in pages and classified pages that need to find their relations. This step can be done parallel with the second step of operation. The last step uses the output of previous steps and automatically annotates each page. With all these steps and components, and with domain ontology and train pages for that domain, the system can automatically learn the ontology parables on the web site and annotate them.

In all these four steps, the classifier could have a very distinct effect on the result of learning. Because its output is the input of other phases and any improvement in this phase could effect on the result of system distinctly. So using classification algorithms with better accuracy could help us in better results in the system. As the most important part of OPLSW, we implemented the classification part and training the classifier model with 3 classification algorithms for Web pages. These are algorithms which used in text mining and with this experience we wanted to consider how proper these algorithms are in web page classification. Our work was on the WEB-KB as a data set of Web to check what algorithms and feature selections can improve the result of this part.

4   Comparing Classification Algorithms for Classifying Web Pages

Classification aids in better information retrieval and knowledge utilization but the size of the Web along with the diversity of the subject matter makes manual classification a tedious and sometimes impractical task. It is highly desirable to be able to perform classification automatically employing computational techniques. Automated categorization is a supervised learning task in which new documents are assigned category labels based on the likelihood suggested by a training set of labeled documents.

For comparing classification on these domain specific data, we use the 3 most important & effective algorithms in text classification. These are Naïve Bays, K-Nearest Neighborhood and SVM. Naive Bays is a wildly used statistical learner known for its simplicity [3], [12]. Support Vector Machines are one of the most accurate classifiers of text currently available. And KNN which used in most of classification works [15], [10].

We use the WEB-KB dataset for our experience. In this dataset, we have 7 classes of pages: course, Department, Faculty, Project, Staff, Student and Other class which all pages that do not belong to the first 6 class will go to this class.

For classifying of Web pages, first of all we should tokenize them. In our tokenizer, we first detect all the tags, eliminate them and make the pages as a pure text. We tokenize the pages by selecting the words between separator marks (like space, commas, quotes and all others) as tokens. After that we tokenize all the pages in each class (according to the train set) and calculate the frequency of each word in that class after checking them with StopWords and stem them. So we eliminate the words in which are StopWords and can not have effect on classifying these pages.

According to the [6], we delete some StopWords from the StopWords list, because they could use as features for classification (for example "I" is a word used in student's pages for several times and it can shows this class pages). After calculating the frequency of each word in these classes, we need a feature selection method to reduce the number of features for classification. According to [14] document frequency (DF) is a reliable measure for selecting information features and it can be used instead of information gain (IG) or $X^2 - test$ (CHI) when the computation of these measures is too expensive. So we use is feature selection but another problem is that how many of the words should be in features. We check this by selecting some thresholds for document frequency of words. By changing the range of threshold we want to find the best features.

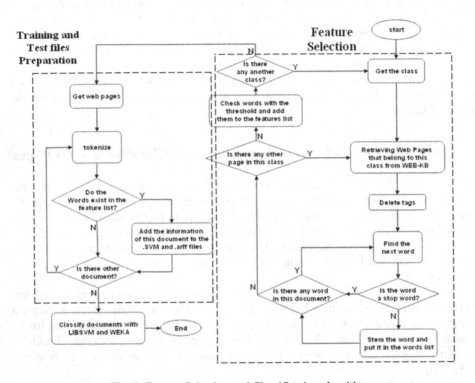

Fig. 2. Feature Selection and Classification algorithm

After selecting each threshold we classified the documents by Naive Bays algorithm. The results have shown in Figure 3. As you can see in the figure, the threshold 30%, have the best accuracy in the classification. After selecting the best threshold, system tokenizes the whole documents to find out the number of times the words repeated in them and then the system write these information in files. After that we use these files as inputs of classification tools (Figure 2).

For NB and KNN, we use WEKA, and for the SVM we use the LIBSVM as tools for classification. The results of our experience have shown in Table 1. Table 1 shows accuracy of different classification algorithms on WEB-KB dataset.

Table 1. Results of different classification algorithms on Web data of a specific domain

Classification Algorithm	NB	KNN	SVM
Accuracy	31.775	76.7829	74.5642

As you can see, because the classes of this domain have common attributes, the accuracy is not high. For example 3 of our classes are subclasses of person class (Student, Faculty and Staff) and the other class has a combination of the other 6 classes. This makes more mistakes in classification of these for the algorithm. For this dataset, KNN have better accuracy than SVM and NB.

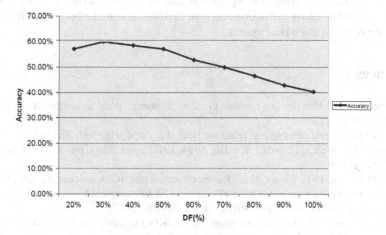

Fig. 3. Accuracy for different thresholds of DF

5 Related Works

[9], Populate an ontology by the use of hand-crafted domain-specific relation patterns. Their algorithm uses parables of some class returned by Google to find parables of other classes. Work on information extraction from large Web sites, so called wrapper algorithms. Brin [2], combines techniques from pattern extraction and wrapper algorithms to extract parables of relations from Web sites. Using Google to identifying relation patterns can be found in [4].

6 Conclusion and Future Works

In this article, we discussed that one of the main challenges in developing the content for Semantic Web is transforming the existing Web content to Semantic Web content by annotating the pages. We proposed a system for changing an existing Website to a Semantic Web Website by using ontology and ontology population on the domain of that Websites. This system helps the designers of Websites with common domain to annotate their websites automatically. Our system operates in training and operation stages. In training the system learn the models and rules of classifiers, attribute extractors and rule extractor. In operation stage, the system uses these rules and models to annotate Web pages. We introduce the classification as the most important step in annotation, because its output is the input of the other 3 steps.

In the last part of our work, we had an experience on classification and we compared the 3 most common classification algorithms for classifying our Web pages and our experience shoed that the KNN algorithm has the best accuracy on our data. The accuracy of all 3 was low because of the common features that the classes in our dataset have. The result of this experience shows that we need more effective algorithm for classifiers that work in Web domain.

For future works we propose implementing this system and work on different parts of this system for improving the algorithms can be used in different parts, such as classification, information extraction and rule extraction to make better results in each part and the result of the whole system.

References

1. Berners-Lee, T., Hendler, J., Lassilo, O.: The Semantic Web. Scientific American 284, 34–43 (2001)
2. Brin, S.: Extracting patterns and relations from the world wide web. In: Atzeni, P., Mendelzon, A.O., Mecca, G. (eds.) WebDB 1998. LNCS, vol. 1590, pp. 172–183. Springer, Heidelberg (1999)
3. Chai, M.K., Ng, H.T., Chieu, H.L.: Bayesian online classifiers for text classification and filtering. In: Proceedings of SIGIR 2007, 25th ACM International Conference on Research and Development in Information Retrieval, pp. 97–104 (2002)
4. Cimiano, P., Staab, S.: Learning by googling. SIGKDD Explorations 6, 24–34 (2009)
5. Ciravegna, F., Wilks, Y.: Designing adaptive information extraction for the semantic web in Amilcare. In: Handschuh, S., Staab, S. (eds.) Annotation for the Semantic Web. ISO Press, Amsterdam (2008)
6. Craven, M., DiPasquo, D., Freitag, D., McCallum, A., Mitchell, T., Nigam, K., Slattery, S.: Learning to construct knowledge bases from the World Wide Web. Artificial Intelligence 118, 69–113 (2005)
7. Davies, J., Studer, R., Warren, P.: Semantic Web technologies: trends and research in ontology-based systems. Wiley Publishing, West Sussex (2006)
8. Ehrig, M., Haase, P., Hefke, M., Stojanovic, N.: Similarity for Ontologies – A Comprehensive Framework. In: 13th European Conference on Information Systems (2009)
9. Geleijnse, G., Korst, J.: Automatic ontology population by googling. In: Proceedings of the Seventeenth Belgium-Netherlands Conference on Artificial Intelligence (BNAIC 2007), pp. 120–126 (2007)
10. Liu, H., Li, J., Wong, L.: A comparative study on feature selection and classification methods using gene expression profies and protein Patterns. In: Genome Informatics 2002: Proceedings of 13th International Conference on Genome Informatics, pp. 51–60. Universal Academy Press, Tokyo (2002)
11. Maedche, A., Staab, S.: Ontology Learning for the Semantic Web. Kluwer Academic Publishers, Dordrecht (2006)

12. McCallum, A., Nigam, K.: A comparison of event models for Naive Bayes text classification. In: Learning for Text Categorization: Papers from the AAAI Workshop, pp. 41–48. AAAI Press, Menlo Park (1998)
13. Motta, E., Vargas-vera, M., Domingue, J., Lanzoni, M., Stutt, A., Cirvegna, F.: MnM: Ontology deriven semi-automatic support for semantic markup. In: Proceedings of the 13th International Conference on Knowledge Engineering and Knowledge Management (EKAW 2007), pp. 379–391 (2007)
14. Yang, Y., Pedersen, J.O.: A comparative study on feature selection in text categorization. In: Proc. of the Fourteenth International Conference of Machine Learning, pp. 412–420 (2002)
15. Yang, Y., Zhang, J., Keseil, B.: A scalability analysis of classifiers in text categorization. In: Proceedings of SIGIR 2008, 26th ACM International Conference on Research and Development in Information Retrieval. ACM Press, New York (2003)

A Study on Vietnamese Prosody

Tang-Ho Lê, Anh-Viêt Nguyen, Hao Vinh Truong, Hien Van Bui, and Dung Lê

Abstract. In the current paper, the Vietnamese prosody was analyzed in order to produce a natural Vietnamese synthesizer. Word duration in texts recorded by various human voices was measured to test several hypotheses for logical segments of the utterances in a sentence. The sound files were analyzed and interpreted by several statistical calculations.

Keywords: Vietnamese, Text to Speech, TTS, Prosody, Ngữ diệu, Ngôn diệu.

1 Introduction

1.1 Naturalness of Voice Synthesizer

Two kinds of speech were distinguished: the natural speech produced by human and the synthesized speech produced by computer. Evidently, an ideally synthesized speech should be natural. This becomes a criterion for the evaluation of a voice synthesizer in the actual literal. In order to reach the naturalness for a synthesizer, the prosody was studied. The basic functions of prosody are to segment and to highlight. The domain variation of prosody extends beyond the phoneme into units of syllables, words, phrases, and sentences [3].

1.2 Prosody Definition and Particular Features of Vietnamese Prosody

In linguistics, prosody is the rhythm, stress, and intonation of connected speech. Prosody may reflect various features of the speaker or the emotional state of a speaker that may not be encoded by grammar or choice of vocabulary [2].

There were two separate steps in this study of Vietnamese prosody. In the first step (presented in this paper), we focused on words, phrases and words groups (usually regarded as "ngữ điệu" in Vietnamese). In essence, this study focused on the local effect of speech. In the later work, the prosody will be examined in the bigger units: the sentences, which involve the emotional expression ("ngôn điệu" in Vietnamese). This will focus on the global effect of speech.

There is a very important difference between the prosody of Vietnamese (as a monosyllabic and tone language) and the one of European languages (polysyllabic languages). The Vietnamese prosody is concerned with rhythm (word's duration) between words in a group of words or in a compound words, while raising the tone (intonation) by augmenting the amplitude and/or the frequency of all words has the global effect on the whole sentence (for example, in a question). The reason is that the Vietnamese speakers cannot change the intonation

N.T. Nguyen et al. (Eds.): New Challenges for Intelligent Information, SCI 351, pp. 63–73.
springerlink.com

(tone) of a word in order to highlight it, because each word has its own meaning thanks to one of the six accents. Moreover, some compound words (or phrases) must be pronounced without emission break between them (a very small silence lesser than 20 milliseconds is considered as no silence). The two phrases "sự mất-mát đáng tiếc" and "chúng ta đã mất đi một số thói quen" are as examples. If an utterance is correctly segmented, it is more natural and comprehensive. For example, the sentence "Học sinh học sinh học" (The student studies biology) cannot be spoken regularly because of the meaning variation. "Học sinh" is "the student" (a noun); "học" is "to study" (a verb); and "sinh học" is "the biology" (a noun). Therefore, a Vietnamese speaks this sentence with two breaks between those words: "Học sinh—học—sinh học".

Another prosody's feature of the Vietnamese is that there is no question to "make accent" on a syllable of a word (as in polysyllabic languages). Because each word has only one syllable, it will be of no sense if one pronounces a word stronger than other, except when one would express a special emotion; e.g. one can speak some words slower and stronger than other ones in a sentence to make them more important.

In order to segment an utterance naturally, a voice synthesizer must take into account the **syntactic feature**: whether the actual word is an article, a preposition or a conjunction; whether it is a compound word, or a word group. Section 3 below has different subtopics for each of these categories. In each subtopic, a hypothesis about the segmentation was presented, and tested through experimentation with sample sentences recorded by different speakers, most of whom were the Vietnamese television broadcasters.

1.3   Some Studies on Vietnamese Prosody

In the actual literal, some recent works and published papers studied the word's duration in Vietnamese. According to Trần [6], the duration of a Vietnamese phonetic unit depends on its position, its pitch, its emphasis and its structure. Other authors [5] stated that the duration of word depends on the structure of the phonetic unit, the silence duration between words, and the structure of the previous word's phonetic unit. Our study presented here examined also this matter with more aspects and quantified measurement.

2   Experimental Description

2.1   Selection of Speakers

For this study of Vietnamese prosody, the audio files were collected via the Internet from the Voice of Vietnamese Broadcasters in several provinces. These data samples include the voices of the north (Ha Noi city), the south (HoChiMinh city) and the central land (Hue city). The original video files were extracted into 36 wave audio files (.wav). There are 20,815 words and 20,779 silences within these audio files. Then, these words and silences were analyzed and measured.

We would like to insist that our test data is not these 36 files (they are not the objects for the test); instead, we aim at the content of these files which are more than 20,000 words' duration and silences between words. This number is very enough for the objectivity of the conclusion or the formulated rules.

2.2 Pre-processing

After collecting the data, the typical audio segments were selected for processing. Only those speak by broadcasters without interviews and without background noise were used, which helped improve the validity of the result.

Fig. 1. A noisy audio segment

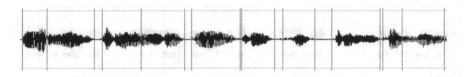

Fig. 2. A non noisy audio segment

2.3 Experiment

The beginning point and the ending point of each word on the spectrum (explained in the section 4 below) were defined. The calculation of the silence between words was automatically conducted with Microsoft Excel work sheets. Some macros in Excel were created to automatically analyze and total up the data in order to prove the hypotheses.

3 Prosodic Hypotheses

Like other languages, Vietnamese has parts of speech such as conjunctions, prepositions, articles, numerals, etc. Also, there are some kinds of phrase (or group of words) called compound words. The distinctive feature about intonation is that each word can be pronounced with two or six tones (*none, high-pitched, low-pitched, long-high-pitched, very-long-high-pitched, and short-high-pitched*). The five following hypotheses about Vietnamese prosody were established.

3.1   Conjunction and Preposition

- Conjunction: using to join the same functional words in a sentence, or to join the main cause with the subordinate cause. There are two types of conjunction: associate conjunction (và, với, hoặc, rồi, lẫn, cùng...) and dependent conjunction (vì, thì, mà, nếu, tuy, mặc dù, vả lại...) [4]

   - Preposition: using to join two words or two phrases having a main-subordinate relation; for example, "của" in "cuốn sách của tôi" (the book of me). Some Vietnamese prepositions are: bằng, bởi, cho, chung quanh, do, dưới, giữa, gần, khỏi, không, lên, qua, ra, thành, trên, tại, tận, tới, từ, vào, về, với, xuống, đến, ở [2].

Hypothesis no.1: **In Vietnamese, a conjunction/preposition and the word followed it must be pronounced without an emission break between them, excepting the preposition at the beginning of a sentence.** For example, "Tuy nhiên" (However) normally followed by a comma signifies an emission break.

   - Let's examine the sentence *"làm đẹp cho Huế, đồng thời tạo ra sân chơi mới cho giới mỹ thuật"*. In this sentence, "cho" is a preposition. Based on the measurement by Sound Forge 9.0c software, the followed silences are 1ms and 2ms, two relatively short breaks.

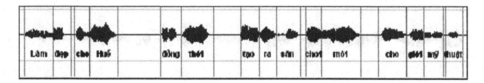

Fig. 3. The spectrum of the sentence with two prepositions ("cho")

3.2   Words Ending with Occlusive Syllables

Occlusive syllables are the phonetic units which terminate with one of the consonants p, t, c, ch (called stop consonants). This means the voice emission is stopped by a vocal physical element such as the lips, the tongue, and the glottis.

   These phonetic units' duration is naturally very short, and correspond only with the accents high-pitched (') and short-high-pitched (.). The latter has shorter duration than the former, e.g. "táp, tạp". (See more explanation in [1]).

Hypothesis no. 2a: **Consequently, all words ending with occlusive have shorter duration than the ones with other accents.** The following spectrum of a sentence is an example: "Ngày mốt là một tháng hai."

   In this example, the word "một" has the shortest duration in spite of different speakers.

Hypothesis no. 2b: **When pronouncing these words, the speakers automatically add a silence to regularize the speaking rhythm (to be in harmony with other words) because the length of those words is very short.**

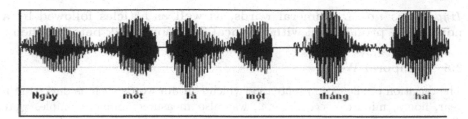

Fig. 4. The spectrum of the sentence
"Ngày mốt là một tháng hai"

3.3 Compound Words and Group of Words

Each Vietnamese word has only one phonetic unit (monosyllabic) and can be a simple word (meaningful) or an element of two or more combined words with a distinctive meaning. Compound words (từ kép) formed by two single words, and group of words (cụm từ) formed by more than two element words. For example, "tôi, bác, nhà, ngựa, cười, đẹp" are single words. "Chợ búa, bếp núc, xe cộ, sân bay, lấp ló, lóng lánh, xanh lè, đỏ rực" are compound words. Finally, "khít khìn khịt, sạch sành sanh, vội vội vàng vàng" are group of words.

Hypothesis no. 3: **To correctly pronounce a compound word or a group of words, speakers do not break the vocal emission between the element words** because this will influence on the meaning, i.e. one can have confusion or an unclear meaning. For example, the sentence "Trời thu *lạnh lẽo* nhưng trong lòng tôi không cảm thấy *lạnh* chút nào" has the following spectrum:

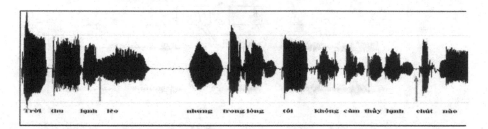

Fig. 5. The spectrum of
"Trời thu lạnh lẽo nhưng trong lòng tôi không cảm thấy lạnh chút nào"

The compound word "lạnh lẽo" is pronounced without silence between these two words, while the next single word "lạnh" has the followed silence before the word "chút".

3.4 Article and Numeral

An article precedes a noun; for example, "cái, con, cục, quyển, sợi, tấm, tờ, hạt, etc." A numeral also precedes a noun to quantify the number of units (e.g. "ba con chó – three dogs").

Hypothesis no. 4: **Numeral words, as well as articles followed by a noun, were pronounced without vocal emission break between them.**

3.5   Temporal Words

The duration of silences in sentences, which contain words such as day, month, year, hours, minutes, seconds, etc., was also measured. Some examples were illustrated:

"*Ngày hai mươi lăm tháng mười hai, ngày hôm qua, ngày mai*" (the twenty fifth of December, yesterday, tomorrow); "*năm nay, năm ngoái*" (this year, last year); "*sáu giờ mười phút*" (six o'clock and ten).

Hypothesis no. 5: **When pronouncing the temporal words, speakers do not break the vocal emission; the silence between them is usually very short.**

4   Measurement

The SoundForge 9.0c was used to define the beginning point (b_i) and the ending point (e_i) of the ith word in a sentence which had n words. By default, the number of silence in this sentence was n-1. The length (l) of ith word was $l_i = e_i - b_i$ and the silence (s) following ith word was $s_i = b_{i+1} - e_i$, where b_{i+1} was the beginning of the next word ($i+1^{th}$ word). For the words with si>200 miliseconds (ms), their silence was adjusted to equal to 200 ms to standardize the speech.

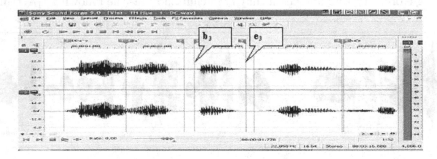

Fig. 6. Spectrum with beginning and ending points of words

After that, the data was transfered to Excel, which was able to automatically calculate the duration, based on the operators and functions established.

The total length of n words in each data file was $L = \sum_{i=1}^{n} l_i$

The total llength of n-1 silence in each data file was $S = \sum_{i=1}^{n-1} s_i$.

From that, the mean was calculated ($\overline{L}$), the standard deviation (StDevL) of the word's length as well as the mean ($\overline{S}$) and the standard deviation (StDevS) of the silence length:

$$\overline{L} = \frac{L}{n} \text{ and } StDev_L = \sqrt{(\frac{1}{n-1}) \sum_{i=1}^{n} (l_i - \overline{L})^2}$$

$$\overline{S} = \frac{S}{n-1} \text{ and } StDev_S = \sqrt{(\frac{1}{n-2}) \sum_{i=1}^{n-1} (s_i - \overline{S})^2}$$

By measuring the silence length which following a conjunction, a preposition, an article, a numeral and a temporal phrase, and then compare to $\overline{S}$ as well as the word's length of words ending with occlusives and short-high-pitched, then compare to $\overline{L}$, we can confirm the established hypotheses above.

5   Self-evaluation of the Current Experiment

The selected and recorded data file contained the voice of the well-trained broadcasters reading news and reports on the local and national broadcasting programs. Therefore, these voices represented the popular voices of Vietnamese, which ensured the correctness and objectivity for the research. The voices of major regions in VietNam were carefully selected including Ha-Noi capital, HCM city, Hue city, Da-Nang city, and Ca-Mau province.

In general, the calculated numbers from the experiment could sufficiently prove the prosody hypotheses. However, there were some exceptions where the broadcasters intentionally highlighted the speech's content, and caused the extension of some silence after the words with stop consonants. More time is necessary to analyze the data files. Further research will surely discover other rules of Vietnamese prosody.

6   Experimental Results

The numbers in table 1, the overall result below, were automatically calculated with Excel. The total number of words in the 36 prosody data files was 20,815 (the total number of silence was 20,815 - 36 = 20,779).

Table 2 indicates that there is not a vocal break (silence) after the conjunctions and prepositions, excepting after the compound conjunction at the beginning of sentence where the silence is equivalent to a comma punctuation mark (178 ms).

Table 3 indicated that the length of words with short-high-pitched & occlusive was generally short (143ms) compared with the mean of all words (184ms). Hence, when a sentence was synthesized by concatenating a word of this category with a followed word, if a short silence is not inserted after them, then the reading was a little odd, and unnatural. Indeed, the words of this category recorded in the sample files had very short length about 100ms to 120ms, because the vocal emission was completely closed. However, when speakers talk, the rhythm was regularized by naturally appending a short silence after them, which was very difficult to be recognized

Table 1. Analysis of silence after conjunctions, prepositions, words with occlusive and compound words

CONTENT	Sum/ Mean	CONTENT	Sum/ Mean
Words		**Compound & Normal words**	
Number of words	20,815	**Compound words**	
Total length of words	3,847.507	Number of silence	774
Mean of a word's length (L_{MEAN})	0.185	Total length of silence	17.893
Standard deviation	0.064	Mean of a silence length	0.023
Silence		Percentage compared to C_{MEAN}	47%
Number of silence	20,779	**Normal words**	
Total length of silence	1,011.617	Number of silence	118
Mean of a silence length (S_{MEAN})	0.049	Total length of silence	5.519
Standard deviation	0.062	Mean of a silence length (C_{MEAN})	0.047
Conjunctions & prepositions		**Other articles**	
Number of silence	2,119	Number of silence	275
Total length of silence	43.082	Total length of silence	5.747
Mean of a silence length	0.020	Mean of a silence length	0.021
Percentage compared to S_{MEAN}	42%	Standard deviation	0.019
Words with Occlusive		Percentage compared to S_{MEAN}	43%
Number of words	1,429		
Total length of words	204.696		
Mean of a word's length	0.143		
Percentage compared to L_{MEAN}	77%		

Table 2. Analysis of silence after conjunctions & prepositions

Measure of	Total number and mean of length	Conjunctions and Prepositions	Conjunctions at the beginning of sentence (e.g. tuy nhiên, tuy vậy, do đó, etc.)
Silence	20,779	2,119	9
Words	20,815	2,119	9
Mean of length	49 ms	20 ms	178 ms

Table 3. Analysis of words with occlusive and stop consonants

Measure of	Total number and other measures	Words with occlusive and stop consonants	Words with short-high-pitched and occlusive			
			"p"	"t"	"c"	"ch"
Words	20,815	1,429	184	517	650	78
Mean of length	184.8	143.2	150.9	151.1	136.5	128.9
Max. duration	565	329	324	329	304	305
Min duration	34	55	66	59	55	74

Table 4. Analysis of silence between compound and simple words

Measure of	Total/ Mean	Compound words	The same words used as simple words
Silence	20,779	774	118
Mean of silence length	49 ms	23 ms	47 ms

Table 5. The mean of the silence length of numerals and temporal words

Type of word	Notation	Mean (ms)
The mean of silence length of article words	Article	21
The mean of silence length of numerals	Num	23
The mean of silence length of temporal words	Temp	18
The mean of silence length in 36 data files	S_{MEAN}	49

Table 4 indicated that the silence between the elements of compound words varied around 23ms, 50% shorter comparing with the mean of all silence (49ms), which meant there were no vocal breaks between the element words. In the last column of this table, the same words used as simple words have the same followed silence as other words in the sentence.

Table 5 compared the mean of the silence length between article words, temporal words or numerals to the mean of the silence length between all words in the 36 data files. The former silence was shorter than the latter (>50%).

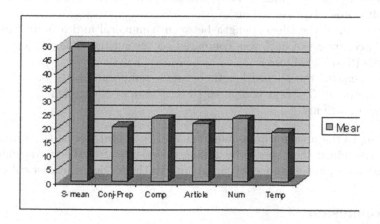

Fig. 7. The spectrum of "Statistical result chart"

Basing on the statistical results (6th column) in figure 7 above, the broadcasters were revealed to automatically adjust the silence length to regularize their speech.

Table 6. The percentage of the silence length between the numerals and the followed nouns of these words compared to the mean of the general silence length (S_{MEAN})

Measure of	không	một	hai	ba	bốn	năm	sáu	bảy	tám	chín	mười
Silence length (ms)	26	51	18	10	14	12	18	7	17	14	9
Compared to $_{SMEAN}$%	54	105	38	20	29	26	37	15	34	30	18

Table 7. The percentage of the silence length of these words within temporal phrases compared to the mean of the general silence length (S_{MEAN})

Measure of	không	một	hai	ba	bốn	năm	sáu	bảy	tám	chín	mười
Silence length (ms)	19	22	4	13	0	8	14	9	15	26	24
Compared to S_{MEAN} (%)	39	45	8	28	0	18	31	18	32	53	51

- The mean of the silence length following a conjunction or a preposition was 20 ms and accounting of 42% was compared to the mean of the silence length in their sentences ($S_{MEAN}=49$ ms)
- The mean of the word length having occlusive and short-high-pitched was 143 ms and accounting of 77% was compared to the mean of all word length ($L_{MEAN}=185$ ms).
- The mean of silence length between a compound words or noun phrase was 23 ms and accounting of 49% was compared to the mean of silence length of normal words ($C_{MEAN}=47$ ms).
- The mean of the silence length between articles words and a followed noun was 19 ms and accounting of 39% was compared to the mean of silence length of all words ($S_{MEAN}=49$ ms).
- The mean of the silence length between a numeral and a noun was 27 ms and accounting of 56% was compared to the mean of silence length of all words ($S_{MEAN}=49$ ms).
- The mean of the silence length between temporal words was 19 ms and accounting of 39% was compared to the mean of silence length of all words ($S_{MEAN}=49$ ms).

Regarding to the article words, numerals and temporal words, the speakers automatically adjust their speech by adding the silence after the words which have the short-high-pitched and stop consonant" in the following statistical tables:

7 Conclusion

Basing on the calculated experimental results on 36 sound files of 20.815 words read by the broadcasters from several distinctive regions in Vietnam, five established hypotheses for Vietnamese prosody were tested and proved as illustrated in table 8 and the comparative chart

By applying these five rules, a speech synthesizer can improve the rhythm of speech to be more natural. To do that, we must integrate a dictionary of

Table 8. Comparison of experimental results

Type of word	Notation	Mean (ms)
The mean of the silence length (S_{MEAN})	S_{MEAN}	49
Mean of the silence length of conjunction/ preposition	Conj-Prep	20
Mean of the silence length of compound words (C_{MEAN})	Comp	23
Mean of the silence length of article words	Article	21
Mean of the silence length of numerals	Num	23
Mean of the silence length of temporal words	Temp	18

articles, conjunctions, prepositions, compound words and group of words into the system. When the program scans the text, it can easily recognize them thank to this dictionary. Then, we have elaborated a sophistical code to reduce, in runtime, the duration of the first element word without influencing its quality. And finally, we added a small silence (21ms) after the ending element word, so that the reader clearly recognized that is a compound word or a group of words. For the temporal word phrase, the program can recognize them by a special algorithm, then add a silence (18ms) between these words.

8   Future Research

Further studies are necessary to understand and ascertain other rules related to the Vietnamese prosody (e.g. rules for rhythm of phrases, idioms, proverbs, etc.). The prosody feature (ngôn điệu) can be considered in the bigger units of the speech (sentences). For instance, the whole sentences involve the emotional expression introduced in the section 1.2 above. The results of this study may be efficiently applied to the second branch of the voice technology: the automated speech recognition, which is just at the initial phase in Vietnam.

References

1. Tang-Hồ, L.: Thử tìm một phương pháp hữu hiệu cho việc dạy đọc và viết tiếng Việt. Huong Viet Review, No. 11, Houston, USA (1999), Khảo cứu ngữ âm Việt Nam on `http://vietsciences1.free.fr/vietscience/vietnam/` `tiengviet/chuongtrinhtiengviet.htm`
2. Bách khoa từ điển. Wikipedia, `www.wikipedia.com`
3. O'Shaughnessey, D.: Speech Communications Human and Machine, 2nd edn. IEEE Press, Los Alamitos (2000)
4. Đức, H.N.: Free Vietnamese Dictionary Project, `http://www.informatik.uni-leipzig.de/~duc/Dict/`
5. Minh, L.H., et al.: Phân tích và tổng hợp đặc tính trường độ của tiếng Việt. Một số vấn đề chọn lọc của CNTT. Thái Nguyên, 29-31 tháng 8 năm (2003)
6. Đạt, T.D.: Analysis and Modeling of Syllable Duration for Vietnamese Speech Synthesis. In: O-COCOSDA 2007 (2007)

Towards Knowledge Acquisition with WiSENet

Dariusz Ceglarek[1], Konstanty Haniewicz[2], and Wojciech Rutkowski[3]

[1] Poznan School of Banking, Poland
dariusz.ceglarek@wsb.poznan.pl
[2] Poznan University of Economics, Poland
konstanty.haniewicz@ue.poznan.pl
[3] CIBER, Poland
wrutkowski@ciber.net

Abstract. This article is a continuation of research work started with an idea of semantic compression. As authors proved that semantic compression is viable concept for English, they decided to focus on potential applications. An algorithm is presented that employing WiSENet allows for knowledge acquisition with flexible rules that yield high precision results. Detailed discussion is given with description of devised algorithm, usage examples and results of experiments.

Keywords: semantic compression, semantic network, WiSENet, knowledge acquisition, natural language processing.

1 Introduction

The aim of this work is to present an application of previously introduced semantic network WiSENet (Wordnet transferred into SenecaNet format). Detailed discussion of various aspects and possible merits of WiSENet were enumerated in previous work [3]. Since earlier publications, developed semantic network has grown taking in account number of concepts. This was necessary action, as most of advanced operations that can be carried with WiSENet cannot function well without extensive concept vocabulary.

Taking into account, that some of readers may not be familiar with specifics of WiSENet a brief summary of its origin and capabilities is given.

To begin with, WiSENet derived its content from Wordnet. The decision was based on overall number of words and potential for further development and restructuring. The most important fact is that, authors had to dismantle synset structure and turn it into a graph where nodes represent concepts and vertices denote relation of hypernymy/hyponymy. This enabled devised algorithms to easily follow relations among particular concepts found in real life textual data. Restructuring was carried out in a lossless manner (algorithm is given in [3]). Additionally, WiSENet proved useful in combination with frequency dictionaries developed for a number of various domains. These frequency dictionaries allow for highly efficient disambiguation of concepts stored in WiSENet. To some point, frequency dictionary coupled with semantic network resembles human cognition

N.T. Nguyen et al. (Eds.): New Challenges for Intelligent Information, SCI 351, pp. 75–84.
springerlink.com   © Springer-Verlag Berlin Heidelberg 2011

when confronted with decisions concerning disambiguation. New structure aided by domain frequency dictionaries proved to work well, results of application of WiSENet to semantic compression for English were highly satisfactory.

Semantic compression is a process throughout which reduction of dimension space (used for indexing) occurs. The reduction entails some information loss, but in general it aims not to degrade quality of results thus every possible improvement is considered in terms of overall impact on the quality. Dimensions' reduction is performed by introduction of descriptors for viable terms. Descriptors are chosen to represent a set of synonyms or hyponyms in the processed passage. Decision is made taking into account relations among the terms and their frequency in context domain.

2   Motivating Scenario for Knowledge Acquisition

As mentioned earlier, it was observed that WiSENet lacks a great number of concepts that are to be met in various textual data. Those most impeding experiments are originating from general culture. Vast majority of identified missing concepts are proper names of various entities. For sake of clarification, by proper names authors understand names of people, organizations and various objects. Wordnet in general does not miss most general categories of entities, yet a lot of highly specialised concepts is not present. As Wordnet was not devised for text processing tasks previous statement is offered not as a criticism but as an observation.

Stating the above, authors decided to invest effort in expanding WiSENet. What is more important, this effort surpasses traditional methods of bulk import of all available resources and their later refactoring to match initial structure of to be extended semantic net.

It was discovered that WiSENet is very useful in discovering concepts that represent some specialisation of other concepts by employing specially prepared rules.

WiSENet can be applied to a set of procedures, that aim to extract information from some textual data. As is well know in the domain of text processing, one can manually prepare a set of rules that trigger when given order of elements is met. A great disadvantage to anyone who has to prepare this set of rules is that he is in need of specifying them in a manner that enumerates every plausible variant of a rule.

If one is to prepare a set of rules that enable him to retrieve information from the data, he should begin with investigation of domain. Let's assume, that the whole process should supply its invoker with new data on people that hold managerial positions at various companies. First of all, one should issue an recognisance query to a search engine of his choice, probing for terms than can denote a managerial position in some company.

It can be easily checked, that querying with search terms such as: chairman, CEO, chief executive officer, managing director, manager; shall bring results similar to following ones:

```
William (Bill) H. Gates is chairman of Microsoft Corporation
Richard K. Matros has been Chairman and CEO of Sun Healthcare
Larry Ellison has been CEO of Oracle Corporation
Jeffrey Epstein as its new chief financial officer
Brian McBride joined Amazon UK as Managing Director
Amazon CEO Jeff Bezos
said Bill McDermott, co-CEO of SAP
Mr Krishan Dhawan, Managing
Director of Oracle India Pvt. Ltd
```

One is ready to observe a vast number of possibilities when it comes to word order in researched material. Furthermore, the given list of search terms is far from completion.

Standard methods of local pattern matching dictate creation of rules that trigger when exact number of tokens of right characteristics is found. Apart from great effort investment spent on rule creation, they are prone to misfiring when slightest change of word order occurs.

Good examples of local pattern matching are regular expressions and text processing automata. While tremendous tools they might induce considerable effort when applied to information extraction. First of all, it was observed that regular expressions tend to fail in information retrieval task, not because their inefficiency but due to users being overwhelmed by their syntax. To exemplify above lets point out that, one has to be an experienced user to produce regular expression that will match more than 99% valid emails. As with practice comes experience, more important issue with regular expression ([10] demonstrated that regular expressions can be converted into non-deterministic finite automata) is its sensitivity to word order permutations.

When one is to consider grammars, one has to remember that they will have to face the challenge of an alphabet that is finite but actual number of symbols cannot be counted a priori. One has to process whole corpora to enumerate all alphabet's symbols. When processing a language such as English this can be troublesome, as there is no known boundaries of resources that should be processed.

Ideal solution to above mentioned issues, shall combine flexibility and ease of use. Flexibility shall be understood as ability to adapt to natural permutations in a word order of processed text. Ease of use shall make user exert the least amount of effort in formulation of his information needs.

3   Application of WiSENet

Coming back to introduced motivation scenario, one can easily observe that given results of recognisance query share common structure. This structure shall be treated as case analysis which leads to introduction of method designed by the authors to automate information retrieval in this specific task.

Every result contains some information on person, its position (managerial one) and some company. Whether there is a task to build a datastore of data

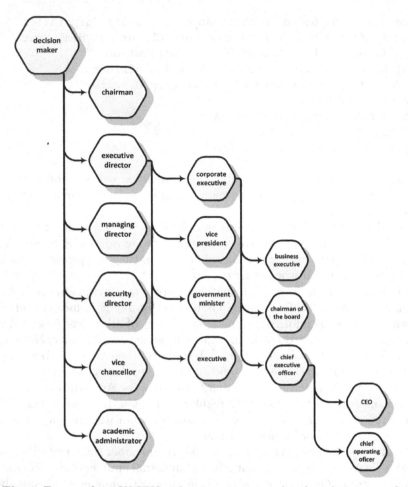

Fig. 1. Excerpt from WiSENet showing concepts related to *decision maker*

on managers in some kind of industry, a method that works with such high level query terms as executive, person and company name will be of tremendous help.

When one is to start with a corpus of some textual data, one can filter it through envisioned method and come up with elements that become candidates to extend current knowledge base. Found elements can be new relations among already stored data, or new more general/specific concepts directly in relation with existing ones. The whole process of acquisition of new concepts and relations bases on WiSENet. Effects of the process are reflected onto it, thus subsequent usage yields better result than previous ones.

WiSENet stores *corporate executive* as a concept. This concept has other concepts in relation, such as its hypernym and various hyponyms. A list of most important is given in figure 1.

Please observe that, given this rather short list, standard local pattern matching would need to include all of them in a OR-like statement or as separate instances of the rule. Using a method that allows for specifying a target concept as a entity from semantic network and informing the system implementing method that it should also take into account all hyponyms and a set of synonyms one is to encode query in a more verbose and flexible matter.

This initial functionality leads to a possibility of semi-automatic semantic network expansion that uses relations and concepts from the to be extended network to gather new concepts.

At this stage of research, one cannot imagine full automation of knowledge acquisition. Authors decided to optimize whole process taking into account two, most important at the moment, parameters. The first parameter is precision of results. Here, precision is treated as a number of text fragments fitting rule with minimum of matching errors. This is especially important as, in order to enhance WiSENet, results have to be evaluated before actual inclusion of newly discovered concepts. The second parameter is postulated flexibility of rule creation. As stated in earlier section, authors are interested in expressive rules that include actual concepts from the network, as this greatly speeds up rule creation. This comes at a price, as all rules have to be processed to enable postulated knowledge acquisition. Details of algorithm that performs actual rule to text data matching are given in separate section.

Exhaustive experiments led to a conclusion, that there has to be seed data to capture potential unknown concepts. By seed data a number of concepts fitting into given rule slot is understood. Key challenge ahead of authors is the bootstrapping nature of all experiments. One wants more exhaustive model. There is no easy way to sidestep investment of time and effort into this task. Ideally, one could use a fairly large body of skilled specialists that extended WiSENet by inclusion of data stored in various resources. Unfortunately, authors cannot afford this, thus methods devised strive for maximum output with minimum input. This can be true even if minimum input in some cases can be as much as 40000 of specific proper name concepts as seed data.

4   Algorithm for Matching Rules

Devised algorithms uses ideas already mentioned in previous publications. All operations are performed with WiSENet as semantic net. The first important step in algorithm is procedure that unwinds rule into all hyponyms stored inside the network. This operation can be of considerable cost in terms of execution as it has to traverse all possible routes from chosen concept to terminal nodes in the network. After completion a list of rules is obtained, listing every possible permutation of concepts from the network. To shorten processing time, one can specify number of levels that procedure shall descend in its course of execution.

Next phase of the algorithm is to step through textual data in order to find matches on computed rules. Stepping through is done by employing bag of concepts approach. Bag of concepts is implemented as a Finite State Automaton with advanced methods for triggering desired actions. At any state, it check whether any of the rules to be matched is completed. Discussion covering details of implementation is beyond the scope of this article. Nevertheless, it can be visualised as a frame passing through textual data. With every shift towards end of text fragment, concepts inside frame are used to check whether they trigger any of the rules obtained in the first phase. Size of the bag is chosen by researcher, yet performed experiments show that best results are obtained for a bag of size 8 to 12 when rules are 2 to 5 concepts long.

Bag of concepts is very important idea, as it tolerates mixins and concept order permutations. All matchings are performed after initial text processing phase is performed. Text processing phase consist of well known procedures such as applying stop list and term normalization.

A mixin is in this case a passage of text that serves some purpose to original text author, yet it separates two or more concepts that exist in one of the computed rules.

Consider following examples:

```
Rule - disease (all hyponyms), therapy (all hyponyms)

Match in: chemotherapy drug finish off remaining cancer
Matched concepts: therapy -> chemotherapy, disease -> cancer
Mixin: drug finish off remaining

Match in: gene therapy development lymphoma say woods
Matched concepts: therapy -> gene therapy, disease -> lymphoma
Mixin: development

Match in: cancer by-bid using surgery chemotherapy
Matched concepts: therapy -> chemotherapy, disease -> cancer
Mixin: by-bid using surgery
```

Examples are taken from one of the experiments performed with biology corpus. It can be observed, that bag of concepts performs well in various cases, it handles long mixins and concept permutation. Additional observation shall be made as concepts being hyponyms to those in the original example rule were matched (as referenced earlier).

All experiments performed took into account possibility of matching more than single rule. Thus a mechanism for triggering a set of rules was devised and was signaled earlier along with bag of concepts.

Procedure matching rules holds internal registers, that store rules that are actively valid with given bag of concepts. To give an example, please consider a set of three rules:

rule 1: university, city (all hyponyms)
rule 2: university, city (all hyponyms), country (all hyponyms)
rule 3: person (all hyponyms), academic

Given exemplary text fragment: SVEN LARS CASPERSEN, Professor of Economics, President of the World Rectors Association, Rector of Aalborg University (Denmark) (1999)

Procedure shall match and matches previously defined rules:

rule number 1 with university → university, city → Aalborg
rule number 2 with university → university, city → Aalborg, country → Denmark
rule number 3 with person → Sven Lars Caspersen, academic → professor

When a complete rule or its part (one can decide whether he is interested in total matches all partial ones) is mapped, it is presented to user to accept match or reject it. When bag of concepts drop earlier concepts and is filled with new ones, rules that were not matched are dropped from register of valid rules.

Algorithm in pseudocode is presented in listing 1

Algorithm 1. Algorithm for matching rules using WiSENet and bag of concepts

SN − *Semantic Network*
R − *semantic relation pattern*

```
//attach rule triggers to concepts in semantic network
mapRulesToSemNet(SN, R[])
for all Rule ∈ R do
   for all Word, Relations ∈ Rule do
      N = SN.getNeighbourhood(Word, Relations)
      for all Word ∈ N do
         SN.createRuleTrigger(Word, Rule)
      end for
   end for
end for
 //text processing: tokenization, phrases, stop list
T = analyzeText(Input)
foreachWordinT
if count(Bag) = size(Bag) then
   //first, deactivate rules hits for a word
   //that drops out from bag of words
   oldWord = pop(Bag)
end if
for all Rule ∈ SN.getTriggers(oldWord) do
   Rule.unhit(Word)
   push(Bag, Word)
   for all Rule ∈ SN.getTriggers(Word) do
      //take all relevant rules and activate word hit
      Rule.hit(Word)
      if Rule.hitCount = Rule.hitRequired then
         //report bag of words when hits reaches required number
         report(Rule, Bag)
      end if
   end for
end for
```

---

5   Experiment

Devised algorithm was used to perform an experiment on biology related data.
Test corpus consisted on 2589 documents. A total number of words in documents
was over 9 million. Authors decided to search for specialists and their affiliations.
This converges with motivating scenario, as WiSENet was enriched by both

Table 1. Sample results of experiments with rules based on WiSENet on corpus of
biology related documents. Discovered concepts are written under matches.

text fragment	match/discovered concept	rule
explain senior author Douglas Smith Md professor department neurosurgery director	Douglas professor department **Douglas Smith**	5
Feb proceedings national academy of sciences researcher University of Illinois entomology professor Charles Whitfield postdoctoral	University of Illinois professor Charles **Charles Whitfield**	1
design function biological network she-bop visiting professor Harvard University Robert Dicke fellow visiting	professor Harvard University Robert **Robert Dicke**	1
modify bacteria Thomas Wood professor –Artie– McFerrin– department chemical engineering have	–Thomas professor department **Thomas Wood**	5
Matthew –Meyerson– professor pathology Dana –Farber– cancer institute senior associate	Matthew professor institute **Matthew Meyerson**	2
an assistant professor medical oncology Dana –Farber– cancer institute researcher broad assistant	professor Dana institute **Dana Farber**	2
sun mat professor emeritus Robert –Hodson– all university Georgia Robert Edwards	professor Robert university **Robert Hodson**	1
vacuole David Russell professor molecular microbiology –Cornell's– college veterinary medicine colleague	David professor college **David Russell**	4
resistant cell professor Peter –Sadler– chairman chemistry department University of Warwick lead research project	professor Peter University of Warwick **Peter Sadler**	1
said first author Quyen Nguyen doctorate assistant professor surgery si tan University of California San Diego school of medicine	Nguyen assistant professor University of California **Quyen Nguyen**	1
scientist –Sirtris– co-author founder prof David –Sinclair– Harvard Medical School published consecutive	prof David Harvard Medical School **David Sinclair**	1

specialists (and their fields of interest), universities, institutes and research centres. Experiment used following rules:

rule 1 first name (all hyponyms), professor (all hyponyms), university (all hyponyms)

rule 2 first name (all hyponyms), professor (all hyponyms), institute (all hyponyms)

rule 3 first name (all hyponyms), professor (all hyponyms), research center (all hyponyms)

rule 4 first name (all hyponyms), professor (all hyponyms), department (all hyponyms)

rule 5 first name (all hyponyms), professor (all hyponyms), college (all hyponyms)

Size of bag of concepts was set at 8 elements. Additionally, all rules were to match exactly all concepts.

Out of 1326 documents where concept professor was found, prepared rules matched 445 text fragments. This gives as recall rate of 33,56%. Precision of results was 84,56%. This level is found to be very satisfactory, especially taking into account that due to algorithm nature there can be duplicates of matched text fragments (due to multiple triggering of rules inside current bag of concepts).

In addition, experiment resulted in 471 concepts that were previously unknown to WiSENet. Context and type of rules that matched text fragments led to extremely efficient updates of the network.

Table 1 demonstrates sample results. Please notice, that match on its own does not discover new concepts. Rules present potential fragments that with high likelihood contain new concepts that can be included into semantic network.

6   Conclusions and Future Work

Work presented in this article continues research efforts started with presentation of SEIPro2S [1]. After realising vision of semantic compression for English and presenting results, authors decided to focus on various applications.

Rules created with WiSENet are interesting application, that has great potential for future development, as it helps to expand body of knowledge represented by WiSENet. Experiments performed with devised algorithm for rule matching showed that envisioned flexibility and precision are available.

As reported in the experiment section, due to reasonably high precision on results, unknown concepts can be easily added, thus realising vision of knowledge acquisition with WiseNet.

Future work will focus on further addition of previously unknown concepts to WiSENet along with restructuring of relations among them. Authors believe that there are even more useful applications of semantic compression and plan to experiment with them and share experiments' results.

References

1. Ceglarek, D., Haniewicz, K., Rutkowski, W.: Semantically enhanced intellectual property protection system - sEIPro2S. In: Nguyen, N.T., Kowalczyk, R., Chen, S.-M. (eds.) ICCCI 2009. LNCS, vol. 5796, pp. 449–459. Springer, Heidelberg (2009)
2. Ceglarek, D., Haniewicz, K., Rutkowski, W.: Semantic compression for specialised information retrieval systems. In: Nguyen, N.T., Katarzyniak, R., Chen, S.-M. (eds.) Advances in Intelligent Information and Database Systems. SCI, vol. 283, pp. 111–121. Springer, Heidelberg (2010)
3. Ceglarek, D., Haniewicz, K., Rutkowski, W.: Quality of semantic compression in classification. In: Pan, J.-S., Chen, S.-M., Nguyen, N.T. (eds.) ICCCI 2010. LNCS, vol. 6421, pp. 162–171. Springer, Heidelberg (2010)
4. Gonzalo, J., et al.: Indexing with WordNet Synsets can improve Text Retrieval (1998)
5. Hotho, A., Staab, S., Stumme, G.: Explaining text clustering results using semantic structures. In: Lavrač, N., Gamberger, D., Todorovski, L., Blockeel, H. (eds.) PKDD 2003. LNCS (LNAI), vol. 2838, pp. 217–228. Springer, Heidelberg (2003)
6. Hotho, A., Maedche, A., Staab, S.: Ontology-based Text Document Clustering. In: Proceedings of the Conference on Intelligent Information Systems, Zakopane. Physica/Springer, Heidelberg (2003)
7. Khan, L., McLeod, D., Hovy, E.: Retrieval effectiveness of an ontology-based model for information selection (2004)
8. Krovetz, R., Croft, W.B.: Lexical Ambiguity and Information Retrieval (1992)
9. Frakes, W.B., Baeza-Yates, R.: Information Retrieval: Data Structures and Algorithms. Prentice-Hall, Englewood Cliffs (1992)
10. McNaughton, R., Yamada, H.: Regular expressions and state graphs for automata. IRE Transactions on Electronic Computers EC-9(1), 39–47 (1960)
11. Fellbaum, C.: WordNet - An Electronic Lexical Database. MIT Press, Cambridge (1998), ISBN:978-0-262-06197-1
12. Zellig, H.: Distributional Structure. Word 10(2/3), 146–162 (1954)
13. Califf, M.E., Mooney, R.J.: Bottom-up relational learning of pattern matching rules for information extraction. J. Mach. Learn. Res. 4, 177–210 (2003)
14. Percova, N.N.: On the types of semantic compression of text. In: COLING 1982 Proceedings of the 9th Conference on Computational Linguistics, Praha, vol. 2 (1982)

Part II
Data Mining and Computational Intelligence

An Approximation Approach for a Real–World Variant of Vehicle Routing Problem

Khoa Trinh, Nguyen Dang, and Tien Dinh

Faculty of Information Technology, University of Science, VNU-HCMC,
227 Nguyen Van Cu, Ho Chi Minh City, Vietnam
ttdkhoa@acm.org, {dttnguyen,dbtien}@fit.hcmus.edu.vn

Abstract. In this paper, we study a real-world variant of the Vehicle Routing Problem, arising from the work of distributing products of an industrial corporation. Given information about vehicles, depots, types of products, and the location of customers, we wish to determine the minimum number of vehicles required to pick up and deliver products to customers within a time limit. Using these vehicles, we are interested in a routing plan that minimizes the total travelling time. Our proposed solution is presented as follows: First, we introduce an exact algorithm using dynamic programming; Second, by relaxing dynamic programming formulation, we give a heuristic algorithm; Third, we provide our experimental results of the heuristic algorithm on real-world datasets provided by the corporation and highlight the effectiveness of the proposed algorithm.

Keywords: Vehicle routing and scheduling, transportation, dynamic programming, heuristics.

1 Introduction

The Vehicle Routing Problem (VRP), which is first proposed by Dantzig and Ramser [9] in 1950s, has proved to be practically useful. In particular, the VRP has a wide variety of applications in transportation and distribution industries [10]. The VRP is long known to be **NP**-hard [8]. In the last fifty years, many different methods have been developed to solve the VRP and its variants, including exact and heuristic algorithms. Laporte et al. published a detailed survey [12] of different approaches for VRP. Dynamic Programming (DP) is among the class of exact algorithms. The use of DP for the classic VRP can be dated back to the work by Eilon, Watson-Gandy and Christofides in [5]. In [3], the authors introduced a method called state-space relaxation which helps prune the state-space associated with the DP formulation, providing a lower bound on optimal solutions for the Capacitated Vehicle Routing Problem. More recently, in [1,4], DP is adopted to give approximate solution to the VRP with stochastic demands. The authors in [2] introduce a DP approach for a variant of Vehicle Routing Problem with Time Windows. However, in most of the real-world case studies,

N.T. Nguyen et al. (Eds.): New Challenges for Intelligent Information, SCI 351, pp. 87–96.
springerlink.com   © Springer-Verlag Berlin Heidelberg 2011

heuristic methods are widely used due to their running time efficiency in solving large-scale problem instances while offering quite good approximate solutions.

In this paper, we consider a real-world variant of the VRP, motivated from the work of distributing dairy products by Vinamilk – the biggest dairy corporation in Vietnam. They have a fleet of 23 vehicles, in which each vehicle has its own capacity. There are two depots where dairy products will be loaded into vehicles. Vinamilk offers two main types of dairy products: milk and ice-cream. Each day, the corporation needs to serve between 100 and 1500 requests from customers in the city. In this problem, each vehicle starts from a garage, comes to depots several times to load products, and delivers them to customers until either all customers are served or the maximum operating time is reached. This means that a vehicle can take more than one route. The main objective is to minimize the required number of vehicles. If there are more than one routing plan using that number of vehicles, we want to minimize the total travelling time of all vehicles.

Moreover, due to the space limitation of the depots, there are only three available parking slots in each depot. In other words, the number of vehicles loading products simultaneously could not be more than three. This suggests a kind of scheduling problem, in which we need to find the order of vehicles entering the depots. A combination of the VRP and scheduling problem like this one often happens in practice. However, to the best of our knowledge, it is just mentioned only one time in [11], in which the number of depots, as well as the number of available parking slots, is fixed to one and the adopted method is the Ant Colony metaheuristics.

2 Problem Formulation

In this section, we will formulate the problem rigorously. Let $N = \{1, 2, \cdots, n\}$, $D = \{1, 2\}$ and $P = \{1, 2\}$ be the sets of customers, depots, and types of products respectively. The customer i requests a quantity q_i of a specific product p_i from some depot d_i. We are given a fleet $F = \{1, 2, \cdots, m\}$ of vehicles, in which the vehicle v can load the product p with the maximum capacity of $C_{v,p}$ (number of boxes.) All vehicles, having the same maximum operating time t_{max}, start at a garage g_0, deliver commodities to customers in several routes and finally go back to this garage. For simplicity of later implementation, we consider g_0 a special customer, requesting a zero quantity of every product type.

In each route, a vehicle needs to go to a depot, and picks up a quantity q of some type of product, which must not exceed its maximum capacity for that type. Note here that, as a constraint of the problem, each vehicle can only carry exactly one type of product in a route. The vehicle then delivers all of the product to appropriate customers before starting a new route. There is a fixed rate of loading and unloading commodity, namely r_{load} and r_{unload}. This means that it takes qr_{load} (or qr_{unload}) time unit to load (or unload) a quantity q of commodity. Besides, there are only n_s "slots," numbered from 1 to n_s, for every depot so that at most n_s vehicles can load products simultaneously at any depot.

We are also given a complete graph $G = (V, E)$, where $V = \{g_0\} \cup N \cup D$. For every pair of vertices $i, j \in V$, we know the travelling time $A_{i,j}$ from i to j. Our objective is to find the minimum number of given vehicles that can satisfy the demand of all customers, breaking tie by minimizing the total travelling time of all vehicles.

3   An Exact Algorithm Using Dynamic Programming

In this algorithm, we iterate through every subset $F' \subseteq F$ of given vehicles and try to find out the minimum total travelling time in order to satisfy all customers by using these vehicles, if possible. For each F', we then break the original problem into many subproblems, which can be characterized by a state. Each state is a 4-tuple $\langle U, l, t, T \rangle$ where

1. $U \subseteq N$ is the set of customers who are to be served.
2. l is a vector containing current locations of all vehicles.
3. t is a vector containing time used by all vehicles.
4. T is a $|D| \times n_s$ matrix, in which $T_{i,j}$ indicates the next free time point of slot j at depot i.

Let $f(U, l, t, T)$ be the minimum total travelling time for serving customers in the set U, given information about locations, time used by vehicles, and the status of slots in depots. Basically, given that l_0 contains g_0's only and there are no elements of t_0 greater than t_{max}, we have $f(\emptyset, l_0, t_0, T_0) = 0$, since there are no customers in this cases, and all vehicles are at garage.

When $U \neq \emptyset$, it is clear that an optimal solution to this subproblem consists of routes of all vehicles. In other words, this problem can be considered a multistage decision problem, in which, at any stage, we decide a route for a vehicle. To be more precise, let $\Re_{i,j} = \{r_{i_1}, r_{i_2}, \cdots \}$ be the set of routes which starts from the slot j at the depot i, in some optimal solution. Let us choose some $\Re_{i',j'} \neq \emptyset$ and a route $r_{opt} \in \Re_{i',j'}$ by some vehicle v with the earliest leaving time. Let X_{opt} be the set of customers who are in the route r_{opt}. According to Bellman's principle of optimality [7], the set remaining routes $\bigcup \Re_{i,j} \setminus r_{opt}$ must form an optimal solution for a smaller subproblem of serving $U \setminus X_{opt}$ customers with the current position of v being the last customer in r_0, the time used by v being the sum of t_v and time spent for the route r_{opt}, and the next free time point $T_{i',j'}$ being the time point when v leaves the depot.

Note that r_{opt} is indeed the shortest Hamiltonian path from some depot d to the last customer in the subgraph $G[X \cup \{d\}]$. Otherwise, we could find a shorter route, which reduces the total travelling time. Since we do not know X_{opt}, we iterate through every nonempty subset X of U, recursively solve the smaller subproblem on $U \setminus X$, combine the cost of this choice with the optimal value of the subproblem, and look for the minimum total cost.

In summary, the dynamic programming formulation is

$$f(U,l,t,T) = \min_{\substack{X \subseteq U, i \in X \cup \{g_0\}, v \in F' \\ d \in D, p \in P, 1 \leq s \leq n_s}} \{f(U \setminus X, l', t', T') + A_{l_v,d} + g(G[X \cup \{d\}], d, i)\},$$

$$(1)$$

with the constraints:

$$\forall i \in X : d_i = d, p_i = p, \tag{2}$$

$$\sum_{j \in X} q_j \leq C_{v,p}, \tag{3}$$

$$t'_v \leq t_{max}. \tag{4}$$

where,

- The function $g(G[X \cup \{d\}], d, i)$ returns the shortest Hamiltonian path from d to i in the induced subgraph $G[X \cup \{d\}]$, where d is the depot that the vehicle started and i is the last customer has been served in the route. This is a classical NP-complete problem, which can also be solved by a slightly modified dynamic programming approach in [6].
- The vectors l', t' and T' take the old values of l, t and T respectively, except the following elements. The new location of v is i. The new time used of v is time point $\max\{A_{l_v,d} + t_v, T_{d,s}\}$, when it was ready load products, plus the total time spent for loading and unloading products, and the travelling time in the route. Similarly, the new next free time point of the slot s at depot d, when v leaves the depot, should be the sum of the time point $\max\{A_{l_v,d} + t_v, T_{d,s}\}$ and the time spent for loading products.

$$l'_v = i, \tag{5}$$

$$t'_v = \max\{A_{l_v,d} + t_v, T_{d,s}\} + (r_{load} + r_{unload}) \sum_{j \in X} q_j + g(G[X \cup \{d\}], s, i), \tag{6}$$

$$T'_{d,s} = \max\{A_{l_v,d} + t_v, T_{d,s}\} + r_{load} \sum_{j \in X} q_j. \tag{7}$$

Note here that the constraint (2) makes sure that every customer in X receives their desired commodity from the right depot. Also (3) and (4) are constraints about the maximum capacity and operating time of vehicles. The solution to the original problem, given a fleet F', is $f(N, 0_{|F'| \times 1}, 0_{|F'| \times 1}, 0_{|D| \times n_s})$.

Although this dynamic programming formulation provides optimal solution to the problem, it is not practical to implement this algorithm in practice since the number of states is exponential to the input size. Moreover, computing the function f for just one state by (1) also requires the access to an exponential number of other states. This phenomenon is known as the "curse of dimensionality." In the next section, we propose a heuristic algorithm based on relaxation of (1).

4   A Heuristic Algorithm

4.1   Description

Since the size of F may be quite large, considering every subset of F, as in the previous section, is not practical. Our approach is to gradually build up a fleet F', and try to satisfy as much customers' demand as possible using this fleet, which we will discuss in the following paragraphs. If F' cannot deliver all of the products, we then greedily add another available vehicle v into F' such that it is capable to carry the largest quantity of products requested by the remaining not served customers, and repeat the process.

Now, assuming that we are given a set of vehicles, we want to plan routes in order to minimize the total remaining demand from customers. Using similar idea as in the above exact algorithm, we start from the initial state $\langle N, 0_{|F'| \times 1}, 0_{|F'| \times 1}, 0_{|D| \times n_s} \rangle$, trying to add a new route in each iteration. Here, we are only interested in "potentially good" states whose ratio of total travelling time by all vehicles and total delivered quantity of products is as small as possible. Intuitively, this ratio is the average time to deliver a unit of product and it tells us how efficient all vehicles has deliver products to the customers so far.

For adding a new route, we first pick up from the current state $\langle U, l, t, T \rangle$ the vehicle v which has the minimum time used t_v, the product p and depot d which maximize the total remaining quantity $\min\{C_{v,p}, \sum_{i \in U, p_i = p, d_i = d} q_i\}$ that v can load, and the slot s at the depot d with the minimum time difference $|T_{d,s} - (t_v + A_{l_v,d})|$. Starting from the depot d with the vehicle v, we construct a new route by using the simple Breadth First Search on customers who requested the product p.

In the same manner as defining states, we characterize each tentative route by a set of customers R in the route, the current position i of the vehicle on the route, the total travelling time t so far and the total quantity q already delivered. Let us denote a tentative route by $\langle R, i, t, q \rangle$. We maintain a queue Q of such routes. We try to extend routes in Q, if possible, and push the new routes into Q'. Next, we assign $Q \leftarrow Q'$, and repeatedly extend routes in Q. Indeed, the size of Q would grow exponentially. Therefore, aiming at a "good" state, we prune Q and only keep at most α routes with the smallest ratio $\frac{t}{q}$. In this paper, we choose $\alpha = O(n)$, and we will provide experimental results on some specific values of α in the next section. For routes which cannot be extended anymore, we push them into a list called $Q_{candidates}$. Finally, we update the current state by adding the route in $Q_{candidates}$ such that the above ratio is minimum.

Fig. 1. illustrates the search tree, in which the root is the depot d. A tentative route is a path from the root to some node. In each iteration, the queue Q contains all nodes which have the same height. All white nodes are pruned and we continue searching from $\alpha = 4$ marked nodes, which indicate routes with the smallest ratio $\frac{t}{q}$.

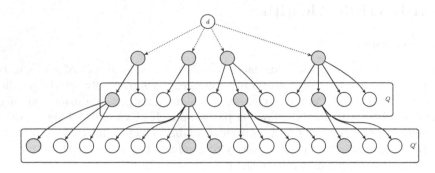

Fig. 1. Illustration of searching for new routes

Assuming that we already chose a vehicle v, the following procedure FIND–ROUTE takes input as the current state and returns the list of routes in $Q_{candidates}$ described above.

Procedure 1. FIND–ROUTE($\langle U, l, t, T \rangle$, v)

1: Choose the product p and the depot d which maximize the total remaining quantity
   $\min\{C_{v,p}, \sum_{i \in U, p_i=p, d_i=d} q_i\}$ that v can load
2: Choose the slot s at the depot d with the minimum time difference $|T_{d,s} - (t_v + A_{l_v,d})|$
3: Push an initial route $\langle \emptyset, d, A_{l_v,d}, 0 \rangle$ into the queue Q
4: $Q_{candidates} \leftarrow \emptyset$
5: **while** $Q \neq \emptyset$ **do**
6:      $Q' \leftarrow \emptyset$
7:      **for all** $\langle R, i, t, q \rangle \in Q$ **do**
8:          $Q'' \leftarrow \emptyset$
9:          **for all** $j \in U \setminus R$ such that $d_j = d, p_j = p$ and $q_j + q \leq C_{v,p}$ **do**
10:              **if** $\max\{T_{d,s}, t_v + c_{l_v,d}\} + (q_j + q)(r_{load} + r_{unload}) + A_{i,j} + A_{j,g_0} \leq t_{max}$
                 then
11:                  Push $\langle R \cup \{j\}, i, t + A_{i,j}, q + q_j \rangle$ into Q''
12:          **if** $Q'' = \emptyset$ **then**
13:              Push $\langle R, i, t, q \rangle$ into $Q_{candidates}$
14:          **else**
15:              Push all tentative routes in Q'' into Q'
16:      Sort all tentative routes in Q' increasing according to the ratio of travelling time used and total amount of commodity delivered.
17:      Push the first α routes in Q' into Q
18: **return** $Q_{candidates}$

The pseudo-code of the heuristic algorithm is shown as follows.

Algorithm 2. HEURISTIC(α)

1: Input $\leftarrow N, D, P, F, G = (V, E), C, r_{load}, r_{unload}, g_0, t_{max}, n_s$
2: $F' = \emptyset, U \leftarrow N$
3: **repeat**
4:     **if** $F = \emptyset$ **then**
5:         **return** NO SOLUTION
6:     Choose a vehicle $v \in F$ such that v is capable to carry the largest quantity of commodity from the remaining customers U
7:     $F' \leftarrow F' \cup \{v\}, F \leftarrow F \setminus \{v\}$
8:     Initialize the state $\langle U, l, t, T \rangle$: $U \leftarrow N$, every vehicles are now at the garage g_0 with $t \leftarrow 0_{|F'| \times 1}, T \leftarrow 0_{|D| \times n_s}$ and the cost $f \leftarrow 0$
9:     **repeat**
10:         Choose the vehicle $v \in F'$ with the minimum t_v
11:         $Q_{candidates} \leftarrow$ FIND–ROUTE($\langle U, l, t, T \rangle, v$)
12:         **if** $Q_{candidates} \neq \emptyset$ **then**
13:             Choose the route $\langle R_0, i_0, t_0, q_0 \rangle$ in $Q_{candidates}$ with the minimum ratio $\frac{t_0}{q_0}$
14:             $f \leftarrow f + t_0, U \leftarrow U \setminus R_0, l_v \leftarrow i_0$
15:             $t_v \leftarrow \max\{T_{d,s}, t_v + A_{l_v,d}\} + q_0(r_{load} + r_{unload}) + t_0$
16:             $T_{d,s} \leftarrow \max\{T_{d,s}, t_v + A_{l_v,d}\} + q_0 r_{load}$
17:     **until** $Q_{candidates} = \emptyset$
18: **until** $U = \emptyset$
19: **for all** vehicle $v \in F'$ **do**
20:     $f \leftarrow f + A_{l_v, g_0}$
21: **return** $\langle F', f \rangle$

4.2   Running Time Analysis

In the procedure FIND–ROUTE, the queue Q contains at most $\alpha = O(n)$ routes so that the two nested for-do loops (lines 7 and 9) run in time $O(n^2)$. After that, sorting all routes in Q' costs $O(n^2 \log n)$ since the size of Q' could contain at most n^2 tentative routes. For each iteration of the while-do loop in line 5, the length of each route increases by one, and every route never contains any vertex more than once, making this loop iterate no more than n times. Therefore, the running time of the FIND–ROUTE subroutine is $O(n^3 \log n)$.

In the main algorithm, the loop in line 3 iterates at most m times since each time the cardinality of F decreases by one. In lines 9..17, after each iteration, we add a new route into our solution and call the procedure FIND–ROUTE. We have indeed the upper-bound of the number of routes is $O(n)$. In summary, the time complexity of the heuristic algorithm is $O(mn^4 \log n)$.

5   Experimental Results

In this section, we provide the experimental results of our heuristic algorithm on datasets provided by Vinamilk Corporation, which will be briefly described as follows. Each dataset records a real-life instance of the problem in one business day, and the maximum operating time of all vehicles t_{max} is equal to 420 minutes (or 7 hours.) There are two types of products, and the number of requests ranges from around 100 to 1500. In addition, the corporation has a fixed fleet of exactly 23 vehicles with different capacity.

Our algorithm is implemented in C++ (compiler GCC 3.4.2) and tested on a personal computer with the Intel Quad-core 2.33GHz processor and 4GB RAM. In the experiment, we choose a specific value $\alpha = 100$ for all test cases. Generally, if α is too small, say less than 20, it may require a larger number of vehicles. On the other hand, if we choose $\alpha > n$, the running time increases, but the total travelling time could decrease. The detailed results are shown in Table 1.

Table 1. Detailed results of the heuristic algorithm

Instance	No. of Customers	Total Demand	No. of Vehicles	Travelling Time (mins)	Running Time (secs)	No. of Vehicles used by Vinamilk
FEB01	821	9861	13	2676	40	20
FEB02	573	7320	10	2200	12	19
FEB03	658	9977	11	2650	22	11
FEB04	582	8376	10	1941	16	20
FEB05	756	9891	12	2337	26	22
FEB06	602	8949	11	2325	18	21
FEB08	939	14009	17	3348	76	20
FEB09	542	11283	12	2499	18	19
FEB10	720	9854	12	2382	27	19
FEB11	482	13379	13	2740	8	16
FEB18	363	5583	7	1557	3	15
FEB19	819	11597	14	2837	69	21
FEB20	510	9561	10	2177	13	23
FEB22*	1513	16483	23	4655	312	23
FEB23	727	13827	17	3306	111	18
FEB24	726	10395	12	2434	52	19
FEB25	678	8464	11	2153	19	14
FEB26	760	10515	12	2341	75	23
FEB27	607	7018	9	1916	21	18

These instances are taken from the work of distributing dairy products of the corporation in February, 2010. We skip several instances which have relatively small number of requests from the customers. The second and third columns describe the size of the input. The next two columns show the output of our algorithm. The last column describes the number of vehicles used by Vinamilk. Currently, the corporation has a team of two people who solves them by hand using greedy techniques and some experienced–based heuristics. Note that the instance FEB22 is very large, in which the corporation had to rent extra vehicles. In most cases, our method can give a solution which uses much smaller number of vehicles. On average, our proposed algorithm helps reduce the required number of vehicles by 33.9%.

We already developed a system which solves and visualizes the planning routes. Fig. 2. shows four routes, which are part of the solution for FEB06, on the real map of Ho Chi Minh City. Customers are numbered increasingly according to the visited order in each route.

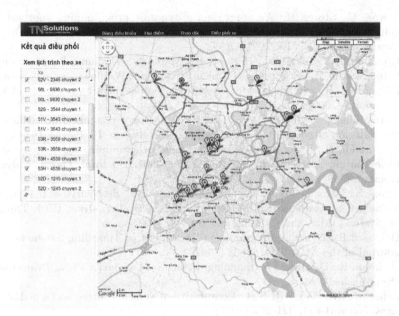

Fig. 2. Illustration of four routes (FEB06) on the real map

6   Conclusions and Future Work

In this work, we have presented a heuristic algorithm to solve a real-world variant of the VRP. In this algorithm, we restrict the state-space, and the solution is built up gradually by adding new routes. The algorithm is then validated experimentally on datasets provided by an industrial corporation. Our experiments show that the proposed method can produce a much smaller required number of vehicles than the current greedy method adopted by the above-mentioned corporation. The running time of the algorithm is also acceptable. Further investigation of metaheuristics on this problem is left open at this point. The literature suggests that metaheuristics such as local search is quite promising to approximate the minimum needed number of vehicles in such a combinatorial optimization problem.

Acknowledgements. We would like to acknowledge the support of Vinamilk Corporation in providing us real data in Feburary 2010 for experimentation and comparison.

This work is a part of the 217/HD-SKHCN project supported by the Department of Science and Technology, Ho Chi Minh City, 2010-2011.

References

1. Novoa, C., Storer, R.: An approximate dynamic programming approach for the vehicle routing problem with stochastic demands. European Journal of Operational Research 196, 509–515 (2009)
2. Leendert, K., Manuel, M., Herbert, K., Marco, J.: Dynamic Programming Algorithm for the Vehicle Routing Problem with Time Windows and EC Social Legislation. Beta Research School for Operations Management and Logistics, University of Twente (2009)
3. Christofides, N., Mingozzi, A., Toth, P.: State-space relaxation procedures for the computation of bounds to routing problems. Networks 11(2), 145–164 (1981)
4. Secomandi, N.: Exact and heuristic dynamic programming algorithms for the vehicle routing problem with stochastic demands. Doctoral Thesis, University of Houston (1998)
5. Eilon, S., Watson-Gandy, C.D.T., Christofides, N., de Neufville, R.: Distribution Management-Mathematical Modelling and Practical Analysis. IEEE Transactions on Systems, Man and Cybernetics, 589–589 (1974)
6. Bellman, R.: Dynamic Programming Treatment of Travelling Salesman Problem. Journal of the ACM 9(1), 61–63 (1962)
7. Bellman, R.: Dynamic Programming. Princeton University Press, Princeton (2003) (Republished)
8. Lenstra, J.K., Kan, A.H.G.R.: Complexity of vehicle routing and scheduling problems. Networks 11, 221–227 (1981)
9. Dantzig, G.B., Ramser, J.H.: The truck dispatching problem. Management Science 6(1), 80–91 (1959)
10. Golden, B.L., Assad, A.A., Wasil, E.A.: Routing vehicles in the real world: applications in the solid waste, beverage, food, dairy, and newspaper industries. In: The Vehicle Routing Problem, Society for Industrial and Applied Mathematics. SIAM Monographs on Discrete Mathematics and Applications, pp. 245–286. SIAM, Philadelphia (2001)
11. Ortega, P., Oliva, C., Ferland, J., Cepeda, M.: Multiple ant colony system for a VRP with time windows and scheduled loading. Ingeniare, Revista chilena de ingenieria 17, 393–403 (2009)
12. Laporte, G.: Fifty years of vehicle routing. Transportation Science 43(4), 408–416 (2009)

CBN: Combining Cognitive Map and General Bayesian Network to Predict Cost of Software Maintenance Effort

Kun Chang Lee[1] and Nam Yong Jo[2,*]

[1] Professor of MIS at SKK Business School
WCU Professor of Creativity Science at Department of Interaction Science
Sungkyunkwan University
Seoul 110-745, Republic of Korea
kunchanglee@gmail.com
[2] PhD Candidate
SKK Business School, Sungkyunkwan University
Seoul 110-745, Republic of Korea
namyong.jo@gmail.com

Abstract. Outsourcing of IT/IS service is now becoming a standard protocol for most of companies. As cost related to maintaining IT/IS service increases quite rapidly due to fierce competition in the market and fast changes in customers' behavior, how to control its cost emerges as the most important problem in the IT/IS service outsourcing industry. It is customary that software maintenance effort (SME) determines cost of IT/IS outsourcing service. Therefore, both IT/IS service providers and demanders have been focused on measuring the cost inflicted by SME, before reaching mutual agreement on price for IT/IS outsourcing service. Problem with this task is that there exist a large number of relevant factors and decision makers ought to be taken all into consideration systematically, which is very hard to do in reality. In this sense, this study proposes a new method called Cognitive Bayesian Network (CBN) in which SME experts first offer a draft causal map for the target SME problem, and the draft cognitive map is translated into the corresponding General Bayesian Network. To determine exact values of conditional probabilities for all of variables and arcs included in CBN, empirical SME data were applied to the CBN. After all CBN showed several merits- (1) it has a flexible structure enough to incorporate relevant variables at any time, and (2) it is capable of producing robust inference results for the given SME problems with rather high accuracy. To prove the validity of the proposed CBN, we interviewed with an expert having more than 15 years of SME experience to draw a draft CBN and than modified it with real SME data, then compared it with other BN models. We found that the performance of CBN is promising and statistically better than other benchmarking BN models.

Keywords: Software maintenance effort, IT/IS outsourcing, Bayesian network, Cognitive map, Cognitive Bayesian network.

---

* Corresponding author.

N.T. Nguyen et al. (Eds.): New Challenges for Intelligent Information, SCI 351, pp. 97–106.
springerlink.com   © Springer-Verlag Berlin Heidelberg 2011

1 Introduction

Since the software crisis argument [3] [6] [8], how to produce software and maintain it within reasonable costs have been debated hotly among researchers and practitioners. As [13] noted, science does not progress continuously, by gradually extending an established paradigm. Rather, science develops as a series of revolutionary paradigm changes [13]. In past several decades, many experts agreed that software crisis had been alleviated to some extents, though not completely reduced, due to dazzling developments in software engineering technologies. However, as the outsourcing of IT/IS service has been established as a standard protocol for most of companies regardless of industry, it is a quite surprising that such outdated software crisis argument is still overshadowing.

Considering the fact that companies are asked vehemently by customers to introduce new products and services very rapidly into the market, developing the related software within reasonable time and cost becomes decisive to maintain competitiveness in the markets [16]. Accordingly, in the IT/IS outsourcing fields, software maintenance efforts (SME) should be measured financially with high accuracy to the extent that both parties, providing companies and demanding clients, can rely on during SME cost negotiation and contract renewal. Nevertheless, it is a hard reality that there is no clear mechanism on exact cost prediction which both IT/IS service providing companies and demanding clients could agree. There exist many reasons underlying this fiasco. First, too many factors are related to SME cost estimation task. Furthermore, from which side you take, interpretation of those factors may differ, adding more dispute and confusion. Second, a set of complicated causal relationships exist among the factors, and they need to be analyzed systematically. Though every practitioners and researchers have been aware of this fact, an explicitly welcomed methodology was never introduced in the field yet.

To fill in this practical need, this study proposes a new paradigm in which combination of cognitive map (CM) and General Bayesian Network (GBN) leads to a new type of Bayesian Network method called Cognitive Bayesian Network (CBN), and rather stable and reliable platform for calculating the SME cost with statistically significant accuracy is created. Main thrust of the proposed CBN is that CM is built based on interview results with the SME expert, and the CM is translated into the corresponding GBN. Logic underlying our proposed CBN is partly backed up by Jørgensen's findings that his efforts of thorough literature review of more than 100 papers from 4 main journals suggested that experts' knowledge is competitive in estimating size of efforts in software development [11]. To prove the validity of our proposed CBN mechanism, real SME data set was applied to the CBN and its results were compared with other benchmarking BN classifiers.

The paper is organized as follows: Section 2 contains a brief overview of related works about SME, CM and BN to introduce the concept of CBN. Section 3 model constructions, performance of the model and simulations with it introduced. Finally, the major contribution of this paper, managerial implications and future researches are discussed.

2 Theoretical Background

2.1 Software Maintenance Effort

2.1.1 Software Maintenance (SM)

The IT/IS outsourcing which providers SM service is now prevailing in all around world. The SM is different from software development (SD). It includes work of a 'short term nature (e.g., fixing critical problems)' as well as of 'a long term nature (e.g., a major enhancement to an existing system)' [1]. Essential difference in features between SM and SD are: (1) the focus of SD is the creation of software, but SM deals mostly with the changes of software (2) the SD is a 'one-of-a-kind project', but the activities of SM with an application usually consist of a large number of small tasks carried out over time in a relatively stable environment [12]. IEEE standard 1219, which is generally adapted to SM outsourcing suppliers to establish quality standard based on it, defines SM as "The modification of a software product after delivery to correct faults, to improve performance or other attributes, or to adapt the product to a modified environment." [15] categorized maintenance activities as four classes: Adaptive-changes in the software environment, Perfective-new user requirements, Corrective-fixing error, Preventive-prevent problems in the future. They insist, after analyze surveyed data, about 75% of the maintenance effort is on first two types, and error correction consumes about 21%.

Table 1. Prior researches on software maintenance effort and cost estimation

Reference	Design of studies	Results
[15]	Surveys on 69 organizations	Maintenance consumes much of the total resources of IS group
[2]	Regression model on a major regional bank	Software maintenance costs are significantly affected by levels of software complexity
[12]	Regression, neural network, pattern recognition model on 109 maintenance tasks in a Norwegian company	Prediction models are instruments to support the expert estimate
[1]	Data Envelopment Analysis(DEA) on software enhancement projects	The presence of significant scale economies are in software maintenance
[19]	Bayesian network on productivity in ERP system localization	BN can be used to support an expert judgment of estimation
[16]	Bayesian network on maintenance project delay	BN can be used for presenting an expert knowledge

2.1.2 SM Effort Prediction

As mentioned prior paragraph, the cost of outsourcing SM is becoming significantly high. Researches have surveyed the proportion of software maintenance costs more than three decades. [5] estimated that more than 90% were spend for maintenance and evolution out of total software costs. [4] surveyed and found Fortune 1,000 companies spent on SM average of 75% in total information system budget. [9] estimated the cost for SM around 60-70% and [21] estimated it 67% out of total IT budget. Like these, the costs of SM and the proportion of them have been increased so far. Thus, for the demander as well as providers of IT/IS outsourcing, the predicting SME has become strategic and economic matters. However, comparing with considerable interests in investigating project scale economies for SD; very little examination of this issue has been done of SM [1]. So it is important to empirically examine SME in the IT/IS outsourcing fields of research. Table 1 shows prior researches of SM cost and effort and design methodologies adapted to the researchers.

2.2 Cognitive Bayesian Network (CBN)

2.2.1 Cognitive Map (CM)

CM is directed graphical representation of a cause and an effect embedded in experts' knowledge [7]. In CM, a node represents certain events or action that eventually leading particular outcomes. The primary emphasis of CM is identifying causal relations between variables in terms of explanation-consequence or means-end as is described in left side of Fig 2. CM has been useful in practice of various decision making environments. The probabilistic inference procedures of BN make it possible to make a prediction with variables in CM. Until recently, the expert based SM predictions doesn't have much attentions from researchers in spite of their high accuracy, because it does not enable an objective and quantitative analysis of what affect the SM productivity [12]. However, it can be overcome when they combined with other statistical models. This study proposes CBN, which is a brand new approach combing CM with BN, to make it possible that usual expert knowledge model can effectively perform various analysis.

2.2.2 Bayesian Network (BN)

BN, also known as a belief network or Bayesian belief network, is a directed acyclic graph which has probabilistic causal relationships and directions. It is recommended for an expert system particularly in domains where uncertainties play vital roles. Nowadays, BNs are used in wide ranges such as a bankruptcy warning, a medical diagnosis, a clinical decision support, a credit-rating of companies, and a reliability analysis of systems etc [18].

BN has several interesting features make it popular in decision supporting model: (1) it can handle incomplete data sets; (2) it shows causal relationship among variables; (3) it can combine expert knowledge with data. The BN's presentation consists of two parts: a qualitative part and a quantitative part [20]:

qualitative: The qualitative part of Bayesian Networks is a graphical representation of variables in the distribution and their relationship. This part takes the form of an directed acyclic graph (DAG) in which each vertex represents the statistical variable that can take one value of a finite set. The arcs represent a direct influential or causal relationship among variables. Fig. 1 shows a digraph in which x, y and z are variables to symbolically represent domain elements and arcs denote the causal dependency among variables. An arc $x \to z$, for example, represent a direct influential relationship between these variables, and the direction of the arc designates z as effect (or consequence) of x. Absence of an arc between two variables means their conditional independence; variables do not influence each other directly. For example, variables y and z are conditionally independent to each other because there is no set of arcs that can connect y to z and vice-versa.

quantitative: For the quantitative part of Bayesian Networks, a set of functions is defined representing numerical quantities of probability distribution over domain variables. Each vertex of the DAG has associated a probability assessment function that describes the influence of each vertex's predecessors to it. The probability assessment functions together (for all vertex) constitute the quantitative part of the Bayesian Network.

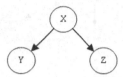

Fig. 1. the BN topology

Thus, the qualitative part is represented by the network structure and the quantitative part is expressed by conditional probability distributions. After those two parts are set up, Bayesian inference can be performed. The inference is the process of calculation the probability of one or more variables X given some evidence *v*. The evidence is expressed as an instantiation of some variables in the BN needs to be calculated. In this paper, we first make a structure of BN by interviewing a SM domain expert, and then to learn conditional probability distributions, we use actual data from an SM outsourcing organization. Then we will make inference to predict SME in the specific situations which variables are represented.

3   Model Construction

There are two basic approaches construct BN models: manual building based purely on human expert knowledge and automatic learning of the structure and the numerical parameter from data. This study suggest CBN model, on which we are going to get CM first then evolve it BN model. It is performed by interviewing domain expert for the qualitative part of BN and then parameter learning is performed for the quantitative part.

3.1   Interaction with Domain Expert for Network Structure

Manual construction a BN model typically involves interaction of the knowledge engineer, who is responsible for building the model, with domain expert. Some times it is recommended to have brainstorming meeting with experts who are not directly involved in building the model [18]. In this study, SM expert was suggested basic variables and given explanations how to draw causal relationship. After he draws the CM then we interview again to modify them. The left side of Fig.2 describes CM after modified after the second interviewing the expert. He finally suggested 11 variables which are causally effect on the target variables; here 'Person-months' is scalable amount of effort which means the number of people needed in a month to maintain certain software system. The right side of Fig.2 describes BN which modified from

CM. The main differences between two are: (1) BN doesn't have four variables that CM has (2) it has CPT which is not represented here but we will learn next paragraph. Some of variable are purposely omitted regretfully because they are not available in the given data for this study.

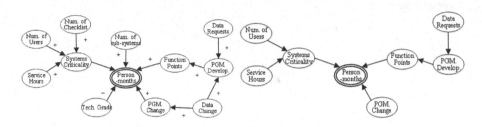

Fig. 2. CM and BN modified from CM(CBN)

3.2 Learning Conditional Probability Distributions

We already know about causal relations among variables from the expert, so we don't have to run further step for make BN structure. Now, parameter learning, which is consists of calculating basic vector of distribution and conditional probability can be followed with data. We donated SME data from one of the Korean IT/IS outsourcing company which has more than 5,000 employees and providing services more than 30 client-companies for academic uses. After cleansing records such as having missing values, we finally got 117 records for experiments. Table 2 describes variables.

Table 2. twelve variables and available values for learning classifiers.

Variables	Explanation of variables	attributes of variables
Num. of User	number of application systems users	<=2461.5, >=2461.6
Num Checklist	number of check-list to monitor system reliability	Not available
Service Hours	application system working hours per week	<=104, >=105
System Criticality	business and technical criticality for customer	A, B, C
Num. of sub-system	number of subsystem subordinated to each application system	Not available
Function Points	size of software system (numbers of program used in this study)	<=2129, >=2130
Date Requests	service request for data per month	<=295, >=296
PGM Develop.	service request for new program development per month	<=23.09, >=23.10
Data Change	service request for data change	Not available
PGM Change	service request for program update per month	<=514, >=252
*Person-month***	*Number of persons estimated to maintain each application software*	*<=4.36, <=8.52, <=12.68, <=16.84, >16.84*

**class node (target variable).

For the sake of discretizing variables and parameter learning we used J Cheng's BNPC[1] and Netica[2]. The class node is discretized by equal width with 5 bins and other variables are discretized with supervised learning by the class. The initial structure of the class node is in Fig. 3. It depicts the causal relationships and basic vectors of distribution which the original data are consisted.

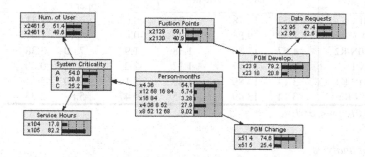

Fig. 3. the initial CBN

3.2.1 Results

(1) Prediction accuracy

For the sake of clear understanding of the experiment performance, we ran 10 times 10 folds iterations to address the accuracy of classification. In addition, benchmarking tests were performed to compare the accuracies of CBN and other popular BN. The resulted CBN accuracy is 68.47 %. Table 3(a) lists the performances of BN classifiers. The prediction accuracies were measured in percent and figures in parenthesis indicate standard deviation. To obtain rigor for each BN classifier's performance, each BN performance was compared with other BN classifiers' performances by using a corrected resample t-test at the 5% significance level. The results in Table 3(b) reassure that (1) CBN, NBN and TAN show the statistically same performance, and (2) GBN-K2 is worst among all of the classifiers. These are meaningful result that knowledge of expert is equal or even outperforms than those of computer aided popular BN structures:

Table 3. Prediction Accuracies of BN Classifiers and Statistical Tests for Comparison

(a) Accuracies for each BN (unit: %)

CBN	NBN	TAN	GBN-K2
68.47(9.53)	66.95(11.97)	66.95(9.95)	64.00(9.91)

Figures in parenthesis indicate standard deviation.

---

[1] http://webdocs.cs.ualberta.ca/~jcheng/bnsoft.htm
[2] http://www.norsys.com

Table 3. *(continued)*

(b) t-test results for BN classifiers

| | Paired Differences | | | | | t | Sig. (2-tailed) |
| | Mean | Std. Dev. | Std. Error Mean | 95% Confidence Interval of the Diff. | | | |
				Lower	Upper		
CBN - NBN	1.52	10.78	1.08	-0.62	3.66	1.41	0.16
CBN - TAN	1.52	9.85	0.98	-0.43	3.48	1.54	0.12
CBN – GBN-K2**	4.67	13.80	1.38	1.93	7.40	3.38	0.00
NBN - TAN	-0.00	11.17	1.12	-2.23	2.22	0.00	1.00
NBN - GBN-K2*	3.14	14.89	1.49	0.19	6.10	2.11	0.04
TAN -GBN-K2*	3.14	13.49	1.35	0.47	5.82	2.33	0.02

** $p < 0.01$, * $p < 0.01$.

(2) What-if analysis

This paper is focused on predicting SME assisted by CBN. The SME can be predicted by performing what-if analyses with CBN structure. Let's investigate what-if results of CBN with Netica software. Fig. 4 shows the prediction and sensitivity of the class node Person-months (PM) when other causally directly related attributes of node are changing.

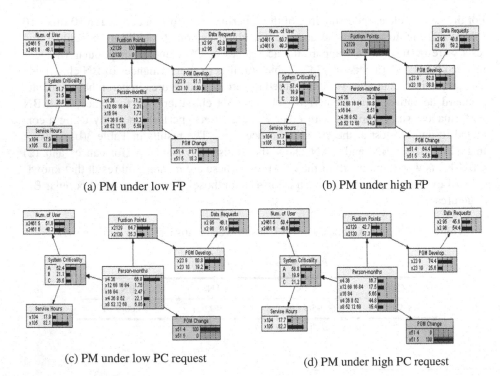

(a) PM under low FP

(b) PM under high FP

(c) PM under low PC request

(d) PM under high PC request

Fig. 4. PM sensitivity under changing value of directly related nodes

Fig. 4(a) and (b) show the effect of Function Point (FP) on PM; when FP is high (>= 2130) the probability of PM between 4.36 and 8.52 is 40.4%. It 21.2% higher than when it is lower (<=2129). In addition, Fig. 4(c) and (d) show the effect of Program Change (PC); PC request is high (>=515) the probability of PM between 4.36 and 8.52 goes up 44.8%, in other words, it is 22.7% higher than when PC is in lower (<=514). Roughly speaking, the sensitivity of PC effect on PM is higher than that of FP. We stochastically predict that 4 to 8 persons month needed for the system SM when PC is more than 514 under other condition are fixed.

4 Conclusion and Future Works

Cost of SME lies at the heart of negotiation between IT/IS service providers and client companies. The problem is how to predict it accurately and objectively. Depend on which side you are the proper cost of SME looks different, which cause serious disputes whenever two parties start negotiation on the IT/IS outsourcing renewal. In this sense, how to predict cost of SME accurately and objectively has been regarded as a long-standing problem in the IT/IS outsourcing industry.

Considering the urgency of SME cost prediction problem, we have proposed a new method called Cognitive Bayesian Network (CBN), a combination of cognitive map and General Bayesian Network. The proposed CBN enables a draft cognitive map suggested by a SME expert to translate into a corresponding BN. In this way, the SME expert's knowledge can be reformulated in a form of BN and simulated with learning and inference capabilities supported by BN. Advantages of CBN, through our experience of empirical tests with it, can be suggested as follows.

First, interested parties regarding the cost of SME can use the CBN as an objective vehicle of compromising their negotiations smoothly, which would have been adrift among endless arguments protecting their positions with their own views and data.

Second, the empirical performance of the CBN is comparable with other BN classifiers such as TAN and GBN. This means that CBN is a better method in reality once its performance is not defeated by others, because CBN can provide more reasonable explanation about a specific SME cost, and various scenarios, from which a set of crucial factors, can be brought to all the interested parties' attention.

Future study issues are as follows. First, more refined method of combining multiple experts' knowledge on SME cost estimation should be suggested. Second, translation from SME expert's cognitive map into a corresponding BN should be written in rigorous algorithmic procedures so that the proposed CBN can be automatically implemented on computer.

Acknowledgments. This research was supported by the World Class University (WCU) program through the National Research Foundation of Korea, funded by the Ministry of Education, Science and Technology, Republic of Korea (Grant No. R31-2008-000-10062-0).

References

1. Banker, R.D., Slaughter, S.A.: A Field Study of Scale Economies in Software Maintenance. Management Science 43, 1709–1725 (1997)
2. Banker, R.D., Datar, S.M., Kemerer, C.F., Zweig, D., Slaughter, S.A.: Software Complexity And Maintenance Costs. Communication of the ACM 36, 81–94 (1993)
3. Dijkstra, E.: On the Cruelty of Really Teaching Computing Science. Communications of the ACM (1989)
4. Eastwood, A.: Firm Fires Shots at Legacy Systems. Computing Canada 19, 17 (1993)
5. Erlikh, L.: Leveraging Legacy System Dollars for E-business. IT Professional 2, 17–23 (2000)
6. Fetzer, H.: Program Verification: The Very Idea. Communications of the ACM 37, 1048–1063 (1988)
7. Fiol, M., Huff, A.S.: Maps for Managers: Where Are We? Where Do We Go From Here? Journal of Management Studies 29, 269–285 (1992)
8. Glass, R.L.: The Standish Report: Does It Really Describe a Software Crisis? Communications of the ACM 49, 15–16 (2006)
9. Huff, S.: Information Systems Maintenance. The Business Quarterly 55, 30–32 (1990)
10. IEEE Std. 1219: Standard for Software Maintenance. IEEE Computer Society Press, Los Alamitos (1993)
11. Jørgensen, M.: A Review of Studies on Expert Estimation of Software Development Effort. Journal of Systems and Software 70, 37–60 (2004)
12. Jørgensen, M.: Experience with the Accuracy of Software Maintenance Task Effort Prediction Models. IEEE Transaction of software engineering 21, 674–681 (1995)
13. Kuhn, T.: The Structure of Scientific Revolutions. The University of Chicago Press, London (1962)
14. Lehman, M.M.: On understanding Laws, Evolution and Conversation in the Large Program Lifecycle. Journal of Software & Systems L, 213–221 (1980)
15. Lientz, B.P., Swanson, E.B., Tompkins, G.E.: Characteristics of Application software maintenance. Communication of the ACM 21, 466–471 (1987)
16. Melo, A.C.V., Sanchez, A.J.: Software Maintenance Project Delays Prediction Using Bayesian Networks. Expert Systems with Applications 34, 908–919 (2008)
17. Nadkarni, S., Shenoy, P.P.: A Bayesian Network Approach to Making Inferences in Causal Maps. European Journal of Operational Research 128, 479–498 (2001)
18. Pourret, O., Naim, P., Marcot, B.: Bayesian Networks: A Practical Guide to Applications. Wiley, Chichester (2008)
19. Stamelos, I., Angelis, L., Dimou, P., Sakellaris, E.: On the use of Bayesian belief networks for the prediction of software productivity. Information And Software Technology 45, 51–60 (2003)
20. van der Gaag, L.C.: Bayesian Belief Networks: Odds and Ends. The Computer Journal 39, 97–113 (1996)
21. Zelkowitz, M., Shaw, A., Gannon, J.: Principles of Software Engineering and Design. Prentice-Hall, New Jersey (1979)

A Central Leader Election Method
Using the Distance Matrix
in Ad Hoc Networks

Mary Wu[1], SeongGwon Cheon[2], and ChongGun Kim[1,*]

[1] Dept. of Computer Engineering,
Yeungnam University, Korea
[2] Dept. of China Business Information,
Catholic Sangji College, Korea
mrwu@ynu.ac.kr, skcheon@angel.csangji.ac.kr, cgkim@yu.ac.kr

Abstract. A leader in a network has important roles to collect data and control
the network. Therefore, when a node with high centrality is selected as the
leader node in the network, the network can be controlled efficiently, and the
amount of resource consumed in controlling can be reduced. This study proposes a central leader election method in ad hoc networks. A representative centrality measure of social network analysis (SNA), degree or closeness is used
for leader selection criteria. Each node's degree or closeness is simply calculated by the proposed method, based on the distance matrix which is generated
based on adjacency matrix. The data-collection process for creating adjacency
matrix applies a tree-based data transmission for the efficient use of the limited
resource in ad hoc networks. Our experiments show that the complexity of the
proposed centrality calculation method is less compared with that of Dijkstra's
algorithm and that of the Bellman–Ford algorithm. The proposed election
method is expected to be appropriate to apply for selecting the leader node on
the limited resource in ad hoc networks.

Keywords: Leader election, Tree-structure, Centrality, Degree, Closeness, Adjacency matrix, Distance matrix.

1 Introduction

A mobile ad hoc network (MANET) is a collection of mobile nodes that can communicate via message passing over wireless links. If they are not within a specified
transmission range, they communicate via other nodes. In ad hoc networks, the leader
election can be employed for gathering and distributing information of the networks.
Many studies for leader election have been proposed [1-9]. Vasudevan et al. [1] propose the election algorithm that can adapt well to topology change, but once a leader
loss is detected, the overall election process is repeated, which greatly increases the
number of messages and time complexity(O). Malpani et al. [2] maintain a directed
acyclic graph with a single sink, which is the leader. However this algorithm is
proved correct for a completely synchronous system with a single link change. The

---

* Corresponding author.

N.T. Nguyen et al. (Eds.): New Challenges for Intelligent Information, SCI 351, pp. 107–116.
springerlink.com   © Springer-Verlag Berlin Heidelberg 2011

algorithms proposed in [3, 4] work only if the topology remains static and hence can not be applied to mobile networks [9].

The selected leader can manage various operations and control many technical problems caused in the networks. For example, in the Core Based Trees protocol (CBT), which provides an ad hoc multicast routing protocol, the whole network is formed as a tree structure centered around one core node and this shared tree is used for multicast routing [10]. In addition, a representative node is used as a core element for providing various services in areas such as key distribution [11], synchronization [12], location-based routing coordination [13] and general control [14].

An election process should select the most proper node among all the nodes of the network as a leader. Values for the leader node election depend on many contexts of the systems. They can be a performance-related metric, e.g., remaining battery life, minimum average distance from other nodes and computation capabilities [1]. The selected leader node performs roles such as exchanging information with other nodes on the network or controlling the entire network. Therefore, when a node is selected with topological centrality, in addition to its performance-related metric, that leader node can manage the network efficiently. For examples, multicast routing protocols construct a single tree that is shared by all members of the group. Multicast traffic for the entire group is delivered along the edges of this unique tree. Therefore, when a "root", or leader is selected which is located in the center of network, network traffic for controlling the network can be reduced [10]. In a mobile wireless environment, user data must be securely multicast from one source to many users; this requires that user data be properly encrypted. When users move in and out of the session, in order to preserve confidentiality, it becomes necessary to rekey each time a user enters or leaves. For secure transmission of keys, key-transmission algorithm is important, but the efficiency of the transmission path is also important. Key transmission from a node which has centrality can provide efficient transmission paths to other nodes [11].

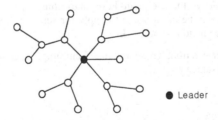

Fig. 1. An example of a central leader in an ad hoc network

Representative measures of centrality are: degree, closeness. These measures have proved of great value in the analysis and understanding of the roles played in social networks as well as in networks of other types, including citation networks and computer networks. Degree is the simplest centrality measure, which is the number of edges. Closeness is the mean geodesic distance of a vertex and all other vertices from it [15-18].

But, more bandwidth to gather the information for each node's centrality calculation is consumed and large overhead due to the complexity of centrality calculation method is required in networks. To solve problems, we propose a tree-based leader election process to minimize the overhead of information collection and a central leader calculation method based adjacency matrix and distance matrix. Degree and closeness as centrality measures for the leader selection process are used. Each node's closeness is calculated by a formula, based on the distance matrix [20] and each node's closeness is simply calculated by a formula, based on the adjacency matrix, respectively. The distance matrix is generated based on the adjacency matrix which presents the neighbor information of each node in the network [9, 19-20].

The remainder of this paper is organized as follows. Section 2 introduces related works. Section 3 introduces the tree-based leader election process and the centrality calculation method based on the adjacency matrix or the distance matrix. Section 4 analyzes the results of the proposed leader election proposed in various scenarios. Finally, in Sect. 5, we show the conclusions of this study.

2   Related Works

2.1   Degree and Closeness

Measures of point centrality based on adjacency of a point have been developed by Shaw (1954), Faucheux and Moscovici (1960), Garrison (1960), Mackenzie (1966), Pitts (1965), Rogers (1974), Czepiel (1974), Nieminen (1973, 1974) and Kajitani and Maruyama (1976). Among them, Nieminen's measure is the number of adjacencies for a point [15]. Nieminen's measure is the counts of the degree or number of adjacencies for a point. Function (1) shows Nieminen's degree measure.

$$Degree(p_k) = \frac{\sum_{i=1}^{n} a(p_i, p_k)}{n-1}, \tag{1}$$

where $a(p_i, p_k) = 1$ if and only if p_i and p_k are connected by a line, otherwise $a(p_i, p_k) = 0$.

Closeness-based measures of point centrality have been developed by Bavelas (1950), Beauchamp (1965), Sabidussi (1966), Moxley (1074) and Rogers (1974). Sabidussi proposed that the centrality of a point be measured by summing the geodesic distance from that point to all other points. This is an inverse centrality since it grows as points are far apart. Beauchamp suggested that the relative point centrality of a point of a point. Beauchamp's measure is the inverse of average distance from a point to other points [15]. Function (2) shows Beauchamp's closeness measure.

$$Closeness(p_k) = \frac{n-1}{\sum_{i=1}^{n} d(p_i, p_k)}, \tag{2}$$

where $d(p_i, p_k)$ is the number of edges in the geodesic linking p_i and p_k.

2.2   Distance Matrix

Distance matrix is created based on the adjacency matrix which presents link connection between nodes. First, an initiate node generates adjacency matrix using the collected neighboring information of each node. For the adjacency matrix A, the ij-th entry is 1 if there is a link connection between node i and node j and all zero entries of the matrix present no link. The adjacency matrix has previously been introduced in [19-20]. The adjacency matrix for fig. 2a) is shown as follows,

$$
A = \begin{bmatrix}
0 & 0 & 0 & 1 & 0 & 0 & 0 & 0 & 0 & 0 & 0 & 0 & 0 \\
0 & 0 & 0 & 1 & 0 & 0 & 0 & 0 & 0 & 0 & 0 & 0 & 0 \\
0 & 0 & 0 & 1 & 0 & 0 & 0 & 0 & 0 & 0 & 0 & 0 & 0 \\
1 & 1 & 1 & 0 & 1 & 1 & 1 & 0 & 0 & 0 & 0 & 0 & 0 \\
0 & 0 & 0 & 1 & 0 & 1 & 0 & 0 & 0 & 0 & 0 & 0 & 0 \\
0 & 0 & 0 & 1 & 1 & 0 & 1 & 1 & 1 & 0 & 0 & 0 & 0 \\
0 & 0 & 0 & 1 & 0 & 1 & 0 & 0 & 0 & 1 & 1 & 1 & 0 \\
0 & 0 & 0 & 0 & 0 & 1 & 0 & 0 & 1 & 0 & 0 & 0 & 0 \\
0 & 0 & 0 & 0 & 0 & 1 & 0 & 1 & 0 & 1 & 0 & 0 & 0 \\
0 & 0 & 0 & 0 & 0 & 0 & 1 & 0 & 1 & 0 & 1 & 1 & 1 \\
0 & 0 & 0 & 0 & 0 & 1 & 1 & 0 & 0 & 1 & 0 & 1 & 1 \\
0 & 0 & 0 & 0 & 0 & 0 & 1 & 0 & 0 & 1 & 1 & 0 & 1 \\
0 & 0 & 0 & 0 & 0 & 0 & 0 & 0 & 0 & 1 & 1 & 1 & 0 \\
\end{bmatrix}
\tag{3}
$$

After establishing the adjacency matrix, the distance matrix can be calculated based on the adjacency matrix. The k-hop matrix $B^{(k)}$ and the shortest k-hop matrix $C^{(k)}$ where k shows the hop counts between node i and node j are matrices for generating the distance matrix. Each entry of matrix $B^{(k)}$ is k or 0 and presents if k-hop path from i to j is present or not. Each entry of matrix $C^{(k)}$ presents the shortest hop counts in the paths from i to j and is calculated by the addition each entry of $B^{(k)}$ and $C^{(k-1)}$.

The ij-th entries of 2-hop matrix $B^{(2)}$ present the hop counts of 2-hop path between nodes and are defined as follows.

$$
b_{ij}^{(2)} = \begin{cases}
0, & i = j \\
0, & a_{ij} = 1 \\
2, & a_{ij}^{(2)} > 0 \text{ and } i \neq j \text{ and } a_{ij} = 0 \\
0, & otherwise
\end{cases}, \quad where \; a_{ij}^{(2)} = \sum_{k=1}^{N} a_{ik} a_{kj}.
\tag{4}
$$

The shortest 2-hop matrix $C^{(2)}$ presents one and two hop counts in the paths from i to j and can be obtained from addition operation based on the 2-hop matrix $B^{(2)}$ and the adjacency matrix A.

$$
c_{ij}^{(2)} = a_{ij} + b_{ij}^{(2)} = c_{ij} + b_{ij}^{(2)}.
\tag{5}
$$

The ij-th entries of 3-hop matrix $B^{(3)}$ present the hop counts of 3-hop path between nodes and are defined as follows.

$$
b_{ij}^{(3)} = \begin{cases}
0, & i = j \\
0, & c_{ij}^{(2)} > 0 \\
3, & \sum_{k=1}^{N} c_{ik}^{(2)} a_{kj} > 0 \text{ and } i \neq j \text{ and } c_{ij}^{(2)} = 0 \\
0, & otherwise
\end{cases}
\tag{6}
$$

The shortest 3-hop matrix $C^{(3)}$ presents one, two, three hop counts in the paths from i to j and can be obtained from addition operation based on the 3-hop matrix $B^{(3)}$ and the shortest 2-hop matrix $C^{(2)}$.

$$
c_{ij}^{(3)} = c_{ij}^{(2)} + b_{ij}^{(3)}.
\tag{7}
$$

By the iterative calculation of the function (8), the distance matrix is generated.

$$c_{ij}^{(m)} = c_{ij}^{(m-1)} + b_{ij}^{(m)}, \quad m \geq 2,$$

$$\text{where } b_{ij}^{(m)} = \begin{cases} 0, & i = j \\ 0, & c_{ij}^{(m-1)} > 0 \\ m, & \sum_{k=1}^{N} c_{ik}^{(m-1)} a_{kj} > 0 \text{ and } i \neq j \text{ and } c_{ij}^{(m-1)} = 0 \\ 0, & otherwise \end{cases} \tag{8}$$

$C^{(2)}$ is created based on A, $C^{(3)}$ is created based on $C^{(2)}$, and $C^{(4)}$ is created based on $C^{(3)}$. The following matrix shows the distance matrix generated in the topology of the fig. 2. The number 4 in the matrix means indicates the maximum hops.

$$C^{(4)} = \begin{bmatrix} 0 & 2 & 2 & 1 & 2 & 2 & 2 & 3 & 3 & 3 & 3 & 3 & 4 \\ 2 & 0 & 2 & 1 & 2 & 2 & 2 & 3 & 3 & 3 & 3 & 3 & 4 \\ 2 & 2 & 0 & 1 & 2 & 2 & 2 & 3 & 3 & 3 & 3 & 3 & 4 \\ 1 & 1 & 1 & 0 & 1 & 1 & 1 & 2 & 2 & 0 & 2 & 2 & 3 \\ 2 & 2 & 2 & 1 & 0 & 1 & 2 & 2 & 2 & 3 & 2 & 3 & 3 \\ 2 & 2 & 2 & 1 & 1 & 0 & 1 & 1 & 1 & 2 & 1 & 2 & 2 \\ 2 & 2 & 2 & 1 & 2 & 1 & 0 & 2 & 2 & 1 & 1 & 1 & 2 \\ 3 & 3 & 3 & 2 & 2 & 1 & 2 & 0 & 1 & 2 & 3 & 3 & 3 \\ 3 & 3 & 3 & 2 & 2 & 1 & 2 & 1 & 0 & 1 & 2 & 2 & 2 \\ 3 & 3 & 3 & 0 & 3 & 2 & 1 & 2 & 1 & 0 & 1 & 1 & 1 \\ 3 & 3 & 3 & 2 & 2 & 1 & 1 & 3 & 2 & 1 & 0 & 1 & 1 \\ 3 & 3 & 3 & 2 & 3 & 2 & 1 & 3 & 2 & 1 & 1 & 0 & 1 \\ 4 & 4 & 4 & 3 & 3 & 2 & 2 & 3 & 2 & 1 & 1 & 1 & 0 \end{bmatrix}. \tag{9}$$

3   Tree-Based Leader Election Process and Centrality Calculation Method Based on the Matrix

3.1   Tree-Based Leader Election Process

In this section, a leader-election algorithm is introduced. The algorithm is designed to send data based on a tree structure, to generate the smallest number of control messages in the ad hoc network which has limited bandwidth.

We assume that each node should periodically broadcast 'Hello' messages in order to check the connectivity with its neighbors and record its own neighbors' information in a neighbor table.

The proposed tree-based leader election algorithm is as follows;

1) **Query phase.** An initiator sends an *Election* message to all its immediate neighbors. Each node designates the neighbour from which it first receives an *Election* message as its parent in the tree. Every node i then propagates the received *Election* message to all of its neighboring nodes except its parent [7]. Fig. 2a) shows *Election* message propagation and fig. 2b) shows the tree which is completed at the end of the Election-message transmission process.

2) **Collection phase.** When any node receives the first *Election* message, it does not immediately return a *Reply* message including the information (node id,

neighbor id). To reduce the amount of control messages in the network, it waits until it has received *Reply* messages from all of its children, before sending a *Reply* message to its parent. This reply process continues until the initiator gets *Reply* messages from all its children.

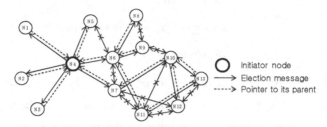

a) Message propagation in the ad hoc network

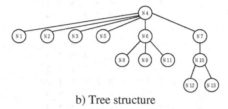

b) Tree structure

Fig. 2. An example of ad hoc networks

3) **Calculation phase.** The initiator generates an adjacency matrix and a distance matrix based on the received information from every node in the network. And then it calculates each node's centrality based on the matrices and elects a node with highest centrality as a leader.

4) **Leader Advertisement phase.** The initiator broadcasts a *Leader* message including the elected leader to announce the leader. The leader message is sent along the tree path.

Election, *Reply* and *Leader* messages are transported to every node. If these messages are transported through flooding, lots of control messages occur on the network. As one might expect, "flooding" results in too many collisions and too much overhead and becomes one of the features degrading performance of the entire network. The tree-based message transmission method which is proposed here minimizes collisions of messages and uses network resources more efficiently.

Leader election process is run periodically and the current leader plays the role as the initiator for the next leader election process.

3.2 Centrality Calculation Method Based on the Matrices

In this section, the calculation methods of each node's degree and closeness centrality a method are introduced. Degree is calculated using a simple formula based on the adjacency matrix and closeness is calculated using a simple formula based on the distance matrix.

In function (10), the degree formula based on the adjacency matrix is presented. It uses the sum of all elements in each row of the adjacency matrix (3) and is normalized by dividing into the sum and the number of neighboring nodes.

$$C_{Degree}(n_i) = \frac{\sum_{k=1}^{n} a_{ik}}{n-1},$$ (10)

where is a_{ik} is the row element of the adjacency matrix and n is the number of nodes.

The closeness formula based on the distance matrix is proposed in function (11). It uses the inverse of the sum of the elements in each row of the distance matrix (9) and is normalized by multiplying the inverse of the sum and the number of neighboring nodes.

$$C_{Closeness}(n_i) = \frac{n-1}{\sum_{k=1}^{n} c_{ik}},$$ (11)

where is c_{ik} is the row element of the distance matrix and n is the number of nodes.

Table 1 shows the results of degree and closeness that are calculated using function (10) and function (11) based on the matrix (3) and (9) respectively.

Table 1. The results of degree and closeness

	N_1	N_2	N_3	N_4	N_5	N_6	N_7	N_8	N_9	N_{10}	N_{11}	N_{12}	N_{13}
EDegree	0.023	0.023	0.023	0.136	0.045	0.136	0.114	0.045	0.068	0.114	0.114	0.091	0.068
ECloseness	40.000	40.000	40.000	63.158	48.000	66.667	63.158	44.444	50.000	52.174	54.545	48.000	40.000

For the central leader election, the degree election weight E_{Degree} is defined by

$$E_{Degree} = C_{Degree} \times S, \text{ where } S \text{ is the value of a node state.}$$ (12)

Depending on the network policies, the value of a node state can be computed capabilities, remaining battery life, and so on.

The closeness election weight $E_{Closeness}$ is defined by multiplying the closeness weight and the value of a node state.

$$E_{Closeness} = C_{Closeness} \times S, \text{ where } S \text{ is the value of a node state.}$$ (13)

Finally, we can consider that the degree election weight E_{Degree} and the closeness election weight $E_{Closeness}$ together for the central leader election. The election weight E is given by $E = \alpha E_{Degree} + \beta E_{Closeness}$. (14)

Fig. 3 show the results of degree and closeness that are calculated using function (12) and function (13) with the fixed value S=1. In fig 3b), when only considering the

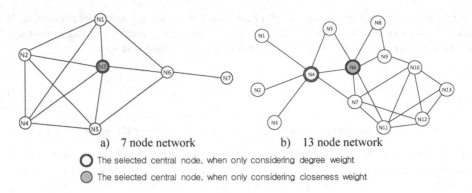

a) 7 node network   b) 13 node network

○ The selected central node, when only considering degree weight

◉ The selected central node, when only considering closeness weight

Fig. 3. The examples of central leader election

degree election weight, N_4 and N_6 has the largest and when only considering the closeness election weight, N_6 has the largest centrality and can be elected as a leader.

4   Experiments and Analysis

4.1   Complexity of Distance Matrix-Based Election Method

We compare the complexities of Dijkstra's algorithm, the Bellman–Ford with that of our proposed methed. To elect a node with the highest centrality, centralities for all nodes in the network have to be calculated. Dijkstra's algorithm is a graph search algorithm that produces a single-source shortest tree. For a given node, the algorithm finds the paths with the lowest cost. The time complexity for a node is $O(N^2)$, where N is the number of nodes. To calculate centralities for all the nodes, the tree calculation of N times is needed. In fig. 3a), there are 7 nodes in the network. Therefore, the centrality calculation of 7 times is required for the spanning trees, with each node as a root. The time complexity of Dijkstra's algorithm to get centralities for all nodes in a network is $O(N^3)$. Bellman–Ford algorithm is similar in its basic structure to Dijkstra's algorithm, but instead of greedily selecting the minimum-weight node, it simply compares all the edges N – 1 times, where N is the number of nodes. The Bellman–Ford algorithm for a source runs in $O(NE)$ time, where N and E are the number of nodes and edges, respectively. Therefore, the time complexity of the Bellman–Ford algorithm to get centralities for all nodes-omit in a network is $O(N^2E)$. The centrality method based on the distance metric requires a complexity of $O(\alpha N^2)$ to calculate a matrix $C^{(\alpha)}$ where α is the maximum hop count in the network topology.

Fig. 4 shows the time complexity depending on the number of nodes with Dijkstra's algorithm, Bellman–Ford algorithm, and the method based on the distance matrix. C is used for experiments, network topology is randomly generated. The number of nodes 50, 55, 60, 65, 70, 75, 80, the number of edges is fixed to 80, and the

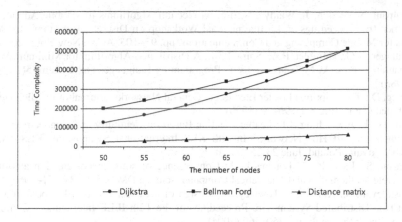

Fig. 4. The amount of time complexity depending on the number of nodes

number of maximum hops is fixed to 10. Fig. 4 shows the average of 10 times for each run. The time complexity of centrality method based on the distance metric is very small compared to the other two methods and increases only slightly, with an increasing number of nodes.

5 Conclusions and Future Works

To use network resources efficiently and help the operation of the entire network smoothly, the centrality of the leader node which controls networks is very important. This study has proposed a novel method for selecting a leader node which has centrality on the network. This research also proposed the tree-based leader message transmission method which minimizes the number of control messages required for information collection for leader election.

For obtaining the centralities of each node in the proposed election method, the simple formulas based on the adjacency matrix and the distance matrix are proposed. The result of the experiment shows the calculation complexity performance of the proposed method is better compared to that of Dijkstra's algorithm and that of Bellman–Ford algorithm. The proposed centrality election method is expected to be good for the efficient use of the limited resource of each node and of the network in ad hoc networks.

Acknowledgement

This research was supported by the yeungman university research grants in 2009.

References

1. Vasudevan, S., Kurose, J., Towsley, D.: Design and Analysis of a Leader Election Algorithm for Mobile Ad Hoc Networks. In: Proceedings of the 12th IEEE International Conference on Network Protocols (ICNP 2004), Berlin, Germany, pp. 350–360. IEEE Computer Society, Los Alamitos (2004)

2. Malpani, N., Welch, J., Waidya, N.: Leader election Algorithms for Mobile Ad Hoc Networks. In: Proceedings of 4th International Workshop on Discrete Algorithms and Methods for Mobile Computing and Communications, pp. 93–103. ACM, New York (2000)
3. Gallager, R.G., Humblet, P.A., Spira, P.M.: A Distributed Algorithm for Minimum Weight Spanning Trees. ACM Transactions on Programming Languages and Systems 5(1), 66–77 (1983)
4. Peleg, D.: Time Optimal Leader Election in General Networks. Journal of Parallel and Distributed Computing 8(1), 96–99 (1990)
5. Wu, M., Kim, C., Jung, J.J.: Leader Election Based on Centrality and Connectivity Measurements in Ad Hoc Networks. In: 4th KES International Symposium, KES-AMSTA 2010, Gdynia, Poland, June 23-25 (2010)
6. Janson, S., Lavault, C., Louchard, G.: Convergence of some leader election algorithms. Discrete Mathematics and Theoretical Computer Science DMTCS 10(3), 171–196 (2008)
7. Vasudevan, S., Kurose, J., Towsley, D.: Design of a Leader Election Protocol in Mobile Ad Hoc Distributed Systems. In: Proceedings of the 12th IEEE International Conference on Network Protocols, pp. 350–360 (2004)
8. Zhang, G., Kuang, X., Chen, J., Zhang, Y.: Design and Implementation of a Leader Election Algorithm in Hierarchy Mobile Ad hoc Network. In: Proceedings of 2009 4th International Conference on Computer Science & Education (2009)
9. Lee, S., Muhammad, R., Kim, C.: A Leader Election Algorithm Within Candidates on Ad Hoc Mobile Networks. In: Lee, Y.-H., Kim, H.-N., Kim, J., Park, Y.W., Yang, L.T., Kim, S.W. (eds.) ICESS 2007. LNCS, vol. 4523, pp. 728–738. Springer, Heidelberg (2007)
10. Gupta, S., Srimani, P.: Core-Based Tree with Forwarding Regions(CBT-FR); A Protocol for Reliable Multicasting in Mobile Ad Hoc Networks. Journal of Parallel and Distributed Computing 61, 1249–1277 (2001)
11. DeCleene, B., Dondeti, L., Griffin, S., Hardjono, T., Kiwior, D., Kurose, J., Towsley, D., Vasudevan, S., Zhang, C.: Secure group communication for Wireless Networks. In: Proceedings of the IEEE Military Communications Conference (MILCOM 2001), pp. 113–117 (2001)
12. Sundararaman, B., Buy, U., Kshemkalyani, A.: Clock Synchronization for Wireless Sensor Networks: A Survey. Ad Hoc Networks 3(3), 281–323 (2005)
13. Blazevic, L., Boudec, J., Giordano, S.: A Location-Based Routing Method for Mobile Ad Hoc Networks. IEEE Transactions On Mobile Computing 4(2) (2005)
14. Bayazit, O.B., Lien, J.M., Amato, N.M.: Better group behaviors in complex environments using global roadmaps. In: Proceedings of 8th International Conference on the Simulation and Synthesis of living systems (Alife 2002), pp. 362–370. MIT Press, Cambridge (2002)
15. Freeman, L.C.: Centrality in Social Networks Conceptual Clarification. Social Networks 1, 215–239 (1978/1979)
16. Newman, M.E.J.: A measure of betweenness centrality based on random walks. Social Networks 27(1), 39–54 (2005)
17. Kleinberg, J.M.: Authoritative sources in a hyperlinked environment. Journal of the ACM (JACM) 46(5), 604–632 (1999)
18. Jung, J.J.: Query Transformation based on Semantic Centrality. Journal of Universal Computer Science 14(7), 1031–1047 (2008)
19. Wu, M., Kim, S.H., Kim, C.G.: A routing method based on cost matrix in ad hoc networks. In: Nguyen, N.T., Katarzyniak, R., Chen, S.-M. (eds.) Advances in Intelligent Information and Database Systems. SCI, vol. 283, pp. 337–347. Springer, Heidelberg (2010)
20. Wu, M., Kim, C.: A cost matrix agent for shortest path routing in ad hoc networks. Journal of Network and Computer Applications (2010), http://dx.doi.org/10.1016/j.jnca.2010.03.013

Facial Soft Tissue Thicknesses Prediction Using Anthropometric Distances

Quang Huy Dinh[1], Thi Chau Ma[1], The Duy Bui[1],
Trong Toan Nguyen[2], and Dinh Tu Nguyen[1]

[1] University of Engineering and Technology,
Vietnam National University, Hanoi, Vietnam
{huy.dq,chaumt,duybt,tund}@vnu.edu.vn
[2] Military Medical Academy, Vietnam

Abstract. Predicting the face of an unidentified individual from its skeletal remains is a difficult matter. Obviously, if the soft tissue thicknesses at every location at the skull are known, we can easily rebuild the face from the skull model. Thus, the problem turns out to be predicting the soft tissue thicknesses for any given skull. With the rapid development of the computer, different techniques are being used in the community for prediction tasks and in recent years the concept of neural networks has emerged as one of them. The principal strength of the neural network is its ability to find patterns and irregularities as well as detecting multi-dimensional non-linear connections in data. In this paper, we propose a method of applying neural networks to predict the soft tissue thicknesses for facial reconstruction. We use the distances between anthropometric locations at the skull as input, and the soft tissue thicknesses as output, as this format is suitable for many machine learning mechanisms. These data is collected and measured from candidates using the Computed Tomography (CT) technique.

Keywords: facial reconstruction, soft tissue thicknesses, data mining, neural networks, machine learning.

1 Introduction

People have been recreating the face of an unidentified individual from their discovered skulls for nearly a hundred years. Traditionally, facial reconstruction is done using clay. This method requires skillful experts who understand the structure of skull and skin very well to attach clay on the skull. The expert skill and amount of time required have motivated researchers to try to computerize the technique. Researchers have pointed that there are some attributes which change the facial shape but not the skull, such as age, weight, and gender. Therefore, a well-designed computer-aided facial reconstruction system has many advantages, including great reduction in time consumption. Using such a system, we can produce several possible facial models from a given skull by using parameters determining the person's age, weight, and gender.

N.T. Nguyen et al. (Eds.): New Challenges for Intelligent Information, SCI 351, pp. 117–126.
springerlink.com

Recently, the rapid development of 3D equipments and technology enable us to advance into this field of research. A lot of computerized methods for 3D facial reconstruction have been proposed and developed, which make use of computer program to transform 3D scanned models of the skull into faces. Most researches try to follow the manual approach, but make use of the computer to fasten the process of reconstruction. In these methods, they have the discovered skull scanned to get its 3D model. After that, they need to calculate the soft tissue thicknesses at every location at the skull surface to build the final facial model and the most critical issue turns out to be discovering these thicknesses. Regards to this problem, there are several approaches. All of them are following the same idea of collecting a database of skin-skull models and then try to discover the relationship between the skin and skull, but differ in the way of collecting and processing. The preview of these related works are described in the next section. Our approach is using the same idea. We consider this problem as the missing data problem, and we also need to obtain a database, but only for the information that is proved to be important by anthropology. This information is the soft tissue thicknesses at the anthropometric locations and the distances between them.

The rest of this paper is organized as follow. Section 2 reviews related work. In Section 3, we describe our approach including the process of obtaining the database, and a method of automatic discovery the skin-skull thicknesses at the anthropometric locations using linear regression and neural networks. Section 4 shows our evaluations. Finally, conclusions and future works are described in Section 5.

2   Related Work

The issue of collecting soft tissue thickness data to clarify the relationship between soft tissue and the underlying bony structure of skull has been discussed by forensic medicine experts for more than a hundred years. In 1883, Welcker [1] obtained a database of soft tissue thicknesses by inserting a thin blade into facial skin of cadavers at selected anatomical landmarks. After that, he measured the depth of the blades penetration. Until middle 1980, all studies that need to collect soft tissue thicknesses data at anatomical landmarks used cadaverous populations and this 'needle technique'. However, this type of approaches has some problems. First of all, a dead person's tissues are not the same as in life due to drying and embalming. Secondly, the skin can be deformed due to the penetration of the needle. Lastly, it is hard to find the landmarks correctly through soft tissue when performing the needle insertion. Since we need to produce a model as accurate as possible, all these matters must be taken into consideration.

The 'needle technique' cannot be used on living subjects, which leads to errors in measurement. After 1980, with the development of technology, non-invasive medical systems become popular. A variety of methods have been used to measure tissue depth in living subjects, including ultrasound, MRI, and CT. In 1987, George [2] used lateral craniographs to record the depths of tissue at the midline anthropometric points. And in 2000, Manhein et. al. [3] used ultrasound to collect information for sample of children and adults of sexes, varying ages and different varieties. El-Mehallawi and Soliman [4] and De Greef et. al. [5] also used ultrasound to conduct study. In 2002, Sahni et. al. [6] used MRI to obtain tissue depth data of Indians. The most accurate measurement can be obtained by using CT. This technique is faster and more accurate as it gives high quality images. With the help of the computer, we can also construct the 3D model from the CT images. In 1996, Phillips and Smuts [7] used CT technique to obtain data of mixed population of South Africa. There are many more related researches that collect soft tissue thicknesses for study. However, most measurements are collected from rather small populations due to the harm it may cause when tests are carried out. Ultrasound techniques seem to be the most accurate and safe as it can be used without any considerable threat for the candidate [8]. MRI has the advantage of collecting data in 3D format. Soft tissue visualization is excellent, but bony tissue is not as well visualized as on the CT scan [9]. In addition, they just gather tissue depths data at anthropometric landmarks, but give no information about any relationship between these depths and the skull shape. Therefore, in most facial reconstruction systems, they just use the average thicknesses that can be calculated from the database for every landmark.

3   Method

Our method consists of the following steps: First, we collect the data from our candidates. After the database is ready, the data processing begins. We use two separate methods, linear regression and neural networks in order to get the prediction model. In this section, we describe these steps in sequence.

3.1   Data Collecting

We obtain our database using the Computed Tomography (CT) technique. This technique is very convenient because of it is fast, accurate, and can produce high quality images. It neglects the disadvantages of [13] because it can capture the skin and skull models of living objects. Besides, we can build 3D models from CT images. Due to some anthropometric researches, the anthropometric landmarks give the majority of information to describe the face. Therefore, we decide to use the thicknesses at the anthropometric landmarks and the distances between them as our features, which makes it easier to implement machine learning algorithms, while maintaining a moderate accuracy. We measure the distances manually by selecting the start point and end point for each distance in the CT image. The image is at the correct ratio with the real person so that the distance can be converted from pixel to milimetre. For each candidate, 9 CT slices are enough to obtain the needed information. From the CT images, we measure the soft tissue thicknesses at anthropometric locations in the cut plane.

Each entry in our database consists of 17 anthropometric distances measured from the skull, along with additional information such as age, gender, weight, and height, coupled with 38 features including soft tissue thicknesses at anthropometric locations in the skull. In the case of facial reconstruction, only these

17 distances are known. In short, each entry in our database is a set of input and output, with input is the 17 distances, and output is the 38 features.

Figure 1 shows an example of our thickness measurement process. Table 1 shows some sample input and output fields.

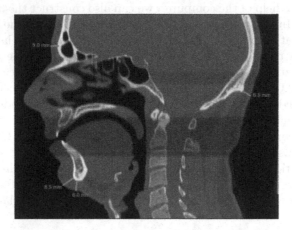

Fig. 1. Soft tissue thicknesses measuarement in sagittal CT image

3.2   Data Processing Using Linear Regression

Linear regression is a method to model the relationship between two variables by fitting a linear equation to the observed data. A linear regression line has an equation of the form Y = a + bX, where X is one of the distances in input set and Y is one of the thicknesses in output. The slope of the line is b, and a is the intercept.

If there is a approximately linear relationship between a distance in input and a thickness in output, linear regression can discover such relationship. For each soft tissue thickness in the output, we apply the linear regression on each distance in the input one by one, to find the one with best performance.

3.3   Data Processing Using Neural Networks

Artificial neural network have seen a rapid increase of interest over the last few years, and are being successfully applied on a wide range of domains as they are capable of solving prediction and classification problems. Multi-Layer Perceptron (MLP) is perhaps the most popular artificial neural network architecture. It is a feedforward artificial neural network model that maps set of input data onto a set of appropriate output. Its units each perform a biased weighted sum of their inputs and pass this activation level through a transfer function to produce their outputs. The MLP consists of three or more layers. Each node in one layer connects to every node in the following layer with a certain weight w_{ij}. We select the two-layer feedforward network, with a tan-sigmoid transfer function

Table 1. Sample Input and Output Fields

Input name	Description
cranial height	distance from the midpoint of the anterior border of the foramen magnum (basion) to the intersection of the coronal and sagittal sutures (bregma).
cranial length	distance from the midsagittal plane from the most anterior point on the frontal (glabella) to the most posterior point on the occipital (opisthocranion).
cranial breadth	greatest width between the parietal eminences (euryon).
facial height	distance from the most anterior inferior point of the mandible in the median sagittal plane (gnathion) to the point of intersection of the internasal suture with the nasofrontal suture (nasion).

Output name	Description
vertex	soft tissue thickness at the vertex.
glabella	soft tissue thickness at the glabella.
nasion	soft tissue thickness at the nasion.
pronasale	soft tissue thickness at the pronasale.
n-pronasale	distance from the nasion to the pronasale.

in the hidden layer and a linear transfer function in the output layer because this structure can represent any functional relationship between inputs and outputs if the hidden layer has enough neurons [14].

For each thickness in output, we calculate the correlation coefficients for the matrix combined of input and the thickness to realize the distances in input that have high correlation with the thickness. After that, the training procedure starts. In this stage, weights are initialised randomly, and the network is trained for several times. After training, the network with the best performance in terms of mean square error on the validation set is selected and applied to the test set to get the final result.

4 Result

We perform the evaluation on the dataset of males which contains 98 samples. In our evaluation, we use the ten-fold cross-validation to compute the output's MSE for the two approaches, linear regression and neural network. As for neural network, the training is done several times, with the number of neurons from 10 to 20 and randomly initialized weights each time. The network with best performance over the validation set is chosen to generate output for the test set.

We then compare these MSE with the 'average method' in which the output thickness for all tests is simply the average of all the output in training set. This 'average method' is what is used in almost every facial reconstruction systems so far.

Table 2. MSE values for 'average method' (AVG), Linear Regression (LR), and Neural Network (NN). The best performance is in boldface.

N#	Output	AVG	LR	NN
1	vertex	1.1914	1.0625	**0.8928**
2	trichion	1.2945	1.0877	**1.0664**
3	glabella	1.2074	**1.0110**	1.0706
4	nasion	0.9699	0.7571	**0.7220**
5	rhinion	0.3886	**0.3400**	0.3797
6	pronasale	7.9621	6.0558	**5.2456**
7	nose length	21.8621	10.8344	**8.7059**
8	subnasale	6.3008	**4.3927**	4.6878
9	upper lip border	4.9468	4.3581	**3.7205**
10	lower lip border	3.1674	2.7312	**2.4167**
11	stomion	2.2193	1.8766	**1.8168**
12	metal	4.1007	3.4298	**3.3625**
13	meton	2.3685	**1.9901**	2.0885
14	opisthooranion	1.8909	1.5124	**1.1001**
15	exocanthion (R)	0.7884	**0.6635**	0.7084
16	exocanthion (L)	0.8609	**0.7121**	0.8459
17	endocanition (R)	2.5804	2.0950	**1.7213**
18	endocantion (L)	2.6779	2.0706	**2.0099**
19	pupil-pupil	10.8380	**4.4587**	4.9687
20	supraobital (R)	0.6689	0.5533	**0.4556**
21	supraobital (L)	0.6859	0.5340	**0.4986**
22	infraobital (R)	1.4038	1.2479	**1.0475**
23	infraobital (L)	1.1147	**0.9573**	1.1920
24	zygomatic arch (R)	0.8485	**0.7432**	1.6805
25	zygomatic arch (L)	0.8857	**0.7400**	0.7982
26	zygomatic (R)	0.8326	0.6982	**0.5635**
27	zygomatic (L)	0.9557	**0.7722**	1.3729
28	porion (R)	3.3546	**2.7241**	2.9786
29	porion (L)	2.5552	2.0471	**1.7367**
30	gonion (R)	1.0521	0.9333	**0.8245**
31	gonion (L)	0.9360	**0.8330**	1.5443
32	alare (R)	2.0965	1.6396	**1.5934**
33	alare (L)	2.0342	1.5304	**1.4494**
34	lateral nasal (R)	1.9751	**1.4220**	1.5541
35	lateral nasal (L)	2.0908	1.3537	**1.3495**
36	nose height	4.1012	**3.5995**	4.5687
37	bucal (R)	13.6992	**11.2034**	12.2837
38	bucal (L)	13.9451	**11.6959**	11.7598

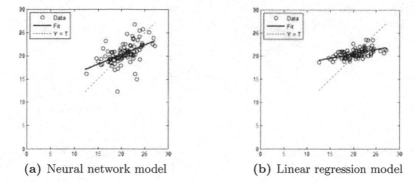

(a) Neural network model   (b) Linear regression model

Fig. 2. Regression results obtained by ten-fold cross validation for **pronasale thickness** using (a) neural network model and (b) linear regression model.

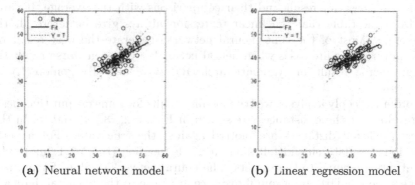

(a) Neural network model   (b) Linear regression model

Fig. 3. Regression results obtained by ten-fold cross validation for **nose length** using (a) neural network model and (b) linear regression model.

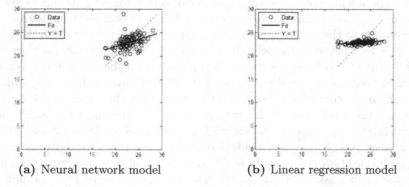

(a) Neural network model   (b) Linear regression model

Fig. 4. Regression results obtained by ten-fold cross validation for **nose height** using (a) neural network model and (b) linear regression model.

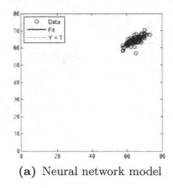

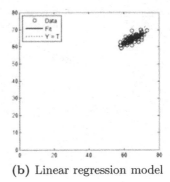

(a) Neural network model   (b) Linear regression model

Fig. 5. Regression results obtained by ten-fold cross validation for **pupil-pupil distance** using (a) neural network model and (b) linear regression model.

Table 2 shows our result and their comparisons with the average. It can be seen from the table that the linear regression always give better result than the average. Most of the time, neural networks generate the best result over all. However, there are cases when neural network gives even worse result than average such as result for zygomatic arch (R), zygomatic (L), gonion (L), and nose height.

In order to deeply analysis, we try plotting results for some output thicknesses. The results over these distances are shown in Figure 2, 3, 4, and 5. In these figures, predicted distances are plotted against the true value. For a perfect prediction, the data should fall along a 45 degree line (the Y=T line), where the outputs are equal to the targets. The neural network's values for pronasale thickness, nose length, pupil-pupil distance are close to the diagonal, indicating the prediction was good. For linear regression, prediction for nose length and pupil-pupil distance seems to have good performance. The other predictions are not as good, but acceptable.

The experimental results show good performance for long distances such as pronasale thickness or nastion-pronasale distance, and bad performance for short distances, due to the error appeared in the measurement process. This is because the longer the distance, the less effect it receives from measurement error. In addition, the thin soft tissues do not depend much on the skull shape. Another thing that needs to be noted is that, in 2009, Pascal Paysan et. al. [15] proposed a method to reconstruct the face from the skull, with the capable of tuning the weight and age attributes. We know that weight and age affect the facial shape. Our candidates' age and weight are within wide range of 18 to 82 and 43kg to 75kg, respectively. Separating candidates into groups is very important because the relationship between features is different from this age and weight range to the others, and missing this step will lead to moderate error in training and validation. However, in our experiment, we could not separate the candidates into groups because the number of entries was not sufficient.

5   Conclusions

Facial reconstruction is an interesting research field as it helps in many cases. Researchers have been developing facial reconstruction systems to fasten the manual process and also to produce better results. In this general problem, one of the most important issues is to determine the soft tissue thicknesses at landmarks on the skull. However, most facial reconstruction systems neglect this issue and use the average thickness for simplicity. Our research has pointed out that this 'average method' has worse performance than our linear regression method in every case, and worse than our neural network method in most cases. Our research also shows that there are relationships between the skull shape and the tissue depths and should people investigate more and more to discover these relationships. However, our research has some limitation which can be improved to obtain better results.

The first possible development is to improve measurement process. As can be seen from the experiments, our results show good performance for long distances such as pronasale thickness or nose length, and bad performance for short distances, due to the error appeared in the measurement process. This is because the longer the distance, the less effect it receives from measurement error. In addition, the thin soft tissues do not depend much on the skull shape, or in other words, they do not have much relationship with the metrics. In addition, to define the landmarks on the CT images depends much on the skill and judgment of the people who perform the measurement, although this technique is the most accurate. This method also requires a lot of time to measure and collect data. We plan to apply image processing to automatic discovery of these metrics. This would save a lot of time in measurement and might give better accuracy.

Secondly, separating candidates into groups is very important because the relationship between features is different from this age and weight range to the others and missing this step will lead to moderate error in training and validation. However, in our experiment, we could not separate the candidates into groups because the number of entries was not sufficient. Separating would give even worse result. In the future, we will collect more data for each group of weight and age. This will improve the prediction performance significantly.

In addition, because our data and problem is straight forward, many other machine learning techniques can be applied such as the decision stump, support vector machines, or boosting. With satisfactory results from neural network approach, it is possibly that better result can be obtained from other techniques. We plan to implement and analyze result using different techniques.

References

1. Welcker, H.: Schiller's schädel und todenmaske, nebst mittheilungen über schädel und todenmaske kants. (1883)
2. George, R.M.: The lateral craniographic method of facial reconstruction. Journal of Forensic Sciences 32, 1305–1330 (1987)

3. Manhein, M.H., Listi, G.A., Barsley, R.E., Musselman, R., Barrow, N.E., Ubelaker, D.H.: In vivo facial tissue depth measurements for children and adults. Journal of Forensic Sciences 45, 48–60 (2000)
4. El-Mehallawi, I.H., Soliman, E.M.: Ultrasonic assessment of facial soft tissue thicknesses in adult egyptians. Journal of Forensic Sciences 117, 99–107 (2001)
5. De Greef, S., Claes, P., Vandermeulen, D., Mollemans, W., Suetens, P., Willems, G.: Large-scale in-vivo caucasian facial soft tissue thickness database for craniofacial reconstruction. Journal of Forensic Sciences 159, 126–146 (2006)
6. Sahni, D., Jit, I., Gupta, M., Singh, P., Suri, S., Sanjeev, Kaur, H.: Preliminary study on facial soft tissue thickness by magnetic resonance imaging in northwest indians. Forensic Science Communications 4 (2002)
7. Phillips, V.M., Smuts, N.A.: Facial reconstruction: Utilization of computerized tomography to measure facial tissue thickness in a mixed racial population. Forensic Sci. Int. 83, 51–59 (1996)
8. Wilkinson, C.: Forensic facial reconstruction. Cambridge University Press, Cambridge (2004)
9. Van der Pluym, J., Shan, W.W., Taher, Z., Beaulieu, C., Plewes, C., Peterson, A.E., Beattie, O.B., Bamforth, J.S.: Use of magnetic resonance imaging to measure facial soft tissue depth. Cleft Palate-Craniofacial Journal 44, 52–57 (2007)
10. Muller, J., Mang, A., Buzug, T.M.: A template-deformation method for facial reproduction. In: Proceedings of the 4th International Symposium on Image and Signal Processing and Analysis, ISPA 2005 , pp. 359–364 (2005)
11. Kähler, K., Haber, J., Seidel, H.P.: Reanimating the dead: reconstruction of expressive faces from skull data. ACM Transactions on Graphics (TOG) 22, 554–561 (2003)
12. Berar, M., Desvignes, M., Bailly, G., Payan, Y.: 3d statistical facial reconstruction. In: Proceedings of the 4th International Symposium on Image and Signal Processing and Analysis, ISPA 2005, pp. 365–370 (2005)
13. Krogman, W.M.: The reconstruction of the living head from the skull. FBI Law Enforcement Bulletin 15, 1–8 (1946)
14. Hagan, M.T., Demuth, H.B., Beale, M.H.: Neural Network Design. PWS Publishing Company, Boston (1996)
15. Paysan, P., Lüthi, M., Albrecht, T., Lerch, A., Amberg, B., Santini, F., Vetter, T.: Face Reconstruction from Skull Shapes and Physical Attributes, ch. 24. In: Denzler, J., Notni, G., Süße, H. (eds.) Pattern Recognition. LNCS, vol. 5748, pp. 232–241. Springer, Heidelberg (2009)

Intelligence Trading System for Thai Stock Index

Monruthai Radeerom[1], Hataitep Wongsuwarn[2], and M.L. Kulthon Kasemsan[1]

[1] Faculty of Information Technology, Rangsit University, Pathumtani, Thailand 12000
[2] ME, Faculty of Engineering at Kasetsart University (Kamphaeng Saen),
Nakhonpathom, Thailand 73140
mradeerom@yahoo.com, fenghtw@ku.ac.th,
kasemsan@rangsit.rsu.ac.th

Abstract. Stock investment has become an important investment activity in Thailand. However, investors often lose money due to unclear investment objectives. Therefore, an investment decision support system to assist investors in making good decisions has become an important research issue. Thus, this paper introduces an intelligent decision-making model, based on the application of Neurofuzzy system (NFs) technology. Our proposed system can decide a trading strategy for each day and produce a high profit for of each stock. Our decision-making model is used to capture the knowledge in technical indicators for making decisions such as buy, hold and sell. Finally, the experimental results have shown higher profits than the Neural Network (NN) and "Buy & Hold" models for each stock index. The results are very encouraging and can be implemented in a Decision- Trading System during the trading day.

Keywords: Computational intelligence, Neuro-Fuzzy System, Stock Index, Decision Making System.

1 Introduction

The prediction of financial market indicators is a topic of considerably practical interest and, if successful, may involve substantial pecuniary rewards. People tend to invest in equity because of its high returns over time. Considerable efforts have been put into the investigation of stock markets. The main objective of the researchers is to create a tool, which could be used for the prediction of stock markets fluctuations; the main motivation for this is financial gain. In the financial marketplace, traders have to be fast and hence the need for powerful tools for decision making in order to work important, efficiently and profit.

The use of Artificial Intelligence (AI) had a big influence on the forecasting and investment decision-making technologies. There are a number of examples using neural networks in equity market applications, which include forecasting the value of a stock index [4,5], recognition of patterns in trading charts[12,17] rating of corporate bonds[8], estimation of the market price of options[11], and the indication of trading signals of selling and buying[3,12].

Even though most people agree on the complex and nonlinear nature of economic systems, there is skepticism as to whether new approaches to nonlinear modeling,

N.T. Nguyen et al. (Eds.): New Challenges for Intelligent Information, SCI 351, pp. 127–136.

such as neural networks, can improve economic and financial forecasts. Some researchers claim that neural networks may not offer any major improvement over conventional linear forecasting approaches [8, 12]. In addition, there is a great variety of neural computing paradigms, involving various architectures, learning rates, etc., and hence, precise and informative comparisons may be difficult to make. In recent years, an increasing amount of research in the emerging and promising field of financial engineering has been incorporating Neurofuzzy approaches [10, 12, 16, 19]. Almost all models are focused on the prediction of stock prices. The difference of our proposed model is that we are focusing on decision-making in stock markets, but not on forecasting in stock markets.

In contrast to our previous work [15], we are not making a direct prediction of stock markets, but we are working on a one-day forward decision-making tool for buying/selling stocks. We are developing a decision-making model which, besides the application of NFs, uses optimization algorithms based on the rate of the return profit of each stock index to construct our NFs Model for a decision support - making system. In this paper, we present a decision-making model which combines technical analysis model and NFs model. The technical analysis model evaluated knowledge about buy, hold and sell strategy from each technique. Our proposed model used result from technical analysis model to input of our NFs. And secondly model, the NFs trading system decides the buy, sell and hold strategy for each stock index. The objective of this model is to analyze the daily stock and to make one day forward decisions related to the purchase of stocks.

The paper is organized as follows: Section 2 presents the background about the neural network and the Neurofuzzy system. Section 3 presents the NFs decision-making model; Sections 4 is devoted to experimental investigations and the evaluation of the decision-making model. This section provides the basis for the selection of different variables used in the model, and models the structure. The main conclusions of the work are presented in Section 5, with remarks on future directions.

2   Computational Intelligence Approaches for the Intelligence Trading System

2.1   Neural Networks (NNs) for Modeling and Identification

The neural networks are used for two main tasks: function approximation and pattern classification. In function approximation, the neural network is trained to approximate a mapping between its inputs and outputs. Many neural network models have been proven as universal approximations, i.e. the network can approximate any continuous arbitrary function well. The pattern classification problem can be regarded as a specific case of the function approximation. The mapping is done from the input space to a finite number of output classes.

For function approximation, a well-known model of NNs is a feed forward multi-layer neural network (MNN). It has one input layer, one output layer and a number of hidden layers between them. For illustration purposes, consider a MNN with one hidden layer (Figure 1). The input-layer neurons do not perform any computations. They merely distribute the inputs to the weights of the hidden layer. In the neurons of

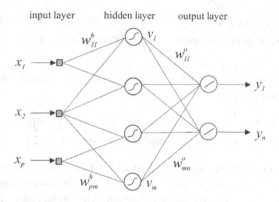

input layer   hidden layer   output layer

Fig. 1. A feedforward neural network with one hidden layer [19]

the hidden layer, first the weighted sum of the inputs is computed. It is then passed through a nonlinear *activation function*, such as the tangent hyperbolic.

Training is the adaptation of weights in a multi-layer network such that the error between the desired output and the network output is minimized. A network with one hidden layer is sufficient for most approximation tasks. More layers can give a better fit, but the training time takes longer. Choosing the right number of neurons in the hidden layer is essential for a good result. Too few neurons give a poor fit, while too many neurons result in overtraining of the net (poor generalization of unseen data). A compromise is usually sought by trial and error methods.

The backpropagation algorithm [16] has emerged as one of the most widely used learning procedures for multi-layer networks. There are many variations of the back propagation algorithm, several of which will be discussed in the next section. The simplest implementation of backpropagation learning updates the network weights and biases in the direction that the performance function decreases most rapidly.

2.2 Neurofuzzy System (NFs) for Modeling and Identification

Both neural networks and the fuzzy system imitate human reasoning process. In fuzzy systems, relationships are represented explicitly in forms of if-then rules. In neural networks, the relations are not explicitly given, but are coded in designed networks and parameters. Neurofuzzy systems combine the semantic transparency of rule-based fuzzy systems with the learning capability of neural networks. Depending on the structure of if-then rules, two main types of fuzzy models are distinguished as mamdani (or linguistic) and takagi-sugeno models [18]. The mamdani model is typically used in knowledge-based (expert) systems, while the takagi-sugeno model is used in data-driven systems

In this paper, we consider only the Takagi - Sugeno-Kang (TSK) model. Takagi, Sugeno and Kang [18] formalized a systematic approach for generating fuzzy rules from an input-output data pairs. The fuzzy if-then rules, for the pure fuzzy inference system, are of the following form:

$$if \ x_1 \ is \ A_1 \ and \ x_2 \ is \ A_2 \ and \ x_N \ is \ A_N \ then \ y = f(x) \qquad (1)$$

Where $x = [x_1, x_2, ..., x_N]^T$, $A_1, A_2, ..., A_N$ fuzzy sets are in the antecedent, while y is a crisp function in the consequent part. The function is a polynomial function of input variables $x_1, x_2, x_3, ..., x_N$.

The first-order TSK fuzzy model could be expressed in a similar fashion. Consider an example with two rules:

$$if \ x_1 \ is \ A_{11} \ and \ x_2 \ is \ A_{21} \ and \ then \ y_1 = p_{11}x_1 + p_{12}x_2 + p_{10}$$
$$if \ x_1 \ is \ A_{12} \ and \ x_2 \ is \ A_{22} \ and \ then \ y_2 = p_{21}x_1 + p_{22}x_2 + p_{20}$$
(2)

Figure 2 shows a network representation of those two rules. The nodes in the first layer compute the membership degree of the inputs in the antecedent fuzzy sets. The product node $\prod$ in the second layer represent the antecedent connective (here the "and" operator). The normalization node N and the summation node $\sum$ realize the fuzzy-mean operator for which the corresponding network is given in Figure 2 Applying fuzzy singleton, a generalized bell function such as membership function and algebraic product aggregation of input variables, at the existence of M rules the Neurofuzzy TSK system output signal upon excitation by the vector, are described by

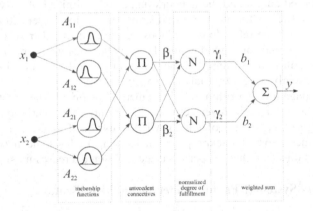

Fig. 2. An example of a first-order TSK fuzzy model with two rules systems [1]

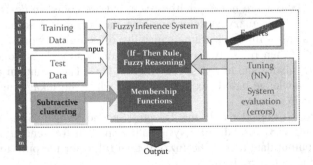

Fig. 3. Constructing Neurofuzzy Networks

In conclusion, Figure 3 summarizes the Neurofuzzy Networks System (NFs). Construction process data called "training data sets," can be used to construct Neurofuzzy systems. We do not need prior knowledge ala "knowledge-based (expert) systems". In this way, the membership functions of input variables are designed by the subtractive clustering method. Fuzzy rules (including the associated parameters) are constructed from scratch by using numerical data. And the parameters of this model (the membership functions, consequent parameters) are then fine-tuned by process data.

3 Methodology for The Intelligence Portfolio Management System

3.1 Decision-Making Model for The Stock Market

Many stock market traders always use conventional statistical techniques for decision-making in purchasing and selling.[7] Popular techniques are used fundamental analysis and technical analysis. They are often net profits on the stock market, but they require a lot of knowledge and experience. Because stock markets are affected by many highly interrelated economic, political and even psychological factors, and these factors interact with each other in a very complex manner, it is, generally, very difficult to forecast the movements of stock markets (see Figure 4). Figure 4 shows Historical Quotes of Bangchak Petroleum public Co., Ltd. (BCP) Stock Prices. It is a high nonlinear system. In this paper, we are working on one day decision making for buying/selling stocks. For that we are developing a decision-making model, besides the application of an intelligence system. We selected a Neurofuzzy system (NFs), which are studied in the emerging and promising field of financial engineering are incorporating Neurofuzzy approaches [2, 8, 10, 14, 19, and 20].

We proposed NFs for the decision making system, called intelligence trading system. The model scenario represents one time calculations made in order to get decisions concerning the purchase of stocks. For this paper, historical data of daily stock returns was used for time interval. At first step of the model realization, technical analysis techniques are used to the decision strategy recommendation. The recommendations (R) represent the relative rank of investment attraction to each stock in the interval [−1, 1]. The values −1, 0, and 1 represent recommendations: Sell, Hold and Buy, respectively. After that, the recommendations are included in the input of a proposed intelligence system. Output intelligence system is the evaluating recommendation based on several recommendations from any technical techniques used by investors. The proposed NNs and NFs for a intelligence trading system are shown in Figure 5.

3.2 Preprocessing of the Proposed NFs Input

Technical analysts usually use indicators to predict the buy and sell signal future. The major types of indicators are Moving Average Convergence/Divergence (MACD), Williams's %R (W), Relative Strength Index (RSI), Exponential Moving Average (EMA), On Balance Volume and etc that corresponding on close price and volume. These indicators can be derived from the real stock composite index. All of indicator is included in input signal for intelligence system. And, the target for training is the buy and sell signal as shown on fig. 5.

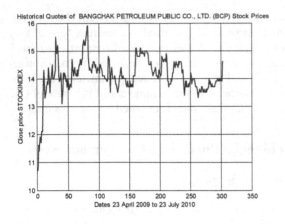

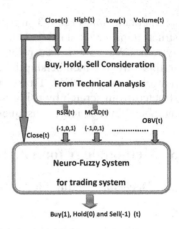

Fig. 4. Historical Quotes of Bangchak Petroleum public Co., Ltd. (BCP) Stock Prices

Fig. 5. The scenario of Intelligence Trading System

For daily data, indicators can help traders identify trends and turning points. The moving average is a popular and simple indicator for trends. Stochastic and RSI are some simple indicators which help traders identify turning points. Some example indicators are defined as follows,

$$RSI = 100 - \frac{100}{1 + \dfrac{\sum (positive\ change)}{\sum (negative\ change)}} \qquad (3)$$

In general, stock price data has a bias due to differences in name and spans. Normalization can be used to reduce the range of the data set to values appropriate for inputs to the activation function being used. The normalization and scaling formula is

$$y = \frac{2x - (\max + \min)}{(\max - \min)}, \qquad (4)$$

Where x is the data before normalizing, y is the data after normalizing.

Because the index prices and moving averages are in the same scale, the same maximum and minimum data are used to normalize them. The max is derived from the maximum value of the linked time series; similarly minimum is derived from the minimum value of the linked time series. The maximum and minimum values are from the training and validation data sets. The outputs of the neural network will be rescaled back to the original value according to the same formula.

For NFs trading system, the expected returns are calculated considering the stock market. That is, the value obtained on the last investigation day is considered the profit. The traders's profit is calculated as

$$\text{Profit(n)} = \text{Stock Value(n)} - \text{Investment value} \qquad (5)$$

Where n is the number of trading days.

And the Rate of Return Profit (RoRP) is

$$RoRP = \frac{Profit(n)}{Investment\ value} \times 100 \tag{6}$$

4 Results and Discussion

The model realization could be run having different groups of stocks (like Banking group, Energy group, etc.), indexes or other groups of securities. For that, we are using market orders as it allows simulating buying stocks, when stock exchange nearly closed market. All the experimental investigations were run according to the above presented scenario and were focused on the estimation of Rate of Return Profit (RoRP). At the beginning of each realization, the start investment is assumed to be 1,000,000 Baht (Approximately USD 29,412). The data set, including the Stock Exchange of Thailand (SET) index, Historical Quotes of Bangchak Petroleum public Co., Ltd. (BCP) Stock Prices, Siam Commercial Bank (SCB) and Petroleum Authority of Thailand (PTT) stock index, has been divided into two different sets: the training data and test data. The stock index data is from April 23, 2009 to July 23, 2010 totaling 304 records. The first 274 records are training data, and the rest of the data, i.e., 30 records, will be test data. Moreover, the data for stock prices includes the buy-sell strategy, closing price and its technical data. Consequently, max-min normalization can be used to reduce the range of the data set to appropriate values for inputs and output used in the training and testing method.

4.1 Input Variables

Technical indexes are calculated from the variation of stock price, trading volumes and time according to a set of formulas to reflect the current tendency of the stock price fluctuations. These indexes can be applied for decision making in evaluating the phenomena of oversold or overbought stock. For input data, several technical indexes are described as shown in Table 1. It is totally 12 inputs. And, Output is only one.

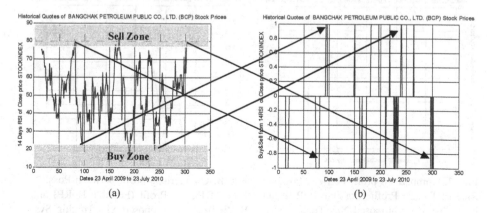

Fig. 6. (a) The 14 periods RSI index (RSI14 (t)) calculated by close price(t)
(b) The Buy (1), Hold (0), Sell (-1) evaluated by RSI14(t)

Table 1. Input and output of Intelligence trading system

NO.	DESCRIPTION
	INPUT
1	Close price(t)
2	Typical price(t)
3	Volume rate of change(t)
4	Price and Volume Trend (t)
5	Price rate of change 12 (t)
6	On-Balance Volume(t)
7	Buy & Sell from RSI 4 Days(t)
8	Buy & Sell from RSI 9 Days(t)
9	Buy & Sell from RSI 14 Days(t)
10	Buy & Sell from William 10 Days(t)
11	Buy & Sell from MACD 10 Days(t)
12	Buy & Sell from EMA 10 and 25 Days(t)
	OUTPUT
1	Buy(1), Hold(0) and Sell(-1)

Table 2. Example of Financial Simulation Model in trading strategy

Possible Buy&Sell			NFs Buy&Sell		
Action	# of Shares	Cash (Baht)	Action	# of Shares	Cash (Baht)
1 STAY	-	1,000,000	1 STAY	-	1,000,000
0 HOLD	-	1,000,000	1 HOLD	-	1,000,000
-1 BUY	13,245	-	0 HOLD	-	1,000,000
0 HOLD	13,245	-	-1 BUY	12,500	-
1 SELL	-	1,066,225	-1 HOLD	12,500	-
0 HOLD	-	1,066,225	-1 HOLD	12,500	-
0 HOLD	-	1,066,225	1 SELL	-	903,125
0 HOLD	-	1,066,225	1 HOLD	-	903,125
-1 BUY	15,178	-	0 HOLD	-	903,125
1 SELL	-	1,107,964	1 SELL	-	903,125

4.2 Evaluating Decision-Making System based on Neurofuzzy Model

We now compare the performance of our proposed NFs with NNs including three types of learning algorithm methods. The learning methods are Batch Gradient Descent (TRAINGD), Scaled Conjugate Gradient (TRAINSCG) and Levenberg-Marquardt (TRAINLM) methods The neural network model has one hidden layer with 30 nodes. And, learning iteration is 10000 epochs. After we trained their learning method, we found scaled conjugate better than other learning methods. Actually, we can conclude that our proposed Neurofuzzy demonstrated four relation types considerably better than a NNs with scaled conjugate gradient learning.

After developing the intelligence trading system, we were given 1,000,000 baht for investment at the beginning of the testing period. The decision for to buy and sell stocks is given by proposed intelligence output. We translated the produced RoRP results that verify the effectiveness of the trading system. Table 2 is show a Financial

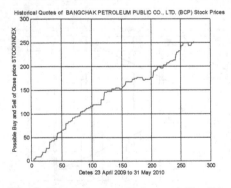

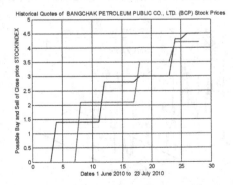

Fig. 7. Comparison profit between Possible Rate of Return Profit (Possible RoRP) and Profit from our proposed NFs Trading System in Training Days

Fig. 8. Comparison profit between Possible Rate of Return Profit (Possible RoRP) and Profit from our proposed NFs Trading System in Testing Days.

Simulation Model for calculating profit in our trading strategy. Results within training days and testing day are shown in Figure 7 and 8, respectively.

Moreover, our proposed decision-making NFs model compared RoRP performance with Buy & Hold Strategy and NNs. The antithesis of buy and hold is the concept of day trading in which money can be made in the short term if an individual tries to short on the peaks, and buy on the lows with greater money coming with greater volatility.

The performance by each stock index is illustrated on Table 3. It reflects the performances of investment strategies in NN, buy and hold and NFs model, respectively. Each line implies the performance of the NFs system in terms of cumulative profit rate of return gained from each stock index. In the case of experimental results, NFs display a greater rate of return than the "buy, sell and hold" model and NN model. The results of difference in the stock index results are small. It is more valuable to calculate the loss and gains in terms of profitability in practice.

Table 3. Rate of Return Profit (RoRP) gained from each trading stock index

STOCK INDEX	STOCK GROUP	Profits of Training Day (274 Days), (%)			
		Possible	NN	NF	Buy & Hold
BCP	Energy	254	240	254	50
PTT	Energy	320	300	320	80
SCB	Banking	180	160	180	70

STOCK INDEX	STOCK GROUP	Profits of Testing Day (28 Days), (%)			
		Possible	NN	NF	Buy & Hold
BCP	Energy	4.5	3.3	4.3	0.8
PTT	Energy	7.1	4.5	6.9	1.5
SCB	Banking	3.2	2.1	2.9	0.5

5 Conclusion

This paper presented the decision-making model based on the application of NFs. The model was applied in order to make a one-step forward decision, considering from historical data of daily stock returns. The experimental investigation has shown, NFs trading system to make a trading strategy, achieving more stable results and higher profits when compared with NNs and Buy and Hold strategy. For future work, several issues could be considered. Other techniques, such as support vector machines, genetic algorithms, etc. can be applied for further comparisons. And other stock index groups, another stock exchange or other industries in addition to electronics one is further considered for comparisons.

Acknowledgment. This piece of work was partly under the Graduate Fund for Ph.D. Student, Rangsit University, Pathumthanee, Thailand.

References

1. Babuska, A.R.: Neuro-fuzzy methods for modeling and identification. In: Recent Advances in intelligent Paradigms and Application, pp. 161–186 (2002)
2. Cardon, O., Herrera, F., Villar, P.: Analysis and Guidelines to Obtain A Good Uniform Fuzzy rule Based System Using simulated Annealing. Int'l J. of Approximated Reason 25(3), 187–215 (2000)

3. Chapman, A.J.: Stock market reading systems through neural networks: developing a model. Int'l J. of Apply Expert Systems 2(2), 88–100 (1994)
4. Chen, A.S., Leuny, M.T., Daoun, H.: Application of Neural Networks to an Emerging Financial Market: Forecasting and Trading The Taiwan Stock Index. Computers and Operations Research 30, 901–902 (2003)
5. Conner, N.O., Madden, M.: A Neural Network Approach to Pre-diction Stock Exchange Movements Using External Factor. Knowledge Based System 19, 371–378 (2006)
6. Liu, J.N.K., Kwong, R.W.M.: Automatic Extraction and Identification of chart Patterns Towards Financial Forecast. Applied soft Computing 1, 1–12 (2006)
7. Doeksen, B., Abraham, A., Thomas, J., Paprzycki, M.: Real Stock Trading Using Soft Computing Models. In: IEEE Int'l Conf. on Information Technology: Coding and Computing, Las Vegas, Nevada, USA, pp. 123–129 (2005)
8. Dutta, S., Shekhar, S.: Bond rating: A non-conservative application of neural networks. In: IEEE Int'l Conf. on Neural Networks, San Diego, CA, USA, pp. 124–130 (1990)
9. Farber, J.D., Sidorowich, J.J.: Can new approaches to nonlinear modeling improve economic forecasts? The Economy As An Evolving Complex System, 99–115 (1988)
10. Hiemstra, Y.: Modeling Structured Nonlinear Knowledge to Predict Stock Markets: Theory. In: Evidena and Applications, Irwin, pp. 163–175 (1995)
11. Hutchinson, J.M., Lo, A., Poggio, T.: A nonparametric approach to pricing and hedging derivative securities via learning networks. Int'l J. of Finance 49, 851–889 (1994)
12. James, N.K., Raymond, W.M., Wong, K.: Automatic Extraction and Identification of chart Patterns towards Financial Forecast. Applied soft Computing 1, 1–12 (2006)
13. LeBaron, B., Weigend, A.S.: Evaluating neural network predictors by bootstrapping. In: Int'l Conf. on Neural Information Process, Seoul, Korea, pp. 1207–1212 (1994)
14. Li, R.-J., Xiong, Z.-B.: Forecasting Stock Market with Fuzzy Neural Network. In: 4th Int'l Conf. on Machine Learning and Cybernetics, Guangzhou, China, pp. 3475–3479 (2005)
15. Radeerom, M., Srisa-an, C.: Prediction Method for Real Thai Stock Index Based on Neurofuzzy Approach. In: Trends in Intelligent Systems and Computer Engineering. LNEE, vol. 6, pp. 327–347 (2008)
16. Refenes, P., Abu-Mustafa, Y., Moody, J.E., Weigend, A.S. (eds.): Neural Networks in Financial Engineering. World Scientific, Singapore (1996)
17. Tanigawa, T., Kamijo, K.: Stock price pattern matching system: dynamic programming neural network approach. In: Int'l J. Conf. on on Neural Networks, vol. 2, pp. 59–69 (1992)
18. Takagi, T., Sugeno, M.: Fuzzy identification of systems and its application to modeling and control. IEEE Trans. on System, Man and Cybernetics 5, 116–132 (1985)
19. Trippi, R., Lee, K.: Artificial Intelligence in Finance & Investing. Chicago, IL, Irwin (1996)
20. Tsaih, R., Hsn, V.R., Lai, C.C.: Forecasting S&P500 Stock Index Future with A Hybrid AI System. Decision Support Systems 23, 161–174 (1998)

Mining for Mining — Application of Data Mining Tools for Coal Postprocessing Modelling

Marcin Michalak[1,2], Sebastian Iwaszenko[1], and Krzysztof Wierzchowski[1]

[1] Central Mining Institute, Plac Gwarkow 1, 40-166 Katowice, Poland
{Marcin.Michalak,Sebastian.Iwaszenko,Krzysztof.Wierzchowski}@gig.eu
[2] Silesian University of Technology, ul. Akademicka 16, 44-100 Gliwice, Poland
Marcin.Michalak@polsl.pl

Abstract. This article deals with the problem of coal postprocessing modelling. The sediment filter centrifuge is the part of the postprocessing that is being modelled with several data mining techniques. In this paper the results of parametrical and nonparametrical models applications are described.

Keywords: data mining, machine learning, parametrical regression, nonparametrical regression, machine modelling.

1 Introduction

Hard coal extractive industry in Europe and particularly in Poland is based on deep underground mines. Along with the mining spoil there is also some amount of barren rock transported to the ground surface. This materials are usually of limited usage and low eceonomic value and often should be treated as a waste. The main goal of the coal postprocesiing is to separate the coal from the barren rock as accureate as it is possible. As the element of this process the sedimental filter centrifuge is used.

In this short article several data mining tools were considered to become the model of centrifuge. It was expected that the quality of process product depends on the feed and working parameters. The final model should describe dependencies between input and output. On the basis of this model it should be possible to match the optimal feed characteristic and machine working parameters that should give the best quality of postprocessed coal.

Firs two parts of this paper describes the sedimental filter centrifuge construction, its operation conditions, working parameters and output product characteristics. Afterwards several regression function estimators are briefly described. Then experiments and their results are presented. The final part of this article contains some conclusions and description of further works.

2 Machine Description

Process of coal cleaning is usually conducted in liquid medium (wet coal cleaning). The coal dewatering is the last step of the process. It is commonly

N.T. Nguyen et al. (Eds.): New Challenges for Intelligent Information, SCI 351, pp. 137–146.
springerlink.com
© Springer-Verlag Berlin Heidelberg 2011

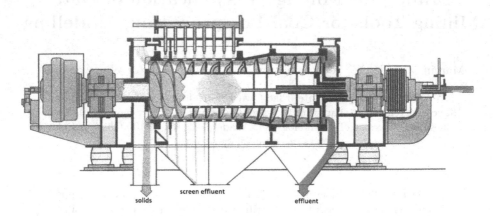

Fig. 1. Sediment filter centrifuge construction.

undertaken by means of centrifuge, which operational parameters are fit to dewatered concentrate grains size. The sediment filter centrifuges went to practical usage for finest coal concentrate grains in the end of last century. The sediment filter centrifuges can be considered as a composition of two types of centrifuges in one apparatus: the sediment centrifuge and filtering centrifuge. The feed is loaded to the installation through the feed pipe into the sediment part of the centrifuge (Fig. 2). Then the feed material is condensed into suspension of very high concentrations of solid parts. Condensed suspension is transported by means of endless screw to filtering part, where the dewatering on filter medium takes place. Dewatered sediment, optionally flashed, leaves the centrifuge through the output pipe. Besides dewatered sediment, the centrifuge produces two kinds of suspensions originating from sediment (effluent) and filtering part (screen effluent). The suspension from filtering part is usually recycled into the feed. The suspension form sediment part is usually treated as a waste. The sediment filter centrifuge has several advantages over the other equipment dedicated to fine grained coal dewatering:

- high efficiency,
- compact construction,
- relatively low moisture content in dewatered product (for selected feed grain size composition),
- loose, easy to homogenize with other concentrate streams output sediment,
- high solid parts recovery ($> 95\%$).

The most important disadvantages of sediment filtering centrifuges are coproduction of very fine and difficult to dewatering and utilization sediment as well as requirements for strict operational conditions.

3 Input and Output Variables

Characteristics of dewatered sediment and sediment part originating suspension are the most interesting aspects of sediment filter centrifuge operation. The characteristics of the filtering part suspension is less important, as it usually is recycled into the feed. The most desirable characteristic of products include the lowest possible moisture content in dewatered sediment and the lowest possible flow and solid part concentration in the sediment part suspension. As is known form previous researches [1,2,5,14] the mentioned characteristics are in relation with the quantitative and qualitative properties of centrifuge feed (the centrifuge working point). Therefore, the dewatering process can be considered as a controlled system. The output variables are quantitative and qualitative properties of dewatered sediment and characteristic of sediment part originating suspension. The quantitative and qualitative properties of feed and centrifuge operational parameters are the input variables. The most important parameters of centrifuge feed are:

- solid particles concentration in suspension,
- suspension grain size composition,
- solid phase mass flow,
- volumetric flow of suspension.

The operational parameters of centrifuge can be characterised as follows:

- rotation parameter — centripetal acceleration measured with gravity acceleration
- height of overflow edge,
- difference between numbers of rotations of endless screw and centrifuge basket.

There were over 130 tests done on the experimental stand built in Central Mining Institute. Different characteristics of feed as well as centrifuge operational states were used. The experimental stand construction is depicted in the Fig. 3. The data gathered during the experiments were used as a basis for investigations of centrifuge model formulation.

4 Analysis Tools

Estimation of the regression function may be realized in one of the two following ways: parametrical or nonparametrical. Parametrical regression is described as the model with the well defined functional form with the finite number of free parameters. It can be written as follows:

$$\widetilde{y} = f(x; \theta) \tag{1}$$

where $\widetilde{y}$ is the estimator of the variable y, x is the independent variable and θ is the vector of the model parameters. Parametrical regression says exactly what is the relation between input and output of the analyzed effect.

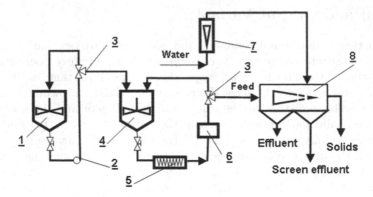

Fig. 2. The schema of experiment stand for coal suspension dewatering in sediment filter centrifuge: 1. Feed suspension tank, 2. Centrifugal pump, 3. Three-way valve, 4. Mixer tank, 5. Screw pump, 6. Flow-meter, 7. Rotameter, 8. Screenbowl centrifuge of type SSC-01.

Nonparametrical regression also describes the dependence between input and output but does not explain its nature. It limits itself to giving the output for a given input. In this short article parametrical and nonparametrical data mining methods were used as prediction tools.

4.1   Parametrical Models

Linear Regression. For a given set of independent variables $\{x_1, x_2, \ldots, x_n\}$ linear regression may be defined as the follows:

$$\widetilde{f}(x_1, x_2, \ldots, x_n) = \alpha_0 + \sum_{i=1}^{n} \alpha_i x_i \tag{2}$$

In the matrix notation for a given set of k observations the linear model may be written as:

$$\begin{bmatrix} y_1 \\ y_2 \\ \vdots \\ y_k \end{bmatrix} = \begin{bmatrix} 1 & x_{1(1)} & \cdots & x_{n(1)} \\ 1 & x_{1(2)} & \cdots & x_{n(2)} \\ \vdots & \vdots & \ddots & \vdots \\ 1 & x_{1(k)} & \cdots & x_{n(k)} \end{bmatrix} \begin{bmatrix} \alpha_0 \\ \alpha_1 \\ \vdots \\ \alpha_n \end{bmatrix} \tag{3}$$

where $x_{v(w)}$ is the value of v–th variable for the w–th object. The solution of the above equation may be found with the least squares method.

Nonlinear Regression. If we assume that the output variable depends from input variables nonlinearly we may define polynomial nonlinear regression. If m is the maximal polynomial degree then nonlinear regression is defined in the following way:

$$\widetilde{f}(x_1, x_2, \ldots, x_n) = \alpha_{00} + \sum_{j=1}^{m} \sum_{i=1}^{n} \alpha_{ij} x_i^j \tag{4}$$

This problem may be also solved with least squares method as the solution of the following matrix equation:

$$
\begin{bmatrix} y_1 \\ y_2 \\ \vdots \\ y_k \end{bmatrix} = \begin{bmatrix} 1 & x_{1(1)} & \cdots & x_{n(1)} & x_{1(1)}^2 & \cdots & x_{n(1)}^2 & \cdots & x_{1(1)}^m & \cdots & x_{n(1)}^m \\ 1 & x_{1(2)} & \cdots & x_{n(2)} & x_{1(2)}^2 & \cdots & x_{n(2)}^2 & \cdots & x_{1(2)}^m & \cdots & x_{n(2)}^m \\ \vdots & \vdots & \ddots & \vdots & \vdots & \ddots & \vdots & \ddots & \vdots & \ddots & \vdots \\ 1 & x_{1(k)} & \cdots & x_{n(k)} & x_{1(k)}^2 & \cdots & x_{n(k)}^2 & \cdots & x_{1(k)}^m & \cdots & x_{n(k)}^m \end{bmatrix} \begin{bmatrix} \alpha_{00} \\ \alpha_{11} \\ \vdots \\ \alpha_{1n} \\ \alpha_{21} \\ \vdots \\ \alpha_{2n} \\ \alpha_{m1} \\ \vdots \\ \alpha_{mn} \end{bmatrix} \tag{5}
$$

4.2 Nonparametrical Models

Support Vector Machines. Support Vector Machines (SVM) were defined in [3] as the classification method and later were adapted into nonparametrical regression task too [4]. In the linear version of this method, the estimated function $f(x)$ is the linear combination of the vector of independent variables:

$$\widetilde{f}(x) = w'x + w_0 \tag{6}$$

A margin ε is a precision of the estimation. Slack variables ξ_i, ξ_i^* are introduced to avoid the overfitting of the regression function. The optimal regression function should minimize the following criterion (n is the number of training objects):

$$J(w, \xi) = \frac{||w||^2}{2} + C \sum_{i=1}^{n} (\xi_i + \xi_i^*) \tag{7}$$

with constraints:

$$
\begin{cases}
y_i - wx_i - w_0 \leq \varepsilon + \xi_i \\
wx_i + w_0 - y_i \leq \varepsilon + \xi_i^* \\
\xi_i, \xi_i^* \quad \geq \quad 0
\end{cases} \tag{8}
$$

The constant $C > 0$ determines the trade-off between the flatness of $\widetilde{f}$ and the amount up to which deviations larger than ε are tolerated [11].

For each train object the pair of the Lagrange multipliers α_i, α_i^* are obtained and the final value of the regression function is calculated as:

$$\widetilde{f}(x) = \sum_{i=1}^{n} wx + w_0 \tag{9}$$

where

$$w = \sum_{i=1}^{n} (\alpha_i - \alpha_i^*) x_i \tag{10}$$

and $w_0 = -w'(x_r + x_s)/2$ where x_r and x_s are support vectors (the notion explained in the next paragraph).

It is worth to notice that not all of the training vectors take part in the evaluation of the regression value. Only vectors x_i which Lagrange parameters $\alpha_i - \alpha_i^* > 0$ influence the result and these vectors are called Support Vectors.

Also the nonlinear version of SVM exists. In this modification the scalar product of two vectors is replaced by the function that performs the corresponding product in higher–dimensional space. Detailed calculations can be found in [11].

Kernel Estimators. Kernel estimators may be described as the simplest and probably the clearest examples of nonparametric estimators. One of the oldest of them, the Nadaraya–Watson estimator [9][13], is defined in the following way:

$$\widetilde{f}(x) = \frac{\sum\limits_{i=1}^{n} y_i K\left(\frac{x - x_i}{h}\right)}{\sum\limits_{i=1}^{n} K\left(\frac{x - x_i}{h}\right)} \tag{11}$$

where $\widetilde{f}(x)$ means the estimator of the $f(x)$ value, n is a number of train pairs (x, y), K is a kernel function and h the smoothing parameter. This estimator assumes that independent (X) and dependent (Y) variables are random variables and the value of the regression function is an approximation of the conditional expected value of the dependent variable Y upon the condition, that the independent variable X took the value x:

$$\widetilde{f}(x) = E(Y|X = x) \tag{12}$$

This estimator may be interpreted as the weighted average of observed values y_i. Similar kernel estimators are: Gasser–Muller [8], Priestley – Chao [12], Stone–Fan [7].

One of the most popular kernel function is the Epanechnikov kernel [6] defined as:

$$K(x) = \frac{3}{4}\left(1 - x^2\right) I(-1 < x < 1) \tag{13}$$

where $I(A)$ means the indicator of the set A. Other popular kernel function are presented in the table 1.

Table 1. Popular kernel functions.

Uniform	$K(x) = \frac{1}{2}I(-1 < x < 1)$
Triangular	$K(x) = (1 - |x|)I(-1 < x < 1)$
Biweight	$K(x) = \frac{15}{16}(1 - u^2)I(-1 < x < 1)$
Gaussian	$K(x) = \frac{1}{\sqrt{2\pi}}\exp -u^2/2$

Radial Basis Functions. Radial Basis Functions [10] belong also to nonparametrical kernel regressors. If we assume that the independent variables space is p–dimensional the estimator takes the form:

$$\widetilde{f}(x) = \alpha + \sum_{j=1}^{r} \beta_j K(||x - x_j||) \tag{14}$$

where α and $\beta_j, j = 1, \ldots, r$ are constant, r is natural ($r \geq 1$), K is the kernel function and x_j are defined points in the independent variables space. These points are called knots and generally do not have anything common with points from train set.

It may be said that continous functions may by approximated with the usage of radial functions with any accuracy. To limit the value of the r parameter keepeing the estimation the following radial functions are used:

$$\widetilde{f}(x) = \alpha + \sum_{j=1}^{r} \beta_j K(A_j ||x - x_j||) \tag{15}$$

where A_j are matrices of constant coefficients. It is very common that diagonal matrices are used with the same value on the main diagonal so the estimator takes the form:

$$\widetilde{f}(x) = \alpha + \sum_{j=1}^{r} \beta_j K \left(\frac{||x - x_j||}{\sigma_j} \right) \tag{16}$$

Coefficients $\alpha, \beta_j, \sigma_j$ and knots x_j are the estimator parameters. Their values may be calculated with the minimizing of the formula:

$$\sum_{i=1}^{n} ||y_i - \widetilde{f}(x_i)||^2 + \lambda \sum_{i=1}^{n} \sum_{j=1}^{n} \left(\frac{\partial^2 \widetilde{f}(x_i)}{\partial x_j^2} \right)^2 \tag{17}$$

With this approach it is assumed that $r = n$ and all training points are considered as knots.

5 Analyzed Data

The analyzed data contains 131 objects, described with 28 variables. One of variables is linearly dependend so it is not considered in the further analysis. One variable describes the origin of the feed as the way of three different streams mixturing. This leads to the definition of three models of independent variables:

model I where feed description was decomposited into three binary variables denoting flot, coarse-grained and raw sludge (29 input variables),
model II where feed description was considered as the discrete variable (27 input variables),
model III where feed description was rejected from the input variables set (26 input variables).

For all models a 10-fold corss-validation was performed. As the prediction error a root of mean squared error was used:

$$err = \sqrt{\frac{\sum_{i=1}^{n}(\widetilde{y}-y)^2}{n}}$$

Names of output variables are encoded as follows:

- **a1** solids: humidity [%]
- **a2** solids: ash content [%]
- **a3** solids: solid phase concentration [%]
- **b1** effluent: ash content [%]
- **b2** effluent: solid phase concentration [%]
- **c1** screen effluent: ash content [%]
- **c2** screen effluent: solid phase concentration [%]

6   Results

The follofwing three tables show the best results of experiments for three models of independent variables. For every model seven variables were predicted (from **a1** to **c2**) with five prediction methods: two parametrical and three nonparametrical. Results of both of groups are separated with the double line. Best results for every predicted variable in the single model are typed with bold font.

Table 2. Best results for model I.

method	predicted variable						
	a1	a2	a3	b1	b2	c1	c2
linear	1,396	0,490	1,565	10,178	1,293	1,134	0,558
nonlinear	**0,687**	0,426	1,291	10,085	1,164	1,134	**0,464**
SVM	1,487	0,952	1,418	12,774	1,218	1,233	0,601
RBF	0,947	0,699	1,384	10,777	1,227	1,329	0,513
NW	0,875	**0,322**	**1,084**	**9,290**	**1,061**	**0,858**	0,551

One may see that the way of feed origin description does not influence in the quality of prediction significantly. The best result for every model are almost equal. Generally, Nadaraya–Watson estimator should be admitted as the best method of centrifuge work prediction.

In the group of parametrical methods nonlinear model occured more accure than linear one. This effect is easy to be explained as nonlinear may describe more complicated dependencies that linear model. Because parametrical models are quite simple we do not observe the "curse of dimensionality" effect for most of predicted variables. It may connects with the models simplicity: every new added variable may provide a bit of new information.

In the group of nonparametrical predictors the simpliest of them gave the best or comparable to the best results.

Table 3. Best results for model II.

method	predicted variable						
	a1	a2	a3	b1	b2	c1	c2
linear	3,237	0,571	1,470	11,534	1,303	1,240	**0,434**
nonlinear	**0,691**	**0,323**	1,316	10,075	1,168	1,140	**0,434**
SVM	1,487	0,952	1,418	12,774	1,218	1,233	0,601
RBF	1,104	0,384	1,404	10,673	1,214	0,947	0,469
NW	0,875	**0,322**	**1,084**	**9,290**	**1,061**	**0,858**	0,551

Table 4. Best results for model III.

method	predicted variable						
	a1	a2	a3	b1	b2	c1	c2
linear	3,121	3,201	4,831	12,553	3,468	0,868	1,393
nonlinear	**0,687**	0,429	1,291	10,085	1,164	0,868	0,464
SVM	1,487	0,952	1,418	12,774	1,218	1,233	0,601
RBF	1,016	0,450	1,432	10,503	1,215	1,106	**0,430**
NW	0,875	**0,322**	**1,084**	**9,290**	**1,061**	**0,858**	0,551

7 Final Conclusions and Further Works

The application of data mining tools for the modelling of the sediment filter centrifuge may be admitted as satisfactory. The results for each of seven predicted variables speak for two models: nonlinear (parametrical) and Nadaraya–Watson (nonparametrical). Each model should be considered as the prediction of different set of predicted variables: nonlinear for solids humidity and solid phase concentration in screen effluent, and Nadaraya–Watson for the other variables. It also worth to notice that Nadaraya–Watson estimator also gave the best or almost the best results amongst all considered predictors.

For the purpose of the centrifuge modelling several data models were used. It turned out that the feed description does not provide any new information that is significant from the predictinal point of view. Decomposition of this variable into three binary values and considering it as the dictionary variable gave the same results as the model that rejects this variable from the set of independent variables. It suggests to use the 3rd model of input variables.

Our further works will focus on the limitation the number of describing variables what should give better prediction results and make models more interpretable.

Acknowledgments

This work was performed at Central Mining Institute (with the cooperation with the Silesian University of Technology) and financed by the Ministry of Science and Higher Education. Publication of this work was financed by the European Community from the European Social Fund.

References

1. Aleksa, H.: Condition and development trends of coal concentrates dewatering technology (in Polish). Mining Review (9), 44–50 (2005)
2. Aleksa, H., Dyduch, F., Piech, E.: Possibility of low ash coal suspends dewatering intensification with the usage of the sedimental filter centrifuge (in Polish). Mining Review (7-8), 41–45 (1994)
3. Boser, B.E., Guyon, I.M., Vapnik, V.N.: A training algorithm for optimal margin classifiers. In: Proc. of the 5th Annu. Workshop on Comput. Learn. Theory, pp. 144–152 (1992)
4. Drucker, H., Burges, C.J.C., Kaufman, L., Smola, A.J., Vapnik, V.N.: Support vector regression machines. In: Adv. in Neural Inf. Process. Syst. IX, pp. 155–161 (1997)
5. Dyduch, F., Aleksa, H.: Impact of hydraulic load of centrifuge and its technical parameters onto size of sorted grain. Prace Naukowe GIG, Mining and Environment 1, 43–51 (2008)
6. Epanechnikov, V.A.: Nonparametric Estimation of a Multivariate Probability Density. Theory of Probab. and its Appl. 14, 153–158 (1969)
7. Fan, J., Gijbels, I.: Variable Bandwidth and Local Linear Regression Smoothers. Ann. of Stat. 20, 2008–2036 (1992)
8. Gasser, T., Muller, H.G.: Estimating Regression Function and Their Derivatives by the Kernel Method. Scand. J. of Stat. 11, 171–185 (1984)
9. Nadaraya, E.A.: On estimating regression. Theory of Probab. and its Appl. 9, 141–142 (1964)
10. Powell, M.D.: Radial Basis Function Approximation to Polynomials. In: Numerical Analysis Proceedings, pp. 223–241 (1987)
11. Smola, A.J., Scholkopf, B.: A tutorial on support vector regression. Statistics and Computing 14(3), 199–222 (2004)
12. Wand, M.P., Jones, M.C.: Kernel Smoothing. Chapman and Hall, Boca Raton (1995)
13. Watson, G.S.: Smooth Regression Analysis. Sankhya - The Indian J. of Stat. 26, 359–372 (1964)
14. Wen, W.W.: An integrated fine coal preparation technology: the GranuFlow Process. Int. J. Miner. Process. 38, 254–265 (2000)

Mining Sequential Rules Based on Prefix-Tree

Thien-Trang Van[1], Bay Vo[1], and Bac Le[2]

[1] Faculty of Information Technology, Ho Chi Minh City University of Technology, Vietnam
[2] Faculty of Information Technology University of Science, Ho Chi Minh, Vietnam
{vtttrang,vdbay}@hcmhutech.edu.vn, lhbac@fit.hcmus.edu.vn

Abstract. We consider the problem of discovering sequential rules between frequent sequences in sequence databases. A sequential rule expresses a relationship of two event series happening one after another. As well as sequential pattern mining, sequential rule mining has broad applications such as the analyses of customer purchases, web log, DNA sequences, and so on. In this paper, for mining sequential rules, we propose two algorithms, *MSR_ImpFull* and *MSR_PreTree*. *MSR_ImpFull* is an improved algorithm of *Full* (David Lo et al., 2009), and *MSR_PreTree* is a new algorithm which generates rules from frequent sequences stored in a prefix-tree structure. Both of them mine the complete set of rules but greatly reduce the number of passes over the set of frequent sequences which lead to reduce the runtime. Experimental results show that the proposed algorithms outperform the previous method in all kinds of databases.

Keywords: frequent sequence, prefix-tree, sequential rule, sequence database.

1 Introduction

Mining sequential patterns from a sequence database was first introduced by Agrawal and Srikant in [1] and has been widely addressed [2-6, 10]. Sequential pattern mining is to find all frequent sequences that satisfy a user-specified threshold called the minimum support (minSup).

Sequential rule mining is trying to find the relationships between occurrences of sequential events. A sequential rule is an expression that has form X→Y, i.e., if X occurs in any sequence of the database then Y also occurs in that sequence following X with high confidence. In sequence databases, there are researches on many kinds of rules, such as recurrent rules [8], sequential classification rules [14], sequential rules [7, 9] and so on. In this paper, we focus on sequential rule mining. In sequential rule mining, there are researches on non-redundant sequential rules but not any real research on mining a full set of sequential rules. If we focus on interestingness measures, such as lift [12], conviction [11], then the approach of non-redundant rule [9] cannot be used. For this reason, we try to address the problem of generating a full set of sequential rules effectively.

Based on description in [7], the authors of paper [9] have generalized an algorithm for mining sequential rules, called *Full*. The key feature of this algorithm is that it requires multiple passes over the full set of frequent sequences. In this paper we

N.T. Nguyen et al. (Eds.): New Challenges for Intelligent Information, SCI 351, pp. 147–156.
springerlink.com   © Springer-Verlag Berlin Heidelberg 2011

present two algorithms: *MSR_ImpFull* and *MSR_PreTree*. The former is an improved algorithm (called *MSR_ImpFull*). The latter is a new algorithm (called *MSR_PreTree*) which effectively mines all sequential rules base on prefix-tree.

The rest of paper is organized as follows: Section 2 presents the related work and Section 3 introduces the basic concepts related to sequences and some definitions used throughout the paper. In section 4, we present the prefix-tree. Two proposed algorithms are presented in Section 5, and experimental results are conducted in Section 6. We summarize our study and discuss some future work in Section 7.

2   Related Work

The study in [7] has proposed generating a full set of sequential rules from a full set of frequent sequences and removing some redundant rules by adding post-mining phase.

Authors in [9] have investigated several rule sets based on composition of different types of sequence sets such as generators, projected-database generators, closed sequences and projected- database closed sequences. Then, they have proposed a compressed set of non-redundant rules which are generated from two types of sequence sets: LS-Closed and CS-Closed. The premise of the rule is a sequence in LS-Closed set and the consequence is a sequence in CS-Closed set. The authors have proved that the compressed set of non-redundant rules is complete and tight, so they have proposed an algorithm for mining this set. For comparison, based on description in [7] they also have generalized *Full* algorithm for mining a full set of sequential rules.

In this paper, we are interested in mining a full set of sequential rules because non-redundant sequential rule mining just minds the confidence, if we use alternative measures then this approach cannot be appropriate as mentioned in Section 1.

3   Preliminary Concepts

Let I be a set of distinct items. An itemset is a subset of items (without loss of generality, we assume that items of an itemset are sorted in lexicographic order). A sequence $s = \langle s_1\ s_2\ ...\ s_n \rangle$ is an ordered list of itemsets. The size of a sequence is the number of itemsets in the sequence. The length of a sequence is the number of items in the sequence. A sequence with length k is called a k-sequence.

A sequence $\beta = \langle b_1\ b_2\ ...b_m \rangle$ is called a subsequence of another sequence $\alpha = \langle a_1\ a_2...\ a_n \rangle$ if there exist integers $1 \le i_1 < i_2 < ... < i_m \le n$ such that $b_1 \subseteq a_{i1}$, $b_2 \subseteq a_{i2}$, ..., $b_m \subseteq a_{im}$. Given a sequence database, the support of a sequence α is defined as the number of sequences in the sequence database that contains α. Given a minSup, we say that a sequence is frequent if its support is greater than or equal to minSup.

Definition 1. *(Prefix, incomplete prefix and postfix).* Given a sequence $\alpha = \langle a_1\ a_2\ ...\ a_n \rangle$ and a sequence $\beta = \langle b_1\ b_2\ ...\ b_m \rangle$ $(m < n)$, (where each a_i, b_i corresponds to an itemset). β is called prefix of α if and only if $b_i = a_i$ for all $1 \le i \le m$. After removing the prefix β of sequence α, the remaining part of α is a postfix of α. Sequence β is called an incomplete prefix of α if and only if $b_i=a_i$ for $1 \le i \le m-1$, $b_m \subset a_m$ and all the items in (a_m-b_m) are lexicographically after those in b_m.

Note that from the above definition, a sequence of size k has $(k-1)$ prefixes. For example, sequence $\langle(A)(BC)(D)\rangle$ have two prefixes: $\langle(A)\rangle$, $\langle(A)(BC)\rangle$. Consequently, $\langle(BC)(D)\rangle$ is the postfix with respect to prefix $\langle(A)\rangle$ and $\langle(D)\rangle$ is the postfix w.r.t prefix $\langle(A)(BC)\rangle$. Neither $\langle(A)(B)\rangle$ nor $\langle(BC)\rangle$ is considered as a prefix of given sequence, however, $\langle(A)(B)\rangle$ is a incomplete prefix of given sequence.

A sequential rule is built by splitting a frequent sequence in two parts: prefix *pre* and postfix *post* (concatenating *pre* with *post*, denoted as *pre++post*, we have the same pattern as before [9]). We denote a sequential rule as $r=pre \rightarrow post$ *(sup, conf)*.

— The support of r: *sup = sup(pre++post)*.

— The confidence of r: *conf = sup(pre++post)/sup(pre)*.

Note that *pre++post* is a frequent sequence, consequently *pre* is also a frequent sequential pattern (by the Apriori principle [1]). For each frequent sequence f of size k, we can possibly form $(k-1)$ rules. For example, if we have a frequent sequence $\langle(A)(BC)(D)\rangle$ which size is 3, then we can generate 2 rules such as $\langle(A)\rangle \rightarrow \langle(BC)(D)\rangle$, $\langle(A)(BC)\rangle \rightarrow \langle(D)\rangle$.

Sequential rule mining is to find out all significant rules that satisfy minSup and minimum confidence (minConf) from the given database. Usually thresholds of support and confidence are predefined by users.

Similar to association rule mining, this problem can be decomposed into two sub problems. The first problem is to find all frequent sequences *(FS)* that satisfy minSup, and the second is to generate sequential rules from those frequent sequences with satisfying minConf.

4   Prefix -Tree

In our approach *(MSR_PreTree)*, the set of rules is generated from frequent sequences, which is organized and stored in a prefix-tree structure as illustrated in Fig. 1. In this section, we briefly outline the prefix-tree (similar to lexicographic tree [5, 6, 10]).

Starting from the root of tree at level 0, root is labeled with a null sequence ϕ. At level k, a node is labeled with a k-sequence. Recursively, we have nodes at the next level $(k+1)$ by extending k-sequence with a frequent item. There are two ways to extend a k-sequence: *sequence extension* and *itemset extension* [6].

In sequence extension, we add an item to the sequence as a new itemset. Consequently, the size of sequence-extended sequences always increases.

Remark 1: In sequence extension, a k-sequence α is a prefix of all sequence-extended sequences. Moreover, α is certainly the prefix of all sub nodes of the nodes which are sequence-extended of α.

In itemset extension, the item is added to the last itemset in the sequence so that the item is greater than all items in the last itemset. So, the size of a itemset-extended sequence does not increase.

Remark 2: In itemset extension, α is an incomplete prefix of all itemset-extended sequences. Moreover, α is an incomplete prefix of all sub nodes of the nodes which are itemset-extended of α.

Based on the above remarks, we develop the algorithm *MSR_PreTree* for generating rules. This algorithm is presented in Section 5.2.

For example, Fig. 1 shows the prefix-tree of frequent sequences. $\langle(A)(A)\rangle$ and $\langle(A)(B)\rangle$ are sequence-extended sequences of $\langle(A)\rangle$, and $\langle(AB)\rangle$ is an itemset-extended sequence of $\langle(A)\rangle$. Sequence $\langle(A)\rangle$ is a prefix of all sequences in T1 and it is an incomplete prefix of sequences in T2.

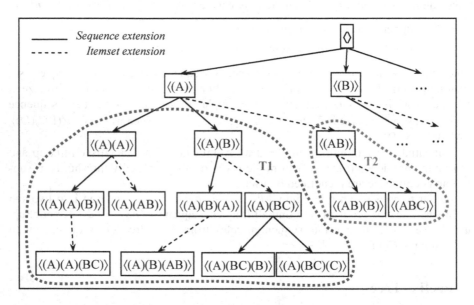

Fig. 1. The Prefix-Tree

5 Proposed Algorithms

The confidence of a rule depends on the support of the prefix. Firstly, for each frequent sequence f, *Full* has to generate all prefixes of f. Then, for each prefix, *Full* has to pass over *FS* to find that prefix's support. Let n be the number of sequences in *FS*, k be the size of the largest sequence in *FS*, the complexity of this algorithm is $O(n*k*n)$ (not to mention the time of generate prefixes). In this section, we proposed two algorithms which effectively reduce the number of *FS* scans.

5.1 MSR_ImpFull

MSR_ImpFull algorithm is given in Fig. 2. It is an improved algorithm of *Full* algorithm. Firstly, the algorithm sorts all sequences in *FS* in ascending order according to their size. As a result, only sequences before X in the list are prefix of X. In detail, for each sequence *f* in *FS*, the algorithm makes a pass over the set of sequences which consist of the sequences following *f* in ordering, we denote this set is *BS*. For each sequence *b* in *BS*, we check whether *b* contains *f* as a prefix or not. If *f* is a prefix of *b* then we try to form a rule *f→post* where *post* is a postfix of *b* with respect to prefix *f*, and compute the confident value of rule (lines 5-8). When the confidence satisfies minConf, we output that rule. If we just mind in the number of FS scans without the time of prefix checking then the complexity of *MSR_ImpFull* is $O(n*n)$.

```
MSR ImpFull algorithm:

Input: Sequence database, minSup, minConf

Output: All significant rules

Method:

   1. Let FS = All sequences with support ≥ minSup

   2. Sort all sequences in FS in ascending order by
      their size

   3. For each sequence f∈FS

   4.    Let BS = All sequences following f in order

   5.    For each sequence b in BS

   6.       If f is a prefix of b

   7.          Let post = postfix of b w.r.t prefix f

   8.          Let r = f→post,

                 sup = sup(b), conf = sup(b)/sup(f)

   9.          If (rconf ≥ minConf)

   10.            Output the rule r(sup, conf)
```

Fig. 2. The *MSR_ImpFull* algorithm

5.2 MSR_PreTree

Although *MSR_ImpFull* algorithm has less numbers of *FS* scans than *Full*, but for each sequence *f*, it scans all sequences following *f* to identify which sequence contains *f* as a prefix. In order to overcome this, we use a prefix-tree structure as mentioned in Section 4. Based on prefix-tree, *MSR_PreTree* generates rules directly because any sequence *α* in the tree (besides the root of tree) is always a prefix of all sequences on the sub trees that their roots are sequence-extended nodes of *α* as Remark 1. We can see that the average number of sequences of those sets approximates *k*. Thus, the complexity of *MSR_PreTRee* is $O(n*k)$.

MSR_PreTree algorithm is shown in Fig. 3. Firstly, we find all frequent sequences using PRISM [6]. As a result, those frequent sequences are stored in the prefix-tree structure (line 1). For each node *r* at level 1, which is the root of a sub tree, we generate rules from each sub tree by calling procedure *Generate-Rule-From-Root(r)* (lines 2-4).

```
MSR PreTree Algorithm:

  Input: Sequence database, minSup, minConf

  Output: All significant rules

  Method:

    1. Let FS = All sequences with support ≥ minSup,
       stored in a prefix-tree (using PRISM [6])

    2. F1 = All nodes at level 1

    3. For each node r in F1

    4.    Generate-Rule-From-Root(r)
```

Fig. 3. The *MSR_PreTree* algorithm

```
Generate-Rule-From-Root(root):

    1. Let Seq-Set = Sequence extensions of root

    2. Let Items-Set = Itemset extensions of root

    3. For each node nseq in Seq-set

    4.    Let Sub-Tree = The tree which rooted at nseq

    5.    Let pre = sequence at root

    6.    Generate-Rules(pre, Sub-Tree)

    7. For each node nseq in Seq-set

    8.    Generate-Rule-From-Root(nseq)

    9. For each node nitem in Items-Set

   10. Generate-Rule-From-Root(nitem)
```

Fig. 4. Generate all rules from a tree

Fig. 4. shows the pseudo-code for the procedure *Generate-Rule-From-Root*. For the root, we have two sets of nodes: a set of sequence-extended nodes and a set of itemset-extended nodes of the root (lines 1, 2). From Remarks 1 and 2, we just generate rules from the sequences on the sub trees that their roots are sequence-extended nodes of the root, because the sequence at the root (denoted as *pre*) is the prefix of these sequences. Hence, for each sub tree, we generate all rules from sequences on the sub tree with respect to prefix *pre*, using procedure *Generate-Rules* (lines 3-6). All extended-nodes of current root will become the prefix for the sub trees at the next level, so we recursively call this procedure for every extended-node of the root (lines 7-10). This recursive process is repeated until the last level of the tree.

Finally, the detail of the procedure *Generate-Rules* is shown in Fig. 5. The input to the procedure is a tree and a sequence *pre* so that *pre* is a prefix of all sequences in that tree. The tree can be traversed in a DFS (Depth First Search) or BFS (Breadth First Search) manner. For each sequence *f* in the tree, we generate the rule *pre→post* which *post* is a postfix of *f* with respect to prefix *pre*.

```
Generate-Rules(pre, Sub-Tree):
  1. For each node n in Prefix-Tree
  2.    Let f = sequence at node n
  3.    Let post = postfix of f w.r.t prefix pre
  4.    Let r=pre→post,
           rsup = sup(f) and rconf = sup(f)/sup(pre)
  5.    if (rconf ≥ minConf)
  6.       Output the rule pre→post(rsup, rconf)
```

Fig. 5. Generate all rules from the sequences on the tree with given prefix

For example, in Fig. 1, in case of generating rules from the root node $\langle(A)\rangle$, we have results as in Table 1.

Table 1. The result of mining sequential rules from the root node $\langle(A)\rangle$

Prefix	Rules
$\langle(A)\rangle$	$\langle(A)\rangle{\rightarrow}\langle(A)\rangle$, $\langle(A)\rangle{\rightarrow}\langle(B)\rangle$, $\langle(A)\rangle{\rightarrow}\langle(A)(B)\rangle$, $\langle(A)\rangle{\rightarrow}\langle(AB)\rangle$, $\langle(A)\rangle{\rightarrow}\langle(B)(A)\rangle$, $\langle(A)\rangle{\rightarrow}\langle(BC)\rangle$, $\langle(A)\rangle{\rightarrow}\langle(A)(BC)\rangle$, $\langle(A)\rangle{\rightarrow}\langle(B)(AB)\rangle$, $\langle(A)\rangle{\rightarrow}\langle(BC)(B)\rangle$, $\langle(A)\rangle{\rightarrow}\langle(BC)(C)\rangle$
$\langle(A)(A)\rangle$	$\langle(A)(A)\rangle{\rightarrow}\langle(B)\rangle$, $\langle(A)(A)\rangle{\rightarrow}\langle(BC)\rangle$
$\langle(A)(B)\rangle$	$\langle(A)(B)\rangle{\rightarrow}\langle(A)\rangle$, $\langle(A)(B)\rangle{\rightarrow}\langle(AB)\rangle$
$\langle(A)(BC)\rangle$	$\langle(A)(BC)\rangle{\rightarrow}\langle(B)\rangle$, $\langle(A)(BC)\rangle{\rightarrow}\langle(C)\rangle$
$\langle(AB)\rangle$	$\langle(AB)\rangle{\rightarrow}\langle(B)\rangle$

In Fig. 1, consider root node $\langle(A)\rangle$, the itemset-extended sequence of $\langle(A)\rangle$ is $\langle(AB)\rangle$ and the sequence-extended sequences of $\langle(A)\rangle$ are $\langle(A)(A)\rangle$ and $\langle(A)(B)\rangle$. Because $\langle(A)\rangle$ is an incomplete prefix of $\langle(AB)\rangle$ and all sub nodes of $\langle(AB)\rangle$, so we don't generate rules from those nodes with prefix $\langle(A)\rangle$. On the contrary, $\langle(A)\rangle$ is a prefix of $\langle(A)(B)\rangle$ and $\langle(A)(C)\rangle$, so we have rules: $\langle(A)\rangle{\rightarrow}\langle(A)\rangle$, $\langle(A)\rangle{\rightarrow}\langle(B)\rangle$. Moreover, $\langle(A)\rangle$

is a prefix of all sub nodes of $\langle(A)(A)\rangle$ and $\langle(A)(B)\rangle$. We also generate rules from those sub nodes with prefix $\langle(A)\rangle$. We repeat the above process for all the sub nodes which are $\langle(A)(A)\rangle$, $\langle(A)(B)\rangle$ and $\langle(AB)\rangle$.

6 Experimental Results

We perform the experiments on two kinds of databases which are in potentially useful areas such as purchase histories, web logs, program execution traces. In the first kind of databases, the size of each itemset of a sequence is greater than or equal to 1. And in the second, all itemsets are of size 1. All experiments are performed on a PC with Intel Core 2 Duo CPU T8100 2x2.10GHz, and 2 GBs memory, running Windows XP Professional.

Synthetic and Real Database: For the first kind of databases, we use synthetic data generator provided by IBM which was used in [1, 6].The generator takes the parameters shown in Table 2. And for the second, we use tests on real database, Gazelle, in which each single item is considered as an itemset. Gazelle was obtained from Blue Martini company, which was also used in KDD-Cup 2000 [13]. This database contains 59602 sequences (i.e., customers), 149639 sessions, and 497 distinct page views. The average sequence length is 2.5 and the maximum sequence length is 267.

Table 2. Parameters for IBM Data Generator

C	Average itemsets per sequence
T	Average items per itemset
S	Average itemsets in maximal sequences
I	Average items in maximal sequences
N	Number of distinct items
D	Number of sequences

Table 3 shows the execution time of three algorithms in two synthetic databases C6T5S4I4N1kD1k and C6T5S4I4N1kD10k with the minSup is decreased from 0.8% to 0.4%. When the support threshold is high, there are only a limited number of frequent sequences, and the length of sequence is short. It is the reason why the number of rules is small, and three algorithms are close in terms of runtime. However, the gaps become clear when the support threshold decreases. Although three algorithms are generated the same rules, both *MSR_ImpFull* and *MSR_PreTree* are more efficient than *Full*. With *MSR_ImpFull* algorithm, runtime is improved up to 2 times and up to 3240 times with *MSR_PreTree* algorithm. So *MSR_PreTree* outperforms both *MSR_ImpFull* and *Full* algorithm. We also test three algorithms on real database (Gazelle) and obtain similar result. With all tests, we set minConf = 50%.

Table 3. The mining time comparison on different synthetic databases

Databases	minSup (%)	#FS	# Rules	Runtime (s)		
				Full	MSR_ImpFull	MSR_PreTree
C6T5S4I4N1kD1k	0.8	11,211	913	11.63	6.17	0.04
	0.7	14,802	1,441	25.86	13.88	0.05
	0.6	20,664	2,302	65.61	34.05	0.07
	0.5	31,311	4,045	157.90	84.33	0.09
	0.4	54,566	9,134	518.40	269.54	0.16
C6T5S4I4N1kD10k	0.8	8,430	248	6.23	4.25	0.02
	0.7	10,480	335	10.44	6.81	0.02
	0.6	13,628	441	19.81	11.33	0.03
	0.5	18,461	613	47.78	24.04	0.04
	0.4	27,168	925	115.77	51.62	0.06
Gazelle	0.11	4,781	4,721	5.59	2.21	0.13
	0.10	5,646	9,527	8.31	3.23	0.23
	0.09	6,883	16,224	14.03	4.88	0.21
	0.08	9,177	37,608	37.93	11.83	0.51
	0.07	15,197	48,672	272.61	49.71	0.77

7 Conclusion and Future Work

In this paper, we have proposed two algorithms named *MSR_ImpFull* and *MSR_PreTree* to mine sequential rules from sequence databases. The key idea is that if given a sequence, we can know which sequences contain a given sequence as a prefix based on sorting the set of frequent sequences; or we can immediately determine which sequences contain a given sequence as a prefix based on prefix-tree. *MSR_PreTree* avoids multiple passes over the full set of frequent sequences in prefix determining process, which can improve the efficiency.

Experimental results show that performance speed up can be obtained in both kinds of databases. Both *MSR_ImpFull* and *MSR_PreTree* are faster than *Full,* and *MSR_PreTree* is also faster than *MSR_ImpFull.* In future, we will apply this approach for generating rules in many kinds of interestingness measures. Besides, based on prefix-tree, we can also mine non-redundant sequential rules.

References

1. Agrawal, R., Srikant, R.: Mining Sequential Patterns. In: Proc. of 11th Int'l Conf. Data Engineering, pp. 3–14 (1995)
2. Srikant, R., Agrawal, R.: Mining Sequential Patterns: Generalizations and Performance Improvements. In: Proc. of 5th Int'l Conf. Extending Database Technology, pp. 3–17 (1996)
3. Zaki, M.J.: SPADE: An Efficient Algorithm for Mining Frequent Sequences. Machine Learning Journal 42(1/2), 31–60 (2000)
4. Pei, J., et al.: Mining Sequential Patterns by Pattern-Growth: The PrefixSpan Approach. IEEE Trans. Knowledge and Data Engineering 16(10), 1424–1440 (2004)

5. Ayres, J., Gehrke, J.E., Yiu, T., Flannick, J.: Sequential Pattern Mining using a Bitmap Representaion. In: SIGKDD Conf., pp. 1–7 (2002)
6. Gouda, K., Hassaan, M., Zaki, M.J.: Prism: A Primal-Encoding Approach for Frequent Sequence Mining. Journal of Computer and System Sciences 76(1), 88–102 (2010)
7. Spiliopoulou, M.: Managing interesting rules in sequence mining. In: Żytkow, J.M., Rauch, J. (eds.) PKDD 1999. LNCS (LNAI), vol. 1704, pp. 554–560. Springer, Heidelberg (1999)
8. Lo, D., Khoo, S.-C., Liu, C.: Efficient Mining of Recurrent Rules from a Sequence Database. In: Haritsa, J.R., Kotagiri, R., Pudi, V. (eds.) DASFAA 2008. LNCS, vol. 4947, pp. 67–83. Springer, Heidelberg (2008)
9. Lo, D., Khoo, S.C., Wong, L.: Non-Redundant Sequential Rules-Theory and Algorithm. Information Systems 34(4-5), 438–453 (2009)
10. Yan, X., Han, J., Afshar, R.: CloSpan: Mining Closed Sequential Patterns in Large Databases. In: SDM 2003, San Francisco, CA, pp. 166–177 (2003)
11. Brin, S., Motwani, R., Ullman, J., Tsur, S.: Dynamic Itemset Counting and Implication Rules for Market Basket Data. In: Proc. of the 1997 ACM-SIGMOD Int'l Conf. on the Management of Data, pp. 255–264 (1997)
12. Berry, M.J., Linoff, G.S.: Data Mining Techniques for Marketing, Sales and Customer Support. John Wiley & Sons, Chichester (1997)
13. Kohavi, R., Brodley, C., Frasca, B., Mason, L., Zheng, Z.: KDD-Cup 2000 Organizers' Report: Peeling the Onion. SIGKDD Explorations 2(2), 86–98 (2000)
14. Baralis, E., Chiusano, S., Dutto, R.: Applying Sequential Rules to Protein Localization Prediction. Computer and Mathematics with Applications 55(5), 867–878 (2008)

A Modified Fuzzy Possibilistic C-Means for Context Data Clustering toward Efficient Context Prediction

Mohamed Fadhel Saad[1], Mohamed Salah[1], Jongyoun Lee[1], and Ohbyung Kwon[2,*]

[1] Department of Informatics
Institute Superior Etudes Technology of Gafsa, Gafsa, Tunisia
fadhel.saad@isetgf.rnu.tn
[2] College of Business,
Professor, Kyung Hee University, Seoul, Korea
obkwon@khu.ac.kr

Abstract. Context prediction is useful for energy saving and hence eco-efficient context-aware service by increasing the interval of context sensing. One way of predicting context is to recognize context patterns in an accurate manner. Traditionally, clustering method has been widely used in pattern recognition, as well as image processing and data analysis. Clustering aims to organize a collection of data items into a specific number of clusters, such that the data items within a cluster are more similar to each other than they are items in the other clusters. In this paper, a modified fuzzy possibilistic clustering algorithm is proposed based on the conventional Fuzzy Possibilistic C-means (FPCM) to obtain better quality clustering results. To show the feasibility and performance of the proposed method, numerical simulation is performed in an actual amusement park setting. The results of the numerical simulation shows that the proposed clustering algorithm gives more accurate clustering results than the FCM and FPCM methods.

Keywords: Context-Aware Service, Clustering, Fuzzy C-Means, Fuzzy Possibilistic C-Means.

1 Introduction

Context aware applications basically run on context sensors and sensory network. The sensors may run all the time, which inevitably needs energy consumption. If the sensing interval increases, then we could save more energy, even though the accuracy of context might decrease. Otherwise, decreasing sensing interval for getting context data in more prompt and accurate manner requires more energy consumption, and hence more costly. For example, Active Badge system usually acquires location data from IR sensors with 10-15 seconds, simply because more frequent infrared sensing needs more grid and hence energy consumption. Even though the trade-offs between energy consumption and context accuracy will be a research challenge for the realization of context aware applications, efforts to resolve these conflicts have been still very few.

---

* Corresponding author.

N.T. Nguyen et al. (Eds.): New Challenges for Intelligent Information, SCI 351, pp. 157–165.
springerlink.com   © Springer-Verlag Berlin Heidelberg 2011

One of the promising methods to avoid energy consumption while preserving context data accuracy is to build up a context prediction method rather than sensing context directly all the time. With this context prediction, estimated context data can be partially inserted to actually sensed context data. As shown in Fig. 1, this method would be called hybrid context sensing, comparing with physical context sensing, which is conventionally used in the context aware applications. As a result, accurate context prediction reduces the interval of physical sensing and hence enables context-aware system which is more efficient in terms of energy savings.

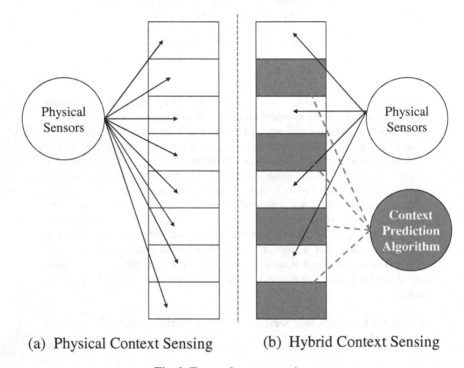

(a) Physical Context Sensing   (b) Hybrid Context Sensing

Fig. 1. Types of context sensing

So far, prediction of future situation which is characterized by a set of context could be performed with several approaches. For example, if some available set of attributes are believed to determine a target context, then regression method would be appropriately applied. Or if we have only serial set of context, then time series such as Box-Jenkins method can be considered. However, the performance of these statistical methods tends to be poor. Alignment prediction method tests if there are any series of context data which are significantly consistent with a sort of rules in rule base. If rules exist, conclusions triggered by the matched rules are adopted as future context [6, 9, 10]. However, alignment prediction basically assumes the availability of prediction rules *a priori*, which would be very difficult or at least very costly. Baysian network algorithm has also been widely used in context prediction, because Baysian network algorithm is excellent in explaining causality between context data. However, this algorithm hardly works well if the prediction should be supported by continuous data acquisition from the sensors.

Hence, in this paper, we will focus on clustering problem for context prediction. Cluster analysis could be used in classifying context data; clustering algorithms find clusters of a data set with most similarity in the same cluster and most dissimilarity between different clusters. Most clustering algorithms do not rely on assumptions like as conventional statistical methods, such as the underlying statistical distribution of data, and therefore they are useful in situations where little prior knowledge exists. The potential of clustering algorithms to reveal the underlying structures in data can be exploited in a wide variety of applications, including classification, image processing, pattern recognition, modeling and identification. The conventional clustering methods put each point of the data set to exactly one cluster. Zadeh proposed fuzzy sets in order to come closer of the physical world [14]. Zadeh introduced the idea of partial memberships described by membership functions. Ruspini first proposed fuzzy c-partitions as a fuzzy approach to clustering [8].

The FCM uses the probabilistic constraint: The memberships of a data point across classes sum to one. While this is useful in creating partitions, the memberships resulting from FCM and its derivatives do not always correspond to the intuitive concept of degree of belongingness. In fuzzy clustering, each point has a degree of belonging to clusters, as in fuzzy logic, rather than belonging completely to just one cluster. Thus, points on the edge of a cluster may be in the cluster to a lesser degree than points in the center of cluster. Moreover, the FCM seems sensitive to noise. To mitigate such an unexpected effect, Krishnapuram and Keller threw away the constraint of memberships in the FCM and propose the Possibilistic C-Means (PCM) algorithm [5]. The advantages of the PCM are that it overcomes the need to specify the number of clusters and is highly robust in a noisy environment. However, some weaknesses still exist in the PCM, i.e., it depends highly on a good initialization and has the undesirable tendency to produce coincident clusters [1].

Pal, founder of the Fuzzy Possibilisitc C-means, deducted that to classify a data point [7]. Therefore, cluster centroid has to be closest to the data point, it is the role of membership. Also for estimating the centroids, the typicality is used for alleviating the undesirable effect of outliers. Indeed, the Fuzzy Possibilistic C-Means combines the characteristics of both fuzzy and possibilistic c-means. At the mercy of the combinatorial characteristics, the algorithm has the advantages rooted from FCM and at the same time PCM. Hence, it is unsupervised method and always converges; the method gives good results despite some noisy points. However, this algorithm also has some disadvantages; it requires a long computational time and is sensitive to the initial guess: Speed and local minima. This concern is more critical when applying to predict context with abundant sensing data. Therefore, even though Fuzzy Possibilistic C-Means gives good performance, the algorithms should be improved in terms of computational time.

Hence, on top of Fuzzy C-means and Possibilistic C-means algorithms, a better classification algorithm which integrates the maximization of the distance between clusters by using a cluster validity index is needed.

2 Fuzzy C-Means Clustering Algorithms

The Fuzzy c-means (FCM) can be seen as the fuzzified version of the k-means algorithm. The k-means algorithm assigns each point to the cluster whose center (also called centroid) is nearest. The center is the average of all points in the cluster. Correspondingly, coordination is the arithmetic mean for each dimension separately over all points in the cluster. The main advantages of this algorithm are its simplicity and speed which allows it to run on larger datasets. However, the main drawback of this algorithm is that it does not yield the same result with each run, since the resulting clusters depend on the initial random assignments. Even though the algorithm can minimize intra-cluster variance, it does not ensure that the result has a global minimum of variance.

The FCM is a method of clustering which allows one piece of data to belong to two or more clusters. This method, which is developed by Dunn in 1973 and modified by Bezdek in 1981, has been widely used in pattern recognition. The algorithm follows iterative clustering method that produces an optimal c partition by minimizing the weighted within group sum of squared error objective function.

The theory of fuzzy logic provides a mathematical environment to capture the uncertainties in much the same human cognition processes. The fuzzy clusters are generated by dividing the training samples in accordance with the membership functions matrix $U = [\mu_{ij}]$. The component μ_{ij} denotes the grade of membership that a training sample belongs to a cluster. Although the FCM is apparently a very useful clustering method, its memberships do not always correspond well to the degree of belonging of the data, and may be inaccurate in a noisy environment, because the real data unavoidably involves some noises.

The FCM algorithms use the probabilistic constraint to enable the memberships of a training sample across clusters to sum up to 1, which means the different grades of a training sample are shared by distinct clusters but not as degrees of typicality. To improve this weakness of FCM, and to produce memberships that have a good explanation for the degree of belonging for the data, Krishnapuram and Keller proposed PCM for unsupervised clustering [5].

Recently, the Fuzzy c-means clustering algorithm has many variations and improvements in terms of stability [4]. The Fuzzy c-means clustering algorithm has adopted in a variety of domains such as image processing and GIS [3, 12].

3 A Proposed Modified Fuzzy Possibilistic Clustering Method

The choice of an appropriate objective function is the key to the success of the cluster analysis and to obtain better quality clustering results; so the clustering optimization is based on objective function [15]. To arrive at a suitable objective function, we started from the following set of requirements: The distance between clusters and the data points assigned to them should be minimized and the distance between clusters should to be maximized [11, 13].

Every point of the data set has a weight in relation to every cluster. Therefore this weight permits to have a better classification especially in the case of noise data. Consequently, the weight is calculated as (1):

$$w_{ji} = exp(\frac{-\|x_j - v_i\|^2}{(\sum_{j=1}^{n} \|x_j - \overline{x}\|^2} \times \frac{n}{c})$$

(1)

where w_{ji} is the weight of the point j in relation to the class i.

Meanwhile, all update methods that were discussed in the previous section are iterative in nature because it is not possible to optimize any of the objective functions reviewed directly. Or to classify a data point, cluster centroid has to be closest to the data point, it is   membership; and for estimating the centroids, the typicality is used for alleviating the undesirable effect of outliers. The objective function is composed of two expressions: The first is the fuzzy function and use a fuzziness weighting exponent, while the second is possibililstic function and use a typical weighting exponent. Note that the two coefficients in the objective function are only used as exhibitor of membership and typicality.

A new relation, lightly different, enabling a more rapid decrease of the function and increase the membership and the typicality when they tend toward 1 and decrease this degree when they tends toward 0. This relation is to add Weighting exponent as exhibitor of distance in the two under objective functions. The objective function of the MFPCM can be formulated as (2):

$$J_{MFPCM}(V, U, T, W, X) = \sum_{i=1}^{c} \sum_{j=1}^{n} (\mu_{ij}^m w_{ji}^m d^{2m}(x_j, v_i) + t_{ij}^\eta w_{ji}^\eta d^{2\eta}(x_j, v_i))$$

(2)

$X = \{x_1, x_2,...,x_n\} \subseteq R^p$ is the data set in the p-dimensional vector space, p is the number of data items, c is the number of clusters with $2 \leq c \leq n-1$. $V = \{v_1, v_2, \ldots, v_c\}$ is the c centers or prototypes of the clusters, v_i is the p-dimension center of the cluster i, and $d^2(x_j, v_i)$ is a distance measure between object x_j and cluster centre v_i.\

U = $\{\mu_{ij}\}$ in (3) represents a fuzzy partition matrix with $u_{ij} = u_i (x_j)$ is the degree of membership of x_j in the ith cluster and all c centroids; x_j is the jth of p-dimensional measured data.

$$\mu_{ij} = [\sum_{k=1}^{c} (\frac{d^2(x_j, v_i)}{d^2(x_j, v_k)})^{\frac{2m}{(m-1)}}]^{-1}$$

(3)

Meanwhile, T = $\{t_{ij}\}$ in (4) represents a typical partition matrix with $t_{ij} = t_i (x_j)$ is the degree of typicality of x_j in the ith cluster and v_i alone.

$$t_{ij} = [\sum_{k=1}^{n} (\frac{d^2(x_j, v_i)}{d^2(x_j, v_k)})^{\frac{2\eta}{(\eta-1)}}]^{-1}$$

(4)

W = $\{w_{ji}\}$ represents a matrix of weight with $w_{ji} = w_i (x_j)$ is the degree of weight of x_j in the ith cluster.

The MFPCM algorithm is summarized as follows:

MFPCM algorithm

- S1:

  Given the data set X, choose the number of clusters $1 < c < N$, the weighting exponent $m > 1, \eta > 1$ and the termination tolerance $\varepsilon > 0$.
  Initialize the c cluster centers v_i randomly.

For $b = 1, 2, \ldots, b_{max}$ do:

- S2:   Compute $W^b = [w_{ji}]$ by using the following formula:

$$w_{ji} = \exp\left(-\frac{\left\|x_j - v_i\right\|^2}{\left(\sum_{j=1}^{n}\left\|x_j - \bar{v}\right\|^2\right) * c/n}\right)$$

- S3:   Update $U^b = [u_{ij}]$ by Eq (15)

  and $T^b = [tij]$ by Eq (16)

- S4:   Modify U by
  $\mu_{ij} = w_{ji} * \mu_{ij}$
  and T by
  $tij = wji * tij$

- S5:   Update the centers vectors Vb=[vi]

- S6:   Compute $E^t = \left| V^t - V^{t-1} \right|$, if $E^t \le \varepsilon$, Stop;

End.

4 Experimental Results

To show the feasibility of the proposed algorithm, we perform some experiments some numerical datasets to compare the performances of these algorithms with those of competing algorithms: FCM and FPC. All competing algorithms are implemented under the same initial values and stopping conditions. The performance evaluations are conducted using MATLAB (Mathworks, Inc., Natick, MA) on an IBM computer with 2.6 GHz Pentium IV processors.

In the experiment, we use actual data set to demonstrate the quality of classification of our approach in relation to FCM and FPCM. Hence, we collected actual context history from the travelers who had actually visited the largest amusement park.

As a result, 400 context records were collected in 2009. Then we randomly selected 70% of the records as training data. The rest are selected as testing data.

The gathered context data set is made up of two subsets: User profile and context. User profile includes gender, ages, preference about experiencing new attractions, and perceived sensitivity on waiting for the service. Meanwhile, we consider visiting time, perceived weather, perceived temperature, perceived average waiting time, attraction visited and attraction to visit next as user context.

In order to assess the quality of a classifier on the validation data, we need a performance index as evaluation measure to know how 'good' it is. Performance index is criteria to choose a good clustering. Partition matrix U such that $U = \{\mu_{ij}\}$ i =1, ..., c, j =1,..., n where μ_{ij} is a numerical value in [0, 1] that tells the degree to which the element x_j belongs to the i-th cluster (with all clusters). Typicality value T such that $T = \{t_{ij}\}$ is a numerical value in [0, 1] that tells the degree to which the element x_j belongs to the i-th cluster (With only cluster i). We adopted Fukuyama and Sugeno index as performance index. Fukuyama and Sugeno modeled the cluster validation by exploiting the compactness and the separation [13].

In below equations of Fukuyama and Sugeno, J_m is a compactness measure, and K_m is a degree of separation between each cluster and the mean ($\overline{v}$) of cluster centroids. Then the performance index (PI) is formed as (5):

$$PI = J_m - K_m \tag{5}$$

where J_m and K_m are described as follows:

$$J_m = \text{Objective function} = J_{FCM}(V, U, X) = \sum_{i=1}^{c}\sum_{j=1}^{n}\mu_{ij}^m d^2(x_j, v_i)$$

$$K_m = \sum_{i=1}^{c}\sum_{j=1}^{n}\mu_{ij}^m d^2(\overline{v}, v_i)$$

Using the above Fukuyama-Sugeno index, we performed experiments with amusement park data set. Table 1 shows the clustering results from running FCM, FPCM and MFPCM. As a result, the clustering results of these algorithms show that our approach, MFPCM, is better than others in terms of Fukuyama-Sugeno index. In this paper, t-test is abbreviated simply because the differences of the performance are big enough. Moreover, as for number of iterations which are proportional to elapsed time, MFPCM (126) is superior to FCM (323) and FPCM (332) apparently. Thus, as expected, MFPCM shows better performance than the clustering competing algorithms in context prediction domain.

Table 1. Performance evaluation with FCM, FPCM and MFPCM

Data set	number of data	number of clusters	Number of data items	Performance Index FCM	Performance Index FPCM	Performance Index MFPCM
Application to Context Prediction (amusement park)	318	5	13	-85,862	-106,500	-171,970
Number of Iterations				323	332	126

5 Conclusions

In this paper we presented a modified fuzzy possibilistic clustering method (MFPCM), which is developed to obtain better quality of clustering results for context prediction. The objective function is based on data attracting cluster centers as well as cluster centers repelling each other and a new weight of data points in relation to every cluster.

The proposed MFPCM is an extension of FPCM which gives better results in context prediction than FPC because the new algorithm adds a new indicator called *weight* for each point in the data set. A new relation is given to enable a more rapid decrease of the objective function and increase of the membership. Moreover, the relation also makes the typicality possible when they tend toward 1 and decrease this degree when they tend toward 0. Hence, a clustering algorithm MFPCM will increase the cluster compactness and the separation between clusters and give better results of clustering.

In particular, cluster compactness is very useful for efficient context prediction, in that compactness positively affects scalability, which in turn helpful for enduring the increase of the number of context sensing events. Moreover, the increase of separation between clusters also contributes to decline erroneous or obscure decision which happens in context prediction. Finally, a numerical example shows that the clustering algorithm gives faster and more accurate clustering results than the FCM and FPCM algorithms. As a result, we could conclude that the proposed algorithm is more useful for better prediction of context.

In this future study, activation sensors will be considered for the empirical test and the proposed algorithm will be used to cluster the behaviors done by humans or smart objects to predict the future actions. Moreover, extensive simulation results with other kinds of real context data are necessary for further proof of the validity.

Acknowledgements

This research is supported by the Ubiquitous Computing and Network (UCN) Project, the Ministry of Knowledge and Economy (MKE) Knowledge and Economy Frontier R&D Program in Korea.

References

1. Barni, M., Cappellini, V., Mecocci, A.: Comments On a Possibilistic Approach to Clustering. IEEE Transactions on Fuzzy Systems 4(3), 393–396 (1996)
2. Bezdek, J.C.: Pattern Recognition with Fuzzy Objective Function Algorithms. Plenum, New York (1981)
3. Chuang, K.S., Tzeng, H.L., Chen, S., Wu, J., Chen, T.J.: Fuzzy C-Means Clustering With Spatial Information For Image Segmentation. Computerized Medical Imaging and Graphics 30(1), 9–15 (2006)
4. Fan, J., Han, M., Wang, J.: Single point iterative weighted fuzzy C-means clustering algorithm for remote sensing image segmentation. Pattern Recognition 42(11), 2527–2540 (2009)
5. Krishnapuram, R., Keller, J.: A Possibilistic Approach to Clustering. IEEE Transactions on Fuzzy Systems 1(2), 98–110 (1993)
6. Loeffler, T., Sigg, S., Haseloff, S., David, K.: The Quick Step to Foxtrot. In: David, K., Droegehorn, O., Haseloff, S. (eds.) Proceedings of the Second Workshop on Context Awareness for Proactive Systems (CAPS 2006). Kassel University Press (2006)
7. Pal, N.R., Pal, K., Bezdek, J.C.: A Mixed C-means Clustering Model. In: Proceedings of the Sixth IEEE International Conference on Fuzzy Systems, vol. 1, pp. 11–21 (1997)
8. Ruspini, E.R.: A New Approach to Clustering. Information Control 15(1), 22–32 (1969)
9. Sigg, S., Haseloff, S., David, K.: Context Prediction by Alignment Methods. Poster Proceedings of the fourth international conference on Mobile Systems, Applications, and Services, MobiSys (2006)
10. Stephan, S., Sandra, H., Klaus, D.: The Impact of the Context Interpretation Error on the Context Prediction Accuracy. In: Third Annual International Conference on Mobile and Ubiquitous Systems: Networking & Services, pp. 1–4 (2006)
11. Timm, H., Borgelt, C., Doring, C., Kruse, R.: An Extension to Possibilistic Fuzzy Cluster Analysis. Fuzzy Sets and Systems 147(1), 3–16 (2004)
12. Wang, J., Kong, J., Lu, Y., Qi, M., Zhang, B.: A Modified FCM Algorithm for MRI Brain Image Segmentation Using Both Local And Non-Local Spatial Constraints. Computerized Medical Imaging and Graphics 32(8), 685–698 (2008)
13. Wu, K.L., Yang, M.S.: A Cluster Validity Index For Fuzzy Clustering. Pattern Recognition Letters 26(9), 1275–1291 (2005)
14. Zadeh, L.A.: Fuzzy Sets. Information Control 8(3), 338–353 (1965)
15. Zhonghang, Y., Yangang, T., Funchun, S., Zengqi, S.: Fuzzy Clustering with Novel Serable Criterion. Tsinghua Science and Technology 11(1), 50–53 (2006)

A New Model of Particle Swarm Optimization for Model Selection of Support Vector Machine

Dang Huu Nghi[1] and Luong Chi Mai[2]

[1] Hanoi University of Mining and Geology, Tu Liem, Ha Noi
dhnghi2005@yahoo.com
[2] Vietnamse Academy of Science and Technology, Institute of Information Technology,
18 Hoang Quoc Viet Road, Ha Noi
lcmai@ioit.ac.vn

Abstract. Support Vector Machine (SVM) has become a popular method in machine learning in recent years. SVM is widely used as a tool in many classifiction areas. The selection optimum values of the parameters (model selection) for SVM is an important step in SVM design. This paper presents the results of research and testing method Particle Swarm Optimization (PSO) in selecting the parameters for SVM. Here we focus on the selection parameter γ of the RBF function and soft margin parameter C of SVM for classification problems. In this paper we propose a Particle Swarm Optimization variation model which combine the cognition-only model with the social-only model and use randomised low discrepancy sequences for initialising particle swarms. We then evaluate the suggested strategies with a series of experiments on 6 benchmark datasets. Experimental results demonstrate the model we proposed better than grid search and full PSO-SVM model.

Keywords: Support vector machine, model selection, particle swarm optimization.

1 Introduction

SVM is a new technique used for regression and classification data. SVM usually depends on some parameters, problem posed is how to choose the parameters of SVM so as to obtain the best results. Here we focus on classification problems and the parameters to be selected are the parameters of the kernel function as well as soft margin parameter C of SVM. The topic of how to select model parameters for SVM is often neglected. In the area of classification, over the last years many model selection approaches have been proposed in the literature such as the gradient descent [10], grid-search [3], genetic algorithms (GA) [2], covariance matrix adaptation evolution strategy (CMA-ES) [5], and more recently Particle Swarm Optimization (PSO) [9]. The original PSO model is not guaranteed to converge on a local (or global) minimiser to the short generation. In this paper we propose a variation to full PSO model which combine the cognition-only model with the social-only model and apply to model selection for SVM. To increase the diversity of population we initialize

N.T. Nguyen et al. (Eds.): New Challenges for Intelligent Information, SCI 351, pp. 167–173.

particle swarm with randomised low-discrepancy sequences and then for the first few generations the velocity are updated by cognition-only model. With the next generation to speed up convergence the velocity are updated by social-only model.

2 Support Vector Machine

Support vector machines are extensively used as a classification tool in a variety of areas. They map the input (x) into a high-dimensional feature space ($z = \phi(x)$) and construct an optimal hyperplane defined by w.z + b = 0 to separate examples from the two classes. For SVMs with L1 soft-margin formulation, this is done by solving the primal problem [11].

$$\min_{w,b,\xi} \quad \frac{1}{2}\|w\|^2 + \frac{1}{2}C\sum_{i=1}^{N}\xi_i \tag{1}$$

With constraints:

$$y_i\left(\langle w.x_i\rangle - b\right) \geq 1 - \xi_i, \quad \xi_i \geq 0, \quad \forall i = 1....N \tag{2}$$

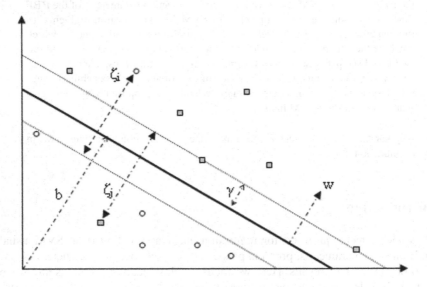

Fig. 1. SVM soft margin and classification problem kernel.

The problem of the Lagrange dual form is:

$$\max \quad \sum_{i=1}^{N}\alpha_i - \frac{1}{2}\sum_{i,j=1}^{N}\alpha_i\alpha_j y_i y_j k\left(x_i, x_j\right) \tag{3}$$

With constraints

$$0 \leq \alpha_i \leq C, \forall i, \quad \sum_{i=1}^{N} y_i \alpha_i = 0 \qquad (4)$$

In which $k(x_i, x_j) = \Phi(x_i). \Phi(x_j)$ is a kernel function of the nonlinear mapping implementation. A number of commonly used kernel functions are:

Gaussian kernel: $\qquad k(x_i, x_j) = \exp\left(-\dfrac{\| x_i - x_j \|^2}{2\sigma^2}\right)$

Polynomial kernel: $\qquad k(x_i, x_j) = (1 + x_i.x_j)^d$

RBF kernel: $\qquad k(x_i, x_j) = \exp(-\gamma \| x_i - x_j \|^2)$

To obtain a good performance, some parameters in SVMs have to be chosen carefully. These parameters include the regularization parameter C, which determines the tradeof between minimizing the training error and minimizing model complexity and parameter (d, σ or γ) of the kernel function.

3 Particle Swarm Optimization

Particle swarm optimization is a non-linear method which falls under the class of evolutionary computation techniques. Particle swarms explore the search space through a population of particles, which adapt by returning to previously successful regions. The movement of the particles is stochastic; however it is influenced by the particle's own memories as well as the memories of its peers. Each particle keeps track of its coordinates in the problem space. PSO also keeps track of the best solution for all the particles (*gbest*) achieved so far, as well as the best solution (*pbest*) achieved so far by each particle. At the end of a training iteration, PSO changes the velocity of each particle toward its *pbest* and the current *gbest* value. The individual velocity is updated by using the following equation:

$$v_i (t + 1) = w \times v_i(t) + c_1 \times r_1 \times (pbest(t)\text{-}x_i (t)) + c_2 \times r_2 \times (gbest(t)\text{-}x_i (t)) \qquad (5)$$

Where $v_i(t)$ is the velocity of particle i at time step t, $x_i(t)$ is the position of particle at time step t, c_1 and c_2 are positive acceleration constants used to scale the contribution of the cognitive and social components respectively and $r_1, r_2 \sim U(0, 1)$ are random values in the range [0, 1], sampled from a uniform distribution. Using the value $c_1 = c_2 = 1.4962$ and w = 0.7298 to ensure the convergence [6].

Kennedy [8] investigated a number of variations to the full PSO models, these models differ in the components included in the velocity equation, and how best positions are determined.

Cognition-Only Model
The cognition-only model excludes the social component from the original velocity equation as given in equation (5). For the cognition-only model, the velocity update changes to

$$v_i (t + 1) = w \times v_i(t) + c_1 \times r_1 \times (pbest(t)\text{-}x_i (t)) \qquad (6)$$

From empirical work, Kennedy reported that the cognition-only model is slightly more vulnerable to failure than the full model. It tends to locally search in areas where particles are initialized. The cognition-only model is slower in the number of iterations it requires to reach a good solution, and fails when velocity clamping and the acceleration coeffcient are small.

Social-Only Model
The social-only model excludes the cognitive component from the velocity equation:

$$v_i (t + 1) = w \times v_i(t) + c_2 \times r_2 \times (gbest(t)\text{-}x_i (t)) \qquad (7)$$

For the social-only model, particles have no tendency to return to previous best positions. All particles are attracted towards the best position of their neighborhood. Kennedy empirically illustrated that the social-only model is faster and more efficient than the full and cognitive models.

4 Implementation

In this report we test the classification using the kernel function RBF so two parameters need to be choice is the γ width of the RBF function and soft margin parameter C of SVM.

Choice of parameters is usually done by minimizing generalization error was estimated as k-fold cross-validation error or leave-one-out error (LOO). With k-fold cross-validation error first the training set is divided into k subsets of the same size then turn evaluated a subset using hyperplane was drawn from the training k-1 subset left.

One method often used to select the parameters is grid search on the log ratio of the parameters associated with cross-validation. Value pairs (C, γ), respectively was assessed using cross-validation and then choose the pair with highest precision. The value of C and γ are increasing exponentially (eg C = 2^{-6}, 2^{-3},......, 2^{14}, $\gamma = 2^{-15}$, 2^{-13},......, 2^{10}) [3][7].

Here we use the particle swarm optimization method to determine value pairs (C, γ) which have the highest accuracy when using cross-validation. Each individual X_i is a vector of two components C and γ. In this paper we propose a model as follows:

To increase the diversity of population we initialize particle swarm with randomised low-discrepancy sequences [12] then first α iteration was implemented to update the location of the individual using the cognition-only model. The next iteration to speed up convergence, the individual position can be updated using the social-only model.

Following algorithm:

```
1:    Randomly initialized population of N individuals
      with randomised low-discrepancy sequences.
2:    t = 0
3:    Repeat
4:      For I = 1 To N
5:        Calculation accuracy of SVM with parameters X_i
          using 5-fold cross-validation
6:        If   accuracy of SVM with parameters X_i greater
          accuracy of SVM with parameters pbest_i Then
7:          Update pbest_i = X_i
8:        If accuracy of  SVM  with  parameters  pbest_i
          greater accuracy of SVM with parameters gbest
          Then
9:            Update gbest = pbest_i
10:     End for
11:   If  t < α  Then
12:     For I = 1 To N
13:        Update X_i using the cognition-only model.
14:     End for
15:   Else
16:     For I = 1 To N
17:        Update X_i using the social-only model.
18:     End for
19:   End if
20:     t = t + 1
21:   Until (number of generations or stop condition
      satisfied)
22:   Return gbest
```

Since doing a complete grid-search may still be time-consuming, in [3] recommend using a coarse grid first. After identifying a "better" region on the grid, a finer grid search on that region can be conducted. For example when searching on the space $C = 2^{-6}, 2^{-4.75}, 2^{-4.50}, ..., 2^{14.75}, 2^{14}$ and $\gamma = 2^{-15}, 2^{-14.75}, ..., 2^{2.75}, 2^{10}$ is need 80 x 100 = 8000 times 5-fold cross-validation. To speed up the implementation in [2] do a search on a coarse grid $C = 2^{-6}, 2^{-3}, ..., 2^{14}$ and $\gamma = 2^{-15}, 2^{-13}, ..., 2^{10}$ to find subdomain which has the best pair value (C, γ) then continue looking glass on this subdomain. For example when rough looking we get pair of values $(2^3, 2^{-5})$ which has best cross-validation value, continue to look fine on the subdomain $C = 2^1, 2^{1.25}, ..., 2^5$ and $\gamma = 2^{-7}, 2^{-6.75}, ..., 2^{-3}$ so the times 5-fold cross-validation is 10 x 12) + (16 x 16) = 376.

When using conventional PSO with a population of 20 individuals, after about 50 generations we obtain accurate results equivalent to the complete grid-search but with 20 x 50 = 1000 times 5-fold cross-validation and when compared the result with coarse then fine search method, the search by PSO method has better results.

With chosen α is 8, experimentally show that the choice of parameters for the SVM by model which we are proposed is equivalent accuracy when using conventional methods PSO but with faster convergence speed. after only 25 to 35 generations we obtain results equivalent conventional PSO method.

To evaluate we test methods on the 6 benchmark datasets was downloaded from http://www.csie.ntu.edu.tw/ ~ cjlin / libsvmtools / datasets

Table 1. Test datasets

Dataset	Training	Test	Features	Classes
Astroparticle	3089	4000	4	2
Vehicle	1243	41	21	2
SpectHeart	80	187	22	2
Adult	3185	29376	123	2
Dna	1400	1186	180	3
Satimage	3104	2000	36	6

Table 2. Accurate results when the parameters are selected using search methods on coarse grid later finer grid search.

Dataset	C	γ	5-fold cross validation rate	Testing
Astroparticle	2.0	2.0	96.8922%	96.875%
Vehicle	128.0	0.125	84.0708%	87.8049%
SpectHeart	4.756825	0.03125	78.7500%	74.3316%
Adult	4.756828	0.03125	83.9873%	84.4533%
Dna	1.414214	0.037163	94.7857%	94.2664%
Satimage	16	0.840896	92.1070%	90.35%

Table 3. Accuracy when the parameters are selected by PSO method.

Dataset	C	γ	5-fold cross validation rate	Testing
Astroparticle	1.814199	4.541019	96.9893%	97.2000%
Vehicle	39.943873	0.166541	84.7949%	85.3659%
SpectHeart	1.181459	0.612769	80%	82.8877%
Adult	207.275585	0.003570	84.0816%	84.2525%
Dna	5984.398006	0.038288	94.9286%	94.3508%
Satimage	22.326996	0.796280	92.2358%	90.1%

Table 4. Compare the number of generations between full PSO model and PSO model proposed

Dataset	Full PSO model		PSO model proposed	
	Generations	5-fold cross validation rate	Generations	5-fold cross validation rate
Astroparticle	50	96.9893%	25	96.9893%
Vehicle	50	84.7949%	33	84.7949%
SpectHeart	50	80%	34	80%
Adult	50	84.0816%	34	84.0816%
Dna	50	94.9286%	35	94.9286%
Satimage	50	92.2358%	35	92.2358%

In many cases show that the parameters chosen by particle swarm optimization method have better results than the grid search method.

5 Conclusion

This paper presents the application of particle swarm optimization method in selecting parameters of SVM simultaneously proposed a new PSO model is a combination of the cognition-only model and the social-only model. To evaluate we compare the results when done by grid search method and fully PSO model method. Results showed that the parameter selection by particle swarm optimization method with model that we propose has better results than searching the grid-search method and rate of convergence is faster than the full PSO model. In the next time we will study the application of particle swarm optimization methods in selecting more than two parameters.

References

1. Engelbrecht, A.P.: Computation Intelligence, ch. 16. John Wiley & Sons, Ltd., Chichester (2007)
2. Chatelain, C., Adam, S., Lecourtier, Y., Heutte, L., Paquet, T.: Multi-objective optimization for SVM model selection. In: Procs. of ICDAR, pp. 427–431 (2007)
3. Hsu, C.-W., Chang, C.-C., Lin, C.-J.: A practical guide to Support Vector Classification, Libsvm: a library for support vector machines (2005)
4. Tu, C.-J., Chuang, L.-Y., Chang, J.-Y., Yang, C.-H.: Feature Selection using PSO-SVM. IAENG International Journal of Computer Science 33 (2007)
5. Friedrichs, F., Igel, C.: Evolutionary tuning of multiple SVM parameters. In: Proceedings of the 12th ESANN, pp. 519–524 (2004)
6. van den Bergh, F.: An Analysis of Particle Swarm Optimization. PhD Dissertation, Faculty of Natural and Agricultural Sciences, University of Pretoria (2002)
7. Frohlich, H., Zell, A.: Efficient parameter selection for support vector machine in classification and regression via model-based global optimization. In: Joint Conf. on Neural Networks (2005)
8. Kennedy, J.: The Particle Swarm: Social Adaptation of Knowledge. In: Proceedings of the IEEE International Conference on Evolutionary Computation, pp. 303–308 (1997)
9. Jiang, M., Yuan, X.: Construction and application of PSO-SVM model for personal credit scoring. In: Procs of the ICCS, pp. 158–161 (2007)
10. Ayat, N., Cheriet, M., Suen, C.: Automatic model selection for the optimization of SVM kernels. Pattern Recognition 38(10), 1733–1745 (2005)
11. Cristianini, N., Shawe-Taylor, J.: An Introduction to Support Vector Machines and Other Kernel-based Learning Methods. Cambridge University Press, Cambridge (2000)
12. Hoai, N.X., Uy, N.Q., Mckay, R.I.: Initialising PSO with Randomised Low-Discrepancy Sequences: Some Preliminary and Comparative Results. To appear in Proceedings of GECCO 2007 (2007)
13. Sivakumari, S., Praveena Priyadarsini, R., Amudha, P.: Performance evaluation of SVM kernels using hybrid PSO-SVM. ICGST-AIML Journal 9(1) (February 2009), ISSN: 1687-4846
14. Wang, Z., Durst, G.L., Eberhart, R.C.: Particle Swarm Optimization and Neural Network Application for QSAR (2002)

Optimization of Short-Term Load Forecasting Based on Fractal Theory

Yongli Wang, Dongxiao Niu, and Ling Ji

School of Economics and Management, North China Electric Power University,
Beijing, China
wyl_2001_ren@163.com, wyl_2001_ren@126.com

Abstract. Power load forecasting is an important part in the planning of power transmission construction. Considering the importance of the peak load to the dispatching and management of the system, the error of peak load is proposed in this paper as criteria to evaluate the effect of the forecasting model. The accuracy of short term load forecasting is directly related to the operation of power generators and grid scheduling. Firstly, the historical load data is preprocessed with vertical and horizontal pretreatment in the paper; Secondly, it takes advantage of fractal and time serial characteristic of load data to design a fractal dimension calculate method for disperse sampling data; Thirdly, the forecasting data image is made by fractal interpolation, the vertical proportion parameter which be used in the interpolation is determined by the similar historical load data, the image can review change condition between load spot. In the view of the nonlinear and complexity in the change of the short-term load, according to current load forecasting technology application and project needs in practice, combined with fractal theory, this paper built a short-term load forecasting model, and obtain good results.

Keywords: fractal theory; data pretreatment; short-term; load forecasting.

1 Introduction

In terms of point of view of macro and micro, changes in power system load are subject to many factors, as a result it has a complex non-linear features[1]. With the advances of modern nonlinear science and technology and higher expect in the forecasting work, the last decade, for all the characteristics of the load forecast, a number of prediction method based on nonlinear theory and its combinations is presented, which has improved the prediction accuracy greatly. Since the concept of fractal proposed in 1970s', it has been got different application in scene modeling, image compression, fractal and chaos in power generation and other aspects. In the power system, fractal has also been applied in power image processing, Fractal characteristics analysis and feature extraction of electrical signals, and other aspects such as experimental data contrast[1]. Paper

[1] This research has been supported by Natural Science Foundation of China(71071052), It also has been supported by Beijing Municipal Commission of Education disciplinary construction and Graduate Education construction projects.

N.T. Nguyen et al. (Eds.): New Challenges for Intelligent Information, SCI 351, pp. 175–186.
springerlink.com   © Springer-Verlag Berlin Heidelberg 2011

[2] has conducted the fractal compression of monitoring image for the substation; the literature [3-5] have extracted the fractal characteristics of power equipment, according to which, PD pattern recognition can be obtained. By calculating the fractal dimension, the statistical self-similarity in the power load has been found in paper [6], besides, the load curves in different or irregular fluctuation extent have different values of fractal dimension.

The basic idea for load forecasting method based on fractal theory in this paper is: at first divide the historical load data into several samples. Based on the fractal collage theorem and the fractal interpolation method introduced by Barnsley, build up an iterated function system (IFS), according to the self-similarity of the historical data on selected similar days. Finally the load curve fitting for the 24h whole point load on forecasting day can be got directly.

2 Data Pretreatment

In the actual load forecasting, the anomaly in the historical data will have a lot of adverse effects, such as Interfering with a correct understanding of the variation, which will lead to the miscarriage of justice of the load forecasting. One of the basic principles of the pretreatment is to reduce the impact of a variety of abnormal data, without losing useful data, as far as possible.

There are many articles proposing different data preprocessing methods for the electric load. For example, literature [7] has used a heuristic approach to process the data, literature [8] has adopted the wavelet analysis method to conversion and reconstruction the load data, and paper [9] filtered out the irregular data by using filtering. Considering the characteristics of the power load data, we provide the following method for missing data pre-processing in this section. Replace the traditional methods with the weighted data on similar days directly, and process the impact and rough load data and other glitches though calculate the deviation rate by mathematical statistics method. Though this way, the regularity of the filled data is stronger, and the data structure processing based on similar days can reflect the characteristics of power load better than the general filtering method.

The mathematical description of power load can be described by the set of state equations as following:

$$\dot{X} = F(X,V,\theta,t) \tag{1}$$
$$Y = G(X,V,\theta,t)$$

Where, X represents a N_x-dimensional state vector, $\dot{X}$ is the derivative of the state vector, Y represents the N_y- dimensional output vector, V is the vector combined effects of the noise and the random noise, θ represents a N_θ-dimensional parameter vector, t means the time, $F(X,V,\theta,t)$ and $G(X,V,\theta,t)$ are both nonlinear functions.

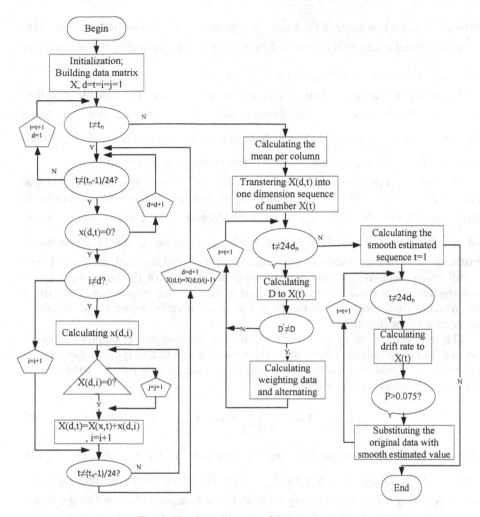

Fig. 1. The flow diagram of data pretreatment

In the calculation of the power load, the load curve is continuous, while the observation samples are discrete, the continuous model above can be written in the type of discrete model as follows:

$$X(k+1) = AX(k) + BV(k)$$
$$Y(k) = CX(k) + V(k)$$

(2)

Where, A, B, C are the parameter matrixes for the state equations.

Since the power load data is 1-dimensional random variable, and its variation can be briefly expressed as:

$$y_i = \bar{y}_i + v_i$$

(3)

Where, y_i is the true value of the load, y_i represents the observations, v_i means the value combined effects of the noise and the random noise. Generally, it is assumed the mean and the variance for v_i is $\bar{v}$ and σ^2 separately.

According to the actual situation in power system, select a suitable threshold δ. When the error between the observed value y_i and the real one y_i is larger than δ, y_i can be considered to be bad data, and further data processing for y_i should be made. While, the error is less than δ, y_i can be taken as the normal.

The entire data pre-processing can be described as follows: input the observed load data, and find the irregular data by comparing the load curve of observed data and the typical load curve, then remove or correct the irregular data. The real load y_i is un-known, but most obtained load data is usually normal. So y_i can be estimated by mathematical statistical methods, then calculate it in accordance with the above ideas.

All in all, data analysis and preprocessing are mainly for the supplement of the missing data and the smoothing of the outliers in the collected information. First, fill up the missing data, then smooth the historical load curve by vertical and horizontal processing of the distortion.

The treatment of missing data can often be done though the supplement of the normal data on the adjacent days. Since the load data on different types of days is quite different, we should use the data on the same type of day and use the following formula for the weighted average processing:

$$x(d,t) = \frac{1}{n} \sum_{i=1}^{n} w_i x(d_i,t) \tag{4}$$

Where, $x(d,t)$ represents the load, and $x(d_i,t)$ is the load at t hour on the d_{ith} day, which is the same type of day with d_{ith}. w_i is the weight of the weighted average, which is usually got from comparing the load data at time point close to t on the similar day with that on the d_{th} day.

The distortion is the abnormal load data generated at a certain time, which is different from the previous load point. It will reduce the similarity of the load curve and have great impact on the forecasting result based on load data of the similar days. However, it's unlike the noise of other continuous functions, the factor for it is usually random, which limits to use the general filtering method, can only consider to use statistical methods to smooth them.

Power load is cyclical, and the 24-hour small cycle should be considered first when pre-processing load data. It means that the load at the same time on different dates should be similar, and the value should be maintained at a certain range. The bad data out of range can be revised though vertical and horizontal processing.

Vertical processing is to compare and analyze the data on the similar days. Similarly, given the load series $x(d,t)$, $t = 1, 2, \cdots, 24$ represent the 24 observation time points. $d = 1, 2, \cdots, N$, means collect the load data on N similar days, though the following formulas we can get the mean $E(t)$ and the variance $V(t)$ of the load at the 24 time points for the N days:

$$E(t) = \frac{1}{N}\sum_{d=1}^{N} x(d,t) \tag{5}$$

$$V(t) = \sigma^2 = \frac{1}{N}\sum_{d=1}^{N}(x(d,t) - E(t))^2$$

Assume the deviation rate of the load at t on the d_{th} day is $\rho(d,t)$, then:

$$\rho(d,t) = \frac{|x(d,t) - E(t)|}{\sigma_t} \tag{6}$$

Suppose η is the permitted deviation rate of the load. And we can control the deviation degree of the revised load, though adjust the size of η. In the actual load data processing, if the actual deviation is more than η, it is the outlier. Otherwise, it's normal. And the outliers can be replaced by the weighted data got from formula (4).

After a longitudinal vertical processing, the missing and abnormal data in the raw data is got pretreatment, as a result, the original series of load become more reasonable, then take further processing by the following method to smooth the load data. Then the hourly load series $x(t) = x(1), x(2), \cdots, x(24)$ is given, use median method to generate a

smooth estimated sequence $\hat{x}(t)$ from the vertical processed load series $x(t)$. Then calculate the deviation rate of the real load series to it, finally conduct a horizontal processing for the hourly load every day.

In order to obtain $\hat{x}(t)$, first generate a new load series $x^{'}(t)$ consisted of the median of the every adjacent five point, then the final load smooth estimated sequence $\hat{x}(t)$ is :

$$\hat{x}(t) = 0.15x^{'}(t-1) + 0.7x^{'} + 0.15x^{'}(x+1); t = 1,2,\cdots,22 \tag{8}$$

$$\hat{x}(t) = x^{'}(t)\text{;}t = 0,23$$

The deviation rate $\rho(t)$ between the real load series $x(t)$ and the smooth estimated sequence $\hat{x}(t)$ is:

$$\rho(t) = \frac{|x(t) - \hat{x}(t)|}{\hat{x}(t)} \tag{9}$$

Similarly, according to the fact situation in power system, given a certain threshold of the deviation rate, we can determine the revise degree of the load curve by adjust the size of the threshold. If the load point is abnormal, we can use the corresponding smooth estimated value $\hat{x}(t)$ to replace $x(t)$.

3 Fractal Theory

3.1 Concepts and Features of the Fractal Theory

Fractal theory is a nonlinear system theory, the study object of which is the rough and non-differentiable geometry generated in the nonlinear system. In this theory, it is considered that any relatively independent part within should be the representation and microcosm of the overall to a certain extent. The "same structure" of the part and whole information is pointed out in this view, therefore the bridge for the transition from part to overall is found. In the nonlinear complex systems, The unique characteristic of fractal is that it describes the entire system though a variable , that is fractal dimension. This means the system turns into the single variable from a number of variables, and the model is reduced from high order, which has great significance in engineering.

It has been confirmed in many work that the load record in power system contains a certain amount of chaotic elements. While the load record are not strictly chaotic, and it contains the development trends, festive trends, cycles, chaos and others. The basic cause of the load uncertainty is that the users are free to the electricity. A large number of different types of loads and too many users result in the complexity of the load behavior, thus the load records are randomly disordered, and the main nature of it is chaotic.

Self-similarity and scale invariance are two basic characteristics of fractal geometry, which is also in line with the scale invariance characteristic of the chaotic system. Analyzing the scale invariance of power load though fractal theory, we can infer the overall load characteristics from local features, such as the load at different times or different regions. Since it is relatively easy to analyze the characteristics of local load, the forecasting based on fractal theory has good prospects.

3.2 Basic Theory of Fractal

3.2.1 Iterated Function Systems and Attract Theorem

So far, the iterated function system (IFS) is the most successful method in the construction and study of the self-similar fractal in the nature. An IFS is consist of a complete metric space (X, d), and a limited set of compression map $W_n : X \to X$ form, the compression factor for W_n is S_n, and $0 \leq S_n < 1$ (n=1,2,…,N). IFS is expressed by $\{X; W_n, n = 1, 2, ..., N\}$, and the compression factor of IFS is $S = \max\{S_n : n = 1, 2, ..., N\}$, the transformation $W : H(X) \to H(X)$ is defined as: $W(B) = \bigcup_{n=1}^{N} W_n(B)$, $\forall B \in H(X)$, the contraction mapping with a compression factor S in the complete metric space (H(X), h(d)), which means if the relationship $h(W(A), W(B)) \leq Sh(A, B)$, $\forall A, B \in H(X)$ holds, then there must be a unique fixed point $P \in H(X)$ meets $P = W(P) = \bigcup_{n=1}^{N} W_n(P)$, And P can be given by: $\forall B \in H(X)$, there is $P = \lim_{n \to \infty} W^n(B)$.

Where: $W^n(B)$ represents the n iterations of transform W, that is $W^n(B) = W(W(...W(B)))$, and the fixed point P is the attractor for the IFS.

3.2.2 Collage Theorem

In a complete metric space (X, d), given $L \in H(X)$ and $\varepsilon > 0$, select an IFS $\{X; W_0, W_1, ..., W_n\}$ with the compression factor S: $0 \leq S < 1$, make

$$h(L, \bigcup_{n=0}^{N} W_n(L)) \leq \varepsilon \tag{10}$$

$$\text{Then}, h(L, P) \leq \varepsilon / (1 - S) \tag{11}$$

Where, P is the attractor for the IFS $\{X; W_0, W_1, ..., W_n\}$; h(L,P) is the Hausdorff metric.

Collage theorem explains that there exists an IFS, whose attractor P approximate to or is similar to a given set of L. Since the fixed point P is constructed from its own transformation W(P), we can make the given set of L a compression transform, then paste them together to reconstruct L.

3.2.3 Fractal Interpolation Algorithm

Fractal interpolation function is first proposed by Barnsley in his paper "functions and interpolation" in 1956, which is generated from a special kind of iterated function systems (IFS). It use the fine self-similar structure presented in many phenomena in nature to fit the volatile curve, and this provides a new way for data fitting. The images of the polynomial interpolation function and spline function are all 1-dimensional smooth curves, therefore, it's appropriate to use them to fit the smooth curves, however, they are not the ideal tool to describe the intense shock curves, such as coastline, electric load curve and so on. While the fractal interpolation functions have great flexibility, as long as adjust its argument appropriately, the dimension of the generated curve can be any value between 1 and 2. Not only can fractal interpolation function be used to fit smooth curves, it has the unique advantages in fitting the rough curves.

Order data set $\{(x_i, y_i): i = 0, 1, ... N\}$, we derive on how to construct an IFS in R^2 as following, its attractor A is the image of the continuous function $F: [x_0, x_N] \rightarrow R$ of interpolated data. Consider the IFS $\{R^2; w_n, n = 1, 2, ... N\}$, where w_n has the affine transformation as following form:

$$w_n \begin{bmatrix} x \\ y \end{bmatrix} = \begin{bmatrix} a_n & 0 \\ c_n & d_n \end{bmatrix} \begin{bmatrix} x \\ y \end{bmatrix} + \begin{bmatrix} e_n \\ f_n \end{bmatrix} \text{ (n=1,2,...,N)} \tag{12}$$

Where, (x, y) is the coordinates for a point; w_i represents the affine transformation; a_i, c_i, d_i are the transformation matrix elements; e_i and f_i are the constant weight for the transformed (x, y).

bn=0 is to ensure that functions between each cell does not overlap, it means that the original vertical line are still vertical line even after the transformation, so as to maintain the transformation vertical invariance.

$$w_n \begin{bmatrix} x_0 \\ y_0 \end{bmatrix} = \begin{bmatrix} x_{n-1} \\ y_{n-1} \end{bmatrix} \tag{13}$$

$$\text{and } w_n \begin{bmatrix} x_N \\ y_N \end{bmatrix} = \begin{bmatrix} x_n \\ y_n \end{bmatrix} \tag{14}$$

This indicates that the left point in the large interval is mapped to the left point of the subinterval, the right end point in the large interval is mapped to the right end point of the subinterval as the same, that is, each transformation must satisfy the following equation:

$$\begin{cases} a_n x_0 + e_n = x_{n-1} \\ a_n x_N + e_n = x_n \\ c_n x_0 + d_n y_0 + f_n = y_{n-1} \\ c_n x_N + d_n y_N + f_n = y_n \end{cases} \tag{15}$$

In fact, the matrix transformation in definition is the elongation transformation, which maps the line parallel to the y-axis to another line, and $|d_n|$ represents the length ratio of two line segments. Therefore, d n is also called as the vertical scaling factor for transform w_n. Order $L = x_N - x_0$, then

$$\begin{cases} a_n = L^{-1}(x_n - x_{n-1}) \\ e_n = L^{-1}(x_N x_{n-1} - x_0 x_n) \\ c_n = L^{-1}[y_n - y_{n-1} - d_n(y_N - y_0)] \\ f_n = L^{-1}[x_N y_{n-1} - x_0 y_n - d_n(x_N y_0 - x_0 y_N)] \end{cases} \tag{16}$$

The parameters got above can determine the ith affine transformation for IFS. After the all parameters in the IFS, we can get the attractor though determined type or random type of iterative algorithm. With the increase in the number of iterations, the fitness between the interpolated curve and the original sample curve increases. After a great number of iterations, a stable constant interpolation curve will be formed at last. Fractal interpolation overcomes the disadvantage in the traditional interpolation method, which is that it can't reflect the local feature between the two adjacent with information. The interpolation curve maintains the most original characteristics of the sampled curve. It's not only by the interpolation points, but also it can display very rich details.

4   Application and Analysis

4.1   Process for Forecasting

In the application of fractal interpolation method for prediction, there are mainly two ways: (1) The fractal extrapolation algorithm based on Extrapolation models of similar days. (2) The fractal interpolation algorithm based on the fitting models of similar days. This paper adopted the latter to study the electricity load forecasting. The fitting model of similar days is to obtain an IFS whose attractor is similar to the historical load data, based on the fractal collage theorem. As the load presents very good periodic, the fitting model of similar day can be considered as the launch of a cycle based on the rich history data.

Take the whole point load forecasting for a local network in Hebei Province on March 25, 2010 as an example, the specific steps are as follows:

1) Take the predicted day as the starting point, and select the same type n days sequential forward as similar days, besides choose one of them as the base date. In this paper, n=3, the day before (March 24) as the base date, and March 22 and 23 as the similar days.

2) Determine the time coordinates (horizontal axis) of the interpolation point set. Analyze the characteristics of the load curve on the base date, and identify the main features points (such as peak and valley points, extreme points, inflection, etc.), then make up the set of interpolation points together, and record their time collections as the basic horizontal axis. In this case the selected interpolation points are the load at 1, 6, 12, 14, 19, 20, 22, and 24 hour, a total of 8 whole point, the corresponding time of which as the basic horizontal axis. In order to ensure the accuracy, we take twice interpolations. The first six interpolation points are the load at 1,6,12,14,19 and 20 hour. And the second ones are the load at 12,14,19,20,22 and 24 hour.

3) Establish the IFS of the load curve on the base date. Establish the iterated function systems according to the interpolation point sets in step 2. in this article, d has been tested as 0.93.since the value of d is determined, we can obtain the affine transformation.

4) Establish the IFS of the load curves on the other similar days. Take the basis time coordinates and the corresponding load value on each similar days as the interpolation point set, set up the iterated function system separately. Where the value of d is unchanged, and the calculation of affine transformation parameters is the same as above.

5) Calculate the average of the corresponding parameters of the obtained three iterated function system, then get the IFS in the statistical sense. According to experience, give higher weight to the parameters on the base date and the closer similar day. In this paper, the weight given to the base date and other similar days are 0.6, 0.25 and 0.15.

6) According to the IFS, start the iteration from any initial point to get the attractors, and the attractor can be considered as the daily load forecasting curve fitted by the historical data. In this case, we select the first point as the initial iteration point.

4.2   Error Analysis

Relative error and root-mean-square relative error are used as the final evaluating indicators:

$$E_r = \frac{x_t - y_t}{x_t} \times 100\% \tag{17}$$

$$RMSRE = \sqrt{\frac{1}{n}\sum_{t=n}^{n}\left(\frac{x_t - y_t}{x_t}\right)^2} \tag{18}$$

The calculation results of the radiation transformation parameters (a, e, c, f) in this case are shown in Table1 and Table 2 as below:

Table 1. The IFS code of the first 6 interpolation points

coefficients	a	e	c	f
W_1	0.2631	0.7368	-28.8695	156.4189
W_2	0.3157	5.6845	21.0464	-154.6795
W_3	0.1052	11.8952	-26.5971	580.1821
W_4	0.2632	13.7374	-5.1509	340.73329
W_5	0.0528	18.9476	-19.9557	545.0029

Table 2. The IFS code of the last 6 interpolation points

coefficients	a	e	c	f
W_1	0.1669	10.0000	1.0539	140.4639
W_2	0.4165	9.0000	35.0098	-485.0076
W_3	0.0831	18.0000	11.5695	-14.2574
W_4	0.1669	18.0000	32.8280	-361.1831
W_5	0.1668	20.0000	-4.8044	253.6931

Use the above IFS codes to calculate the attractor though iteration, the result of the load forecasting for the whole point of 24 hours can be seen Table 3. In this numerical study, the predicted result is close to the actual value, and the relative error is relatively small, which means the obtained result is satisfactory.

Table 3. Load forecasting for a local network in Hebei Province on March 25, 2010

Time point	Actual load/MW	Forecasting load/MW	Relative mistake %	Time point	Actual load/MW	Forecasting load/MW	Relative mistake %
1	2568.28	2595.8781	1.074	13	2977.215	3096.3651	4.002
2	2422.6695	2518.2282	3.944	14	2782.5306	2874.0549	3.289
3	2352.9285	2385.6465	1.391	15	2880.2418	2828.3604	-1.801
4	2341.4403	2278.7472	-2.678	16	2899.7742	2978.0268	2.699
5	2262.2529	2217.4563	-1.980	17	2972.2827	2950.7085	-0.726
6	2204.1723	2287.7016	3.790	18	3041.6793	3118.8987	2.539
7	2314.7493	2268.858	-1.983	19	3099.0957	2986.0587	-3.647
8	2532.8775	2466.8019	-2.609	20	2961.6309	3037.1898	2.551
9	2651.0067	2575.4109	-2.852	21	3072.2571	3178.3077	3.452
10	2908.3227	2916.4161	0.278	22	3154.8639	3188.8119	1.076
11	3039.2562	3113.8926	2.456	23	3055.2216	3096.8817	1.364
12	3119.0094	3156.4014	1.199	24	2798.4591	2872.3698	2.641
--	--	--	--	RMSRE	--	--	2.55

Fractal interpolation functions utilize the similar structure characteristic of the time series in the power load system to construct a special iteration function system to fit the highly fluctuate irregular random load curve. It can be seen from the above figure that the forecast point is almost consistent with the actual one, which means the prediction is accurate.

5 Conclusion

Fractal interpolation function is different from the traditional function interpolation, because it constructs an entire iterated function system rather than a function, and the IFS can reflect more characteristics of the known adjacent information points, as a result, the accuracy of load forecasting can be improved greatly. Fractal interpolation functions and fractal collage principle for power load forecasting is feasible, besides, it's rapid, has high precision and no convergence issues, the data is easy to collect, so it has good practicability. However, the method is entirely based on historical data, and gives different weight coefficients to the factors which impact the load changes, there are some inevitable defects caused by human factors in the trend forecast. Considering the power system is a multi-factor interaction system, how to determine the non-mathematical experience quantify relationship between the each factor and the weight coefficient is the key in further practice, which requires in-depth study for the system.

References

1. Li, X., Guan, Y., Qiao, Y.: Fractal theory based durability analysis and forecasting of power load. Power System Technology 30(16), 84–88 (2006)
2. Qu, W., Zhu, J.: A low bit fractal compression and coding in image surveillance system of substations. Proceedings of the EPSA 15(2), 54–57 (2003)
3. Li, Y., Chen, Z., Lü, F., et al.: Pattern recognition of transformer partial discharge based on acoustic method. Proceedings of the CSEE 23(2), 108–111 (2003)
4. Gao, K., Tan, K., Li, F., et al.: Pattern recognition of partial discharges based on fractal features of the scatter set. Proceedings of the CSEE 22(5), 22–26 (2002)
5. Cheng, Y., Xie, X., Chen, Y., et al.: Study on the fractal characteristics of ultra-wideband partial discharge in gas-insulated system (GIS) with typical defects. Proceedings of the CSEE 24(8), 99–102 (2004)
6. Hu, P., Bo, J., Lan, H., et al.: Analysis of the power load's character based on fractal theory. Journal of Northeast China Institute of Electric Power Engineering (4), 45–52 (2002)
7. Kiartzis, S.J., Zoumas, C.E., Theocharis, J.B., et al.: Short-term load forecasting in an autonomous power system using artificial neural Networks. IEEE Trans. on Power System 18(2), 673–679 (2003)
8. Xu, F.: Application of Wavelet and Fractal Theory on Data Treatment of short-time Load Forecasting. Jiangsu Electrical Engineering 25(3), 37–38 (2005)
9. Connor, J.T.: A robust neural network filter for electricity demand prediction. Forecast 15(6), 437–458 (1996)
10. Li, S.: Fractal. Higher Education Press, Beijing (2004)

11. Fan, F., Liang, P.: Forecasting about national electric consumption and its constitution based on the fractal. Proceedings of the CSEE 24(11), 91–94 (2004)
12. Li, T., Liu, Z.: The chaotic property of power load and its forecasting. Proceedings of the CSEE 20(11), 36–40 (2000)
13. Barnsley, M.F.: Fractals Everywhere. Academic Press, Boston (1988)
14. Xue, W.-l., Yu, J.-l.: Application of Fractal Extrapolation Algorithm in Load Forecasting. Power System Technology 30(13), 49–54 (2006)
15. Xin, H.: Fractal theory and its applications. China University of Technology Press, Beijing (1993)
16. Xie, H.: Mathematical foundations and methods in fractal applications. Science Press, Beijing (1998)

A Performance Evaluation Study of Three Heuristics for Generalized Assignment Problem

Tomasz Kolasa[1] and Dariusz Król[2]

[1] Faculty of Computer Science and Management,
Wrocław University of Technology, Poland
Kolasa.Tomek@gmail.com
[2] Institute of Informatics,
Wrocław University of Technology, Poland
Dariusz.Krol@pwr.wroc.pl

Abstract. The classical generalized assignment problem (GAP) has applications that include resource allocation, staff and job scheduling, network routing, decision making and many others. In this paper, principles of three heuristics are identified. The detailed study on common algorithms like genetic algorithm, ant colony optimization, tabu search and their combinations is made and their performance for paper-reviewer assignment problem is evaluated. The computational experiments have shown that all selected algorithms have determined good results.

Keywords: paper-reviewer assignment problem, genetic algorithm, tabu search, ant colony optimization, hybrid algorithm.

1 Introduction

This paper presents a study on practical application of three artificial intelligence algorithms for generalized assignment problem [1]. Proposed solutions and their evaluation can be a valuable reference to solve similar NP-hard problems [2] including paper-reviewer assignment problem, resource allocation, staff and job scheduling, network and vehicle routing, decision making and others. The following artificial intelligence algorithms have been selected: genetic algorithm (GA), ant colony optimization (ACO) and tabu search (TS). GA [3] is a population-based general-purpose technique used to find approximate solutions to optimization and search problems (especially in poorly understood, irregular spaces). Genetic algorithm is inspired by natural phenomena of genetic inheritance and survival competition of species. The second algorithm, ACO [4] is a probabilistic search technique used to solve computational problems that can be reduced to finding an optimal path in a graph. Ant colony optimization algorithm is based on the observations of real ants and their ability to find shortest paths between their nest and food sources. In nature, ants leave a chemical pheromone trail as they move and they tend to choose the paths marked by the strongest pheromone smell. Thus while searching for the food, an ant will choose the path that other ants extensively use – probably the shortest one. The last algorithm, TS is a local search technique that enhances the performance of searching by using memory structures – tabu list [5]. Once

N.T. Nguyen et al. (Eds.): New Challenges for Intelligent Information, SCI 351, pp. 187–196.
springerlink.com   © Springer-Verlag Berlin Heidelberg 2011

a candidate solution has been found, it is marked as "tabu" so that the search does not take into account that possibility repeatedly. Tabu search is designed to prevent other methods from getting trapped in local minima.

Additionally, this work presents and evaluates following hybrids of selected algorithms: genetic-ant colony optimization hybrid (ACO-GA) and genetic-tabu search hybrid (GA-TS). The motivation in developing a hybrid system is to achieve better performance for large instances. The first method combines advantages of both used algorithms: the ability to cooperatively explore the search space and to avoid premature convergence to quickly find good solutions within a small region of the search space. These two algorithms complement each other in order to find good assignments. In each iteration of ACO-GA algorithm, the best solution of the two algorithms is selected. The second hybrid uses the tabu search method as local heuristic search for selected chromosomes in genetic algorithm population. In each iteration of GA a tabu search algorithm is applied on the best solutions in the population.

The approach followed in this paper provides proposition of tuning parameters for optimization of selected algorithms. In this context, a new genetic algorithm crossover operator — the "random or" crossover operator is introduced. Also, ACO adaptation to search for optimal path in complete bipartite graph (or biclique) is examined. The algorithms have been implemented for solving the paper-reviewer assignment problem [6] and evaluated on real-world conference data set (KNS). This real-world data set comes from the academic conference KNS held in 2009 at Wroclaw University of Technology. The main characteristics are as follows: number of papers - 179, number of reviewers - 64, number of conference keywords - 80, maximal number of keywords per paper - 7, maximal number of keywords per reviewer - 15 and best possible fitness - 0.560. All computations where performed on IBM BladeCenter HS21 XM with 5GB RAM and Linux CentOS release 5.3.

2 Genetic Algorithm Evaluation

First, we focus on the genetic algorithm. The experiments mainly concern the novel "random or" crossover operator, initial population generation heuristic and mutation implementation variants, for details see [7]. GA has the following initial parameters: crossover type - one-point, crossover probability - $p_c = 0.8$, mutation type - switch, mutation probability - $p_m = 0.01$, initial population - B added, population size - 100, generations - 10000. Following modifications of GA are tested: population initialization with "best possible" fitness F_B of solution B or not, crossover operator type (one-point, two-point, uniform and "random or"), mutation operator type (switch and random), crossover probability p_c and mutation probability p_m.

Fig. 1 illustrates mean values of GA computations with different variations of initial population, crossover and mutation operators. Based on these computational data the following conclusions can be drawn:

Initial population: When the "best possible" fitness solution B is added, the GA visibly surpasses the one with purely random initial population, giving better result in a slightly less time. Adding precomputed partly-feasible solution B to initial population helps the genetic algorithm to converge faster, basically it can be said that the GA fixes

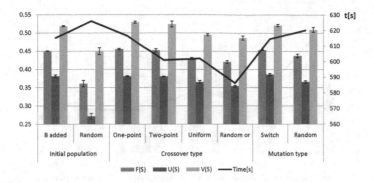

Fig. 1. Results of initial population, crossover and mutation operators experiment ($F(S)$ - overall fitness of solution S, $U(S)$ - average degree of common paper's and reviewer's keywords, $V(S)$ - average degree of paper's keywords covered by assigned reviewers' keywords, for details see [6]).

the solution computed by simple iterative algorithm. Using heuristics and problem-specific knowledge strategies in GA is a common practice in evolutionary computation community [3].

Crossover operator: The one-point and two-point operators, when used as GA crossover operator, equally produce better results than two other operators: the uniform operator and proposed "random or" operator. However, more experiments with results given in Fig. 4 have shown that it is not always the case. Moreover, the GA with "random or" operator runs noticeably faster.

Mutation operator: Basically the "switch" mutation operator overperforms the "random" operator, but not in a great manner. Moreover, it always produce a feasible chromosome, so there is no need to apply the fix procedure after the mutation, like in "random" mutation operator.

The additional detailed study on GA classic crossover operators and proposed "random or" operator was made. The chart in Fig. 2 depicts the computational experiment results of crossover operators performance for different mutation probability p_m. It can be easily seen that in view of the solution quality, the "random or" crossover operator overperforms others. Only in case when $p_m = 0.01$ it produces significantly worse solution. What is more, for all values of mutation probability p_m the "random or" crossover operator runs faster then other crossover operators. The solution fitness grow for all crossover operators is proportional to mutation probability p_m. The fact that the solution quality increase with the growth of mutation probability p_m is confirmed by mutation probability p_m computational experiment with results demonstrated in Fig. 3. The time performance and solution quality obtained by of one-point and two-point operators is rather equal and in most cases is greater than the uniform crossover operator. Thus, in case of the *KNS* input data and used parameters this implementation of genetic algorithm performs better with the one-point or two-point crossover operator, than when the uniform operator is used. On the other hand, the work [3] have presented the results

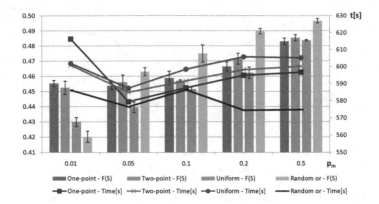

Fig. 2. The GA operators experiment results for different mutation probability p_m.

when the uniform crossover operator overperforms the one-point and two-point operators. Such studies are input data and problem depended, so it is hard to propose general hypothesis.

Although, the detailed theoretical and practical study on mutation p_m and crossover p_c probability parameters had been made during the years [8], there is unfortunately always a need to tune the algorithm parameters up. Even though, that there are suggestions about optimal values of such parameters, they are always problem depended.

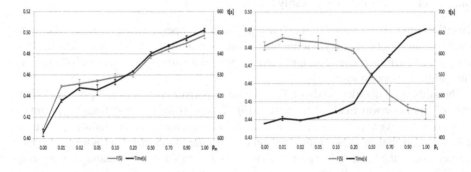

Fig. 3. The experiment results of mutation probability p_m parameter.

Fig. 4. The experiment results of crossover probability p_c parameter.

The quality of GA solution results based on mutation probability p_m parameter is plotted in Fig. 3. The experiment computations of various values of probability p_m parameter have shown that the quality of obtained solution is proportional to mutation probability grow. Normally a rather small values of mutation probability p_m are advised, because the large values could destroy the population, so in the best case (when population is elitist) change the genetic algorithm into a random search algorithm. However in this case, the chromosome representation is crucial. Because of problem characteristic and the use of binary two-dimensional chromosome representation, the mutation of one or even few bits in the individual will not change its fitness value dramatically. The chromosome fitness values is a mean of all assignments finesses (when $n = 200$, $m = 100$ and $k = 2$ there are 40 000 assignments). In problems where the position of bit in chromosome representation does matter (for example when using binary encoded numbers representations) the fitness value of mutated individual can largely differ from the original.

Corresponding to previous diagram, Fig. 4 provides the quality of GA solution results based on crossover probability p_c parameter. It shows that the most optimal crossover probability p_c parameter values are in range $[0, 0.2]$. What is more, when $p_c \in [0, 0.2]$ the computations are also less time consuming. The work [8] presents optimal values of crossover probability for few example problems: 0.32, 0.95, $[0.25, 0.95]$. Thus again the genetic algorithm parameter optimal value can differ significantly when algorithm is applied to different problems.

3   Ant Colony Optimization Evaluation

In this section we are interested in the ant colony optimization algorithm. ACO depends on the number of initial parameters: heuristic information importance - $\beta = 1.0$, pheromone information importance - $\alpha = 1.0$, pheromone evaporation rate - $\rho = 0.01$, number of ants used - $g = 5$, iterations - 1000. In the experiments we are taken into account the following: heuristic information importance α, pheromone information importance β, pheromone evaporation rate ρ, number of ants (g) used.

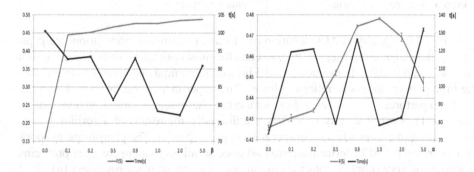

Fig. 5. The experiment results of heuristic importance β parameter.

Fig. 6. The experiment results of pheromone importance α parameter.

Based on the results depicted in Fig. 5 the the fitness values of ACO solution results is proportional to the heuristic information importance β parameter grow. If $\beta = 0$, only pheromone is used, without any heuristic information. This generally leads to rather poor solutions and stagnation situation when all ants follow the same path. The work [4] have presented the heuristic importance optimal values - $\beta \in [2, 5]$. Thus it confirms that achieved results for β parameter optimization are reasonable ones. The time performance is directly connected with the Java `Math.pow(a,b)` method implementation, which computes the a^b operation. The performance of `Math.pow` method is lower for $\alpha \in \{0.1, 0.2, 0.9\}$, due to computation of fractional powers (the computation for $\alpha = 0.5$ is faster probably because $a^{0.5} = \sqrt{a}$).

The dependence of quality of ant colony optimization solution results on pheromone information importance α parameter is given in Fig. 6. It shows that the most optimal pheromone importance parameter value is $\alpha = 1$. The work [4] have also found the $\alpha = 1$ value as optimal. Considering the experiments results for α and β parameters and the literature, the values $\alpha = 1$ and $\beta = 5$ can be recognized as optimal. Similarly to experiment results of heuristic information importance β parameter, the time performance of pheromone importance α parameter is directly connected with the Java Math.pow(a,b) method implementation.

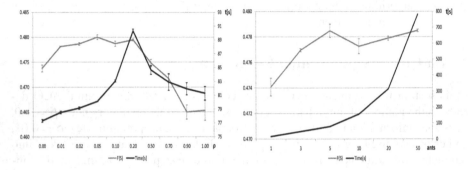

Fig. 7. The experiment results of pheromone evaporation rate ρ parameter.

Fig. 8. The experiment results of number of ants g parameter.

In Fig. 7 the quality of ACO solution results and pheromone evaporation rate ρ parameter dependence is presented. It shows that the most optimal pheromone evaporation rate value is $\rho = 0.05$. The value $\rho = 0.05$ is close to optimal one; however, in this case the time performance is significantly lower. The experimental results have shown that the ACO performs worse when it does not use the pheromone information (evaporation rate $\rho = 1$) or use it slightly ($\rho = 0.9$). Additionally, the results of algorithm performance are also poorer when no evaporation is used ($\rho = 0$). The pheromone evaporation improves the ACO performance to find good solutions to more complex problems. However, on very simple graphs the pheromone evaporation is not necessary [4].

The quality of ant colony optimization solution results, based on number of ants g parameter is aggregated and depicted in Fig. 8. The results of number of ants g parameter optimization have shown that ACO performs well when $g \in \{5, 20, 50\}$. The time needed for an algorithm to end grows exponentially with the number of ants used. Thus, the number of ants $g = 5$ is selected as an optimal parameter.

4 Tabu Search Evaluation

In this section we report the computational results on the performance of TS algorithm. Two parameters are specified: neighborhood size $|N(S)|$, tabu tenure T. The initial parameters are: neighborhood size - $|N(S)| = 50$, tabu tenure - $T_n = 10$, iterations - 10000.

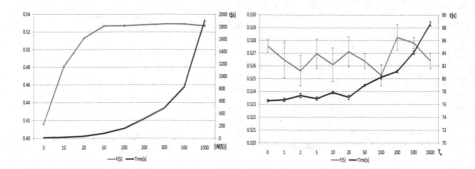

Fig. 9. The experiment results of neighborhood size $|N(S)|$ parameter.

Fig. 10. The experiment results of tabu tenure T_n parameter.

In order to consider the TS performance dependence on the neighborhood size $|N(S)|$ parameter, the result are shown in Fig. 9. The experimental computations of various values of neighborhood size $|N(S)|$ parameter have shown that the quality of obtained solution grows logarithmically with the neighborhood size grow. However, the time needed for an algorithm to finish grows exponentially with neighborhood size. Thus, the reasonable value for neighborhood size parameter is $|N(S)| = 50$.

Fig. 10 presents the dependence of quality of tabu search solution results on tabu tenure T_n parameter. The experiment results of tabu tenure T_n have shown that there is probably no dependence over solution quality and tenure T_n parameter. Furthermore, the experiments have shown that the tabu list have no significant importance in paper-reviewer assignment problem and KNS input data scope.

5   ACO-GA Evaluation

In ACO-GA algorithm the genetic algorithm and ant colony optimization algorithm complement each other in optimal solution search. Each time the ACO-GA algorithm iteration ends the better solution of two algorithms is selected. So naturally the ACO and GA optimal parameters should also be the same for ACO-GA and parameters optimization experiments should not be needed. Nevertheless, the one thing about this is worth checking, namely which one of component algorithms solutions, and in what extent, is chosen more frequently. The ACO-GA experimental results have shown that the ant colony optimization solution is chosen roughly over 100 times more often then genetic algorithm solution. Thus in fact, the ACO-GA algorithm can be understood as an enhanced form of ant colony optimization algorithm. Due to the fact that ACO solutions are selected more often the GA solution, the genetic algorithm solution chosen rate is established as the quotient between the number of times the GA solution was chosen and the ACO-GA iterations.

Fig. 11 shows the genetic algorithm chosen rate dependence over the GA mutation probability p_m and ACO pheromone evaporation rate ρ. The algorithm was run with optimal parameters for ACO and GA algorithms, the number of iterations was 1000. Interpretation of this chart is straightforward, when the ACO performance is lower

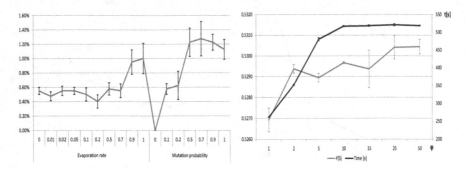

Fig. 11. The experiment results of GA chosen rate.

Fig. 12. The experiment results of maximum modifications φ parameter.

(it is the case for big values of pheromone evaporation rate) the GA solution chosen rate grows. Similarly, when the GA performance is bigger (when mutation probability grows), then the genetic algorithm solutions are chosen more often.

6 GA-TS Evaluation

The proposed tabu search and genetic algorithm hybrid is investigated in this section. Similarly to ACO-GA hybrid algorithm, the optimal parameters for both tabu search and genetic algorithm should also be optimal for GA-TS algorithm. However, the way of unification of TS and GA algorithms bases on different principles than those in ACO-GA hybrid algorithm. The GA-TS algorithm uses additional parameters that should be studied: number of local TS iterations LTS_{iter}, elite closeness selection criterion ε, maximum number of modifications φ. Basically, those parameters control the number of TS invocations, for details see [7]. We assume the following initial values for the parameters: local TS iterations - $LTS_{iter} = 5$, elite closeness criterion - $\varepsilon = 0.01$, maximum modifications - $\varphi = 10$, neighborhood size - $|N(S)| = 10$, tabu tenure - $T_n = 10$, crossover type - random or, crossover probability - $p_c = 0.1$, mutation type - switch, mutation probability - $p_m = 0.5$, initial population - B added, population size - 50, generations - 5000.

The results of maximum modifications φ parameter are depicted in Fig. 12. It explains that the maximum number of modifications φ parameter has not got significant impact on quality of GA-TS solutions. The fitness values of algorithm solutions slightly grow with the maximum modifications φ parameter. The little changes of fitness values are probably caused by the fact that algorithm returns near optimal solutions and it is not possible to enhance the solution quality further. The time needed for the algorithm to end grows logarithmically with the maximum modifications φ parameter. It is reasonable to use the parameter's value from range $\varphi \in [2,5]$.

The dependence of quality of GA-TS solution results on the number of local TS iterations LTS_{iter} parameter and on the elite closeness selection criterion ε parameter are summarized in Fig. 13 and Fig. 14 respectively. Similarly to maximum number of modifications φ parameter experiment results, the change of parameter values does not give

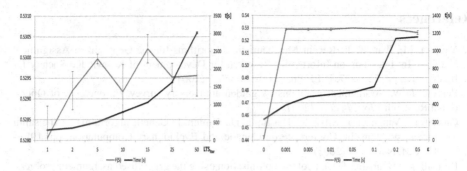

Fig. 13. The experiment results of local TS iterations LTS_{iter} parameter.

Fig. 14. The experiment results of elite closeness selection criterion ε parameter.

big impact on quality of GA-TS solutions. Due to exponential dependence of running time on local TS iterations LTS_{iter} parameter values, rather small values of parameter are encouraged $LTS_{iter} \in [2, 10]$. Similarly to previous GA-TS parameter experiments results, the change of elite closeness selection criterion ε parameter values does not give big impact on quality of GA-TS solutions. However, the difference between the use of local tabu search algorithm ($\varepsilon > 0$) and plain GA ($\varepsilon = 0$) can be easily seen. Although the more detailed study should be made, the GA-TS hybrid algorithm seems to overperform the plain genetic algorithm. Additionally, due to exponential dependence of running time on elite closeness selection criterion ε parameter values, rather small values of parameter are encouraged $\varepsilon \in [0.001, 0.01]$.

7   Conclusions

This work evaluates an artificial intelligence approach to generalized assignment problem, especially to paper-reviewer assignment problem. The parameters optimization experiments have disclosed the superiority of proposed genetic algorithm "random or" crossover operator. Moreover, the usability of heuristic initialization of GA population have been confirmed. Furthermore, based on the genetic algorithm experiments, changes to the algorithm mutations operators can be considered. As a result of optimization experiments, a fairly good parameters values were found for all tested algorithms. However, parameters optimization experiments presented here are mostly one dimensional and do not check possibilities of all cross dependencies.

An interesting finding was made during ACO-GA experiments. Tests have shown that GA gives about 100 time less contribution to ACO-GA final solution than ACO algorithm. However, it helps the ACO algorithm to avoid premature convergence. The hybridization mechanism has great potential to provide even more improvements and still requires further work.

Acknowledgments. This research was partially supported by Grant no. N N519 407437 funded by Polish Ministry of Science and Higher Education (2009-2012).

References

1. Yagiura, M., Ibaraki, T.: Recent Metaheuristic Algorithms for the Generalized Assignment Problem. In: Int. Conf. on Informatics Research for Development of Knowledge Society Infrastructure, pp. 229–237. IEEE Press, Los Alamitos (2004)
2. Pentico, D.W.: Assignment problems: A golden anniversary survey. European J. of Operational Research 176, 774–793 (2007)
3. Garrett, D., Vannucci, J., Silva, R., Dasgupta, D., Simien, J.: Genetic algorithms for the sailor assignment problem. In: Conference on Genetic and Evolutionary Computation, pp. 1921–1928. ACM, New York (2005)
4. Randall, M.: Heuristics for ant colony optimisation using the generalised assignment problem. In: Congress on Evolutionary Comput., pp. 1916–1923. IEEE Press, Los Alamitos (2004)
5. Wu, T.H., Yeh, J.Y., Syau, Y.R.: A Tabu Search Approach to the Generalized Assignment Problem. J. of the Chinese Inst. of Industrial Eng. 21(3), 301–311 (2004)
6. Kolasa, T., Król, D.: ACO-GA approach to paper-reviewer assignment problem in CMS. In: Jędrzejowicz, P., Nguyen, N.T., Howlet, R.J., Jain, L.C. (eds.) KES-AMSTA 2010. LNCS, vol. 6071, pp. 360–369. Springer, Heidelberg (2010)
7. Kolasa, T., Król, D.: A Survey of Algorithms for Paper-Reviewer Assignment Problem. IETE Tech. Rev. 28 (2011) (in Press)
8. Feltl, H., Raidl, G.R.: An improved hybrid genetic algorithm for the generalized assignment problem. In: Symp. on Applied Computing, pp. 990–995. ACM, New York (2004)

Smart User Adaptive System for Intelligent Object Recognizing

Zdenek Machacek, Radim Hercik, and Roman Slaby

Department of Measurement and Control, FEECS, VSB – Technical University of Ostrava,
Ostrava, Czech Republic
{zdenek.machacek,radim.hercik,roman.slaby}@vsb.cz

Abstract. This paper focuses on the intelligent adaptive user system for intelligent object recognizing of traffic signs for car applications with wide setting of parameters and limits, which depends on user type and requirements. The designed smart user adaptive system gives uninterrupted attention and intervention of the service for user by the process control. That is the reason why the control system destined for scanning and processing of the image is entered into the control applications. The advantages of these systems are preventing of possible errors made by human inattention. This developed adaptive system is usable to assist the car driving. The paper paragraphs describe an algorithms, methods and analyses of designed adaptive system. The developed model is implemented in the PC, ARM processor and PDA device.

Keywords: user modeling, adaptation, recognition, image processing, detection.

1 Introduction

There is presented user smart adaptive system for intelligent object recognizing for car driving system upgrading. The developed recognition system is adaptable for various parameters and situation cases. The algorithm parts of whole adaptive recognition system are simulated as particular algorithms in Matlab mathematical environment and realized in Visual Studio programming environment.

The recognition driver system was developed for traffic signs detection. The system processes real-time image captured by a camera. Implemented algorithms are based on the idea of standardized form and appearance of traffic signs. The algorithms search for traffic signs and then transmit this information to the dilution, or an airborne vehicle computer. Information about traffic signs is reducing the risk of an inadvertent violation of traffic rules, which is particularly high for drivers who drive in an unfamiliar area or passing through sections of the ongoing construction activities. Algorithms are proposed primarily to identify traffic signs in the Czech Republic, but it can be easily extended for use in all states that use standardized forms of road signs.

The developed system for image processing meets the requirements for exactness, rate and reliability. These three basic requirements are closely connected with each other and it is necessary to consider them by designing of such system. According to

N.T. Nguyen et al. (Eds.): New Challenges for Intelligent Information, SCI 351, pp. 197–206.
springerlink.com   © Springer-Verlag Berlin Heidelberg 2011

wide range of possibilities of use, each application requires different outputs from the image processing system. It depends on what image scanner is used and what are the parameters of the environment. This is the reason why it is necessary to design the system with wide range of modularity. The developed application shows examples of user adaptive recognition system for traffic signs with modifiable target as operating system Windows CE, Windows 7, Windows Mobile 6 in Fig.1, Fig.9. [1],[2].

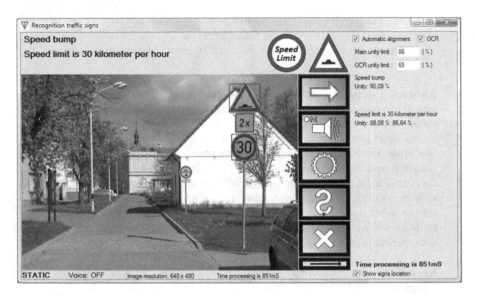

Fig. 1. The user adaptive application in device with operating system Windows CE/Windows 7

2   Intelligent Recognizing System for Target Modifications

An application is created and modified for these operating systems: Linux, Windows CE, Windows Mobile. Implementable hardware target is a development kit the 32-bit multimedia applications processor iMX31 based on ARM11 ™ core, which contains CMOS / CCD input, 2D/3D graphics accelerator and VGA MPEG-4 hardware encoder. The development kit i.MX LITEKIT of Zoom ™. The development kit has a serial port RS232, RJ45 connector for Ethernet connection, number of USB ports and integrated 60-pin connector for the Touch panel, TFT display connection. User interfaces for these types of embedded systems are derived from the function they perform. Scope of user interface varies from none to full desktop-like interface of PC. The developed application allows you to fully visualize and control the intelligent recognizing by the embedded system through the touch screen LCD. [7].

The application and implemented algorithms are suitable for wide range of applications in existing PDA devices and mobile phones. These devices are limited by low frequency computing power and memory space compared to being commonly used in desktop computers. The application was tested in devices E-TEN Glofiish X650. This device contains a Samsung S3C2442 processor 500 MHz, 128 MB Flash ROM, 64 MB RAM and a VGA TFT display 2.8-inch.

The developed intelligent recognition system is applicable in the classical desktop PC. The developing was implemented for these computers same as algorithms modeling in mathematical environment Matlab. The tests of algorithms executing are evaluated for chosen desktop PCs and the results are explained in text below. The developed application is suitable and prepare for modification for other named device targets.

Video input of applications determined to image recognition, the camera with VGA resolution is enough. Higher resolution would extend the process time because each point of image matrix is being worked up by recognition process. Web camera with VGA resolution based on CMOS technology is used for image recording. The advantage of these cameras is a low price and very good availability. The disadvantage of this camera is the undesirable noise, which can be cleared up by software device. The developed system for intelligent recognition was tested with input signal from Webcams Logitech with resolution 352 * 288 pixels. Image quality reflects necessary quality of traffic signs identification. Low image resolution of traffic signs causes that recognition success will decrease to 90%. It is about 5% less than if the original high resolution image is evaluated. Application speed is convenient, as it is able to process more than 10 frames per second with presented camera resolution. [3],[5].

3   Software Implementation and Algorithms Testing

The recognition algorithms are implemented in Matlab and C# programming language. The developed Matlab mathematical model is used to verify the functionality of the detection algorithms and for testing the system behavior for different conditions and environments. The C# application program is designed for Windows XP, Vista/7, real-time operating Windows CE, PDA devices using Windows Mobile 6.0.

The algorithms were first implemented in the Matlab mathematical modeling language to confirm design assumptions and to simulate the various functions and function blocks. The model in Matlab allowed easy debugging of code that was later implemented in other programming languages. After modeling the recognition algorithm was implemented in C#. Programming C# allows you to create applications for Windows platforms.

The Matlab mathematical code is written by m-Files, which contain the source codes. The recognition algorithms are divided into logical units and functions. Such a separation increases reliability and increases the modularity of the developed intelligent recognition system. Each part of code contains results plot of modified figure after algorithm execution.

The application for recognition of traffic signs depends on computer performance time recognition of road signs. This time is given by parameters of input image size and number of objects that are inside image, the performance of computer and others.

The processing time of implemented code in C# on a desktop PC was tested on six different computers whose hardware configuration is shown in Fig.2. [4].

There was tested image example, which contains traffic signs, for comparison of measured computer performance. To test the application for recognition of traffic signs have been selected image with VGA resolution, taken from real environment. Resolution size affects the overall accuracy for detection and time, but for this test is quite sufficient VGA resolution. The time performance test of recognition traffic

Fig. 2. The time performance test example of recognition traffic signs code with results

signs code was measured as time from the beginning of image analysis to identify the various traffic signs. As shown in Fig.2, the time results are the best for PC with dual-core Intel Core 2 Duo T9300, the worst for Intel Celeron M 420, followed by a PC with AMD Sempron 3400 +. Testing has shown that reduced performance of the device has affected algorithms time consumption. There is important to more efficient resource devices utilization and memory management optimization for sufficient algorithms acceleration of application. [6].

4 Structure and Algorithms of Intelligent Recognizing

The recognition system of traffic signs processes in real-time images captured by camera. System is based on an algorithm which is based on the idea of a standardized form and appearance of traffic signs. Parameters are defining in the Czech Republic a notice by the Ministry of Transport č.30/2001 Sb. The algorithm consists of three basic parts. The algorithm structure of intelligent recognizing system is shown in Fig.3. The first part realizes the input image adjustment. This meant brightness correction and image thresholding. The second part realizes image segmentation and the third part realize find pattern and the actuator transmission or display the result. Adaptability is necessary for traffic signs recognition system, because of variable input image parameters as a wide range of values, as on the scene, the location, the size of objects on the image, contrast, intensity of light at the time of camera image loading. The first step of recognition algorithm is to standardize the input image, which is performed by image thresholding. After image adaptation and modification, there is possible to search for continuous areas and segments in image, which may search traffic signs on the image.

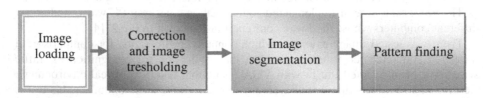

Fig. 3. The structure of intelligent recognizing system basic algorithms.

From knowledge of continuous areas and segments and the known properties of signs, there could be process the analysis by pattern comparison. Evaluation of the specific traffic sign shall be chosen by continuous correlation algorithm with the file of patterns of traffic signs. Before applying the correlation function, the image must be standardizing to defined size of focused area. Application of the correlation function is not efficient in the case of traffic signs with the speed limit. The best solution in these cases is the OCR algorithm choice. The implemented OCR code is implemented for objects located within the selected discrete area. In many cases, traffic signs are not placed in a perfectly vertical position. The given properties knowledge of the traffic signs enables to adjust the angle of the selected segmented areas before applying the correlation function. This implemented mathematical algorithm assures enough increasing the reliability of the correlation function and recognition results accuracy.

An example of individual results plots from implemented recognition process is shown in Fig.4. The Figure presents input image from camera, image after brightness correction and image thresholding, image after segmentation algorithm, which is prepared to pattern traffic signs comparison.

Fig. 4. Example of input, image after correction and thresholding, image after segmentation

The turning algorithm can correct skewing of the traffic sign up to 30 degrees over the vertical position. In the case, the traffic signs without vertical displacement have been recognized with approximately 95% accuracy. The traffic signs with vertical displacement distortion in the sense of vertical displacement were detected with approximately 88% accuracy. The shown figure Fig.5 presents camera image containing the traffic sign with or without vertical displacement. In the case, the algorithm for automatic rotation is switched off; the recognition algorithm doesn't successfully detect any traffic signs on image. Therefore the rotation method has a positive effect on the recognition system.

The defined patterns of the traffic signs are arranged side by side in one line. There are stored in *. bmp files with bit depths 1, what means monochromatic color of image. The dimension of each defined pattern is a standardized traffic sign to the size of 100x100 pixels. Thus, the selected size is a good compromise between the quality of the representation of traffic signs and the number of pixels by which the marking is represented. Because of the processing time saving, the implemented algorithms of the correlation function enable to minimize the number of process pixels, which are comparing with the traffic sign patterns. The pattern matrix of the traffic signs is

Fig. 5. Example of image containing the traffic sign with or without vertical displacement

divided into two separate *.bmp files, because of the character parameters of each traffic sign image. The one matrix of the patterns includes the traffic signs, which are consisted only of a single continuous area. The last member of this matrix is a special case, that is identified in the case of traffic signs "no stopping" or "no standing". In this case, identified object is recognized by additional algorithms, which decide the relevant object to one of these two traffic signs. The matrix of the patterns with only one single continuous area is shown in Fig.6.

Fig. 6. The matrix of the traffic sign patterns composed of a single continuous area.

The second matrix of the patterns includes the traffic signs, which are consisted of more than one continuous area. The correlation and other algorithms usable for traffic signs recognition is more difficult and accuracy needful compare to matrix with only of a single continuous area in traffic sign. The matrix of the patterns with more then one continuous area is shown in Fig.7.

There are four basic types of traffic signs, if we are speaking about their size parameters. Each of those four types of traffic signs is defined by a fixed ratio of own height h and width w. These properties of their relative size in the image are usable as the base for the structural filter. This filter enables to ignore objects, which can not be searched traffic signs, and deletes all objects that can not even be associated with the searched traffic signs. The basic types of traffic signs, which are different by their size parameters, are shown in Fig.8.

The implemented OCR algorithms are usable for traffic signs with the speed limit, which means for recognition process of numeral symbols inside the traffic sign. The possibility of OCR algorithms is not always capable to surely identify all the characters of a word or phrase. Therefore, the OCR code contains additional correctional algorithms, which compare the results with the numeral symbols dictionary. Moreover, there is implemented analysis for logical evaluation, for example the searching

Fig. 7. The matrix of the traffic sign patterns composed of more than one continuous area.

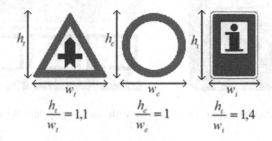

$$\frac{h_t}{w_t} = 1,1 \qquad \frac{h_c}{w_c} = 1 \qquad \frac{h_i}{w_i} = 1,4$$

Fig. 8. The basic types of traffic signs, which are different by their size parameters

speed limit should be always divisible by value equal to 5. All implemented algorithms in specific combination bring together the adaptive intelligent recognition system of traffic signs with the sufficient accuracy of detection.

5   The Adaptive User Interface Description

The smart user adaptive system for intelligent object recognizing is realized as variable application with changeable parameters and visually face depends on operating system - Windows XP/Vista/7, Windows Mobile 6.0, Windows CE and hardware platform. The implemented user interface and the system design are dependent on the possibility of the speed, accuracy, reliability and display ability given by chosen platform for object (traffic signs) recognition. There are presented the realized applications for various platforms. The main application user windows are shown in Fig.1, Fig.9. The application has two execution modes of processing. The one mode (static mode) is the static images recognition, which is stored on hard drive. The second mode (online mode) is the live video stream processing, which is received from the connected video equipment.

The processing information about chosen mode, information of the analyzed image and processing time are displayed in the status bar windows of the main application window. The recognized traffic signs, it means results, are displayed at the top bar window of the main application window, which are shown in the symbols and text form. The recognized objects – traffic signs are presented not only by image displaying, but by voice output too. In the extended application mode, there is viewable also percentage of each successful recognized object, which can be helpful for users to recognition parameters setting. In the case of speed limit traffic signs recognition,

Fig. 9. The user adaptive application in device with operating system Windows Mobile 6

there is displayed the percentage of each successful recognized numeral symbol. The application is designed for touch user control realized on the devices equipped with the touch panel. [8].

The application allows performing a variety of parameter settings. The user can switch the voice output module by icons on the taskbar application window. The choice of function mode is available by the settings button and it is easily changeable via three touches. The parameters setting of the detection algorithm is implemented in the system setting window for users, which is shown in Fig.10. The system settings menu contains text and voice help description for each parameter for intelligent object recognizing.

The parameters for intelligent object recognizing, which affects recognition algorithms, are settable by users. The life time parameter for online mode gives the value of time in seconds, when the traffic sign is displayed on the display from the moment of the recognition.

The diagonal r and length v parameters are usable for setting of threshold boundaries. The red diagonal r parameter is used for the threshold secondary function, which reflects the red color in the image. Maximal threshold area and the minimal threshold area indicate the relative size of the object in the image area in percentage.

These values for analysis function define removing areas represented continues areas with noise elimination from image. The maximal ratio y/x and the minimal ratio y/x parameters represent the dispersion of searching height and width ratio of continuous areas. The maximal and the minimal angle of rotation are used in auto traffic signs angle correction.

The unity limit for main recognition algorithm specifies the minimum degree of identity with the patterns of traffic signs. The unity limit for OCR recognition algorithm specifies the minimum degree of consistency compared to patterns which is recognized as successful.

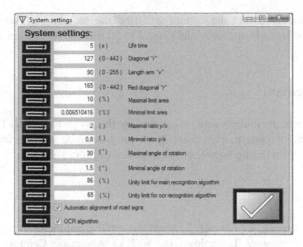

Fig. 10. The system setting menu of developed smart user adaptive application

There are configurable algorithm functions by user, which enable automatic alignment of Road Signs and OCR algorithm processing. These functions are settable by checkboxes, which are accessible in the system settings window.

6 Conclusion

The purpose of the paper was to verify the possibility of user adaptability in the developed application for intelligent recognition system of traffic signs. The adaptable realization is usable for automotive and other applications. The particular parts of algorithms and structure were developed in theoretical knowledge base. The basic innovation is that the smart user adaptive system is variable for actual situations and video image recognition.

The main goal of this paper is to show development of adaptive automotive system for object recognition. The basic system architecture and algorithms for traffic signs recognition are presented here. The problem of image processing and object recognition was discussed in this contribution.

The suitable tool in developed system for image processing seems to be the i.MX31 multimedia application processor with support of the operating system windows embedded CE and PDA device with the operating system Windows Mobile. System built in this way offers sufficient modularity needed for developments, processing and operation of complicated evaluating algorithm.

The verification of the developed software was accomplished approximately on 60 traffic signs. The test results described implemented algorithms dependency on system performance and traffic signs were detected correctly in almost cases, but rarely there was not detected any traffic sign. This error was caused by damage to traffic signs, the excessive pollution or poor light conditions. Traffic signs were successfully recognized with patterns in range from 90 to 95%.

During the recognition process, the distance of traffic signs from the camera plays a detectable role. Due to a small number of pixels of traffic sign that is located too far, it cannot be guaranteed that the system will find the required consensus. Limit for identifying traffic signs is approximately equal to 50 meters, if the camera zoom value is equal to one.

Acknowledgement

The work and the contribution were supported by the project FR-TI2/273 – Research and development of progressive methods of long-distance monitoring of physico-mechanic quantity including wire-less transmission of processed data.

References

1. Nevriva, P., Machacek, Z., Krnavek, J.: Simulation of Thermal Fields of Sensors Supported by an Image Processing Technology. In: WSEAS Automatic Control, Modelling and Simulation, 7p (2008)
2. Ozana, S., Machacek, Z.: Implementation of the Mathematical Model of a Generating Block in Matlab&Simulink Using S-functions. In: The Second International Conference on Computer and Electrical Engineering ICCEE, Session 8, pp. 431–435 (2009)
3. Krejcar, O.: Problem Solving of Low Data Throughput on Mobile Devices by Artefacts Pre-buffering. EURASIP Journal on Wireless Communications and Networking, 8 (2009), doi:10.1109/EURCON.2009.5167783
4. Tutsch, M., Machacek, Z., Krejcar, O., Konarik, P.: Development Methods for Low Cost Industrial Control by WinPAC Controller and Measurement Cards in Matlab Simulink. In: Proceedings of Second International Conference on Computer Engineering and Applications, pp. 444–448 (2010), doi:10.1109/ICCEA.2010.235
5. Lacko, L.: Programujeme mobilni aplikace ve Visual Studiu .NET. 19.8.2004, p. 470. Computer Press, Praha (2004), ISBN 80-251-0176-2, 9788025101766
6. Kotzian, J., Konecny, J., Krejcar, O., Lippa, T., Prokop, H., Kuruc, M.: The Indoor Orientation of Autonomous Mobile Devices for Increasing Human Perspective. In: Mobilight Workshops 2010, Barcelona,Spain. LNCIST. Springer, Heidelberg (2010), ISBN 978-3-642-16643-3
7. Sridhar, L.: Design Embedded Communications Software. CMP Books, San Francisco (2003), ISBN 1-57820-125-X
8. Srovnal Jr., V., Machacek, Z., Srovnal, V.: Wireless Communication for Mobile Robotics and Industrial Embedded Devices. In: Proceedings ICN 2009 8th International Conference on Networks, Cancun, Mexico, pp. 253–258 (2009), ISBN 978-0-7695-3552-4

Student Modeling by Data Mining

Alejandro Peña-Ayala[1,2,3,4] and Riichiro Mizoguchi[4]

[1] WOLNM, [2] ESIME-Z-IPN, [3] CIC-IPN, [4] Osaka University
31 Julio 1859 # 1099B Leyes Reforma DF 09310 Mexico
apenaa@ipn.mx, miz@ei.sanken.osaka-u.ac.jp
http://www.wolnm.org/apa, http://www.ei.sanken.osaka-u.ac.jp/

Abstract. This work pursues to find out patterns of characteristics and behaviors of students. Thus, it is presented an approach to mine repositories of student models (SM). The source information embraces students' personal information and assessment of the use of a Web-based educational system (WBES) by students. In addition, the repositories reveal a profile composed by personal attributes, cognitive skills, learning preferences, and personality traits of a sample of students. The approach mines such repositories and produces several clusters. One cluster represents volunteers who tend to abandon. Another group clusters people who fulfill their commitments. It is concluded that: educational data mining (EDM) produces some findings to depict students that could be considered for authoring content and sequencing teaching-learning experiences.

1 Introduction

Witten and Frank state data mining (DM) as: 'The process to seek understandable and useful knowledge, previously unknown, from huge heterogeneous repositories" [10]. Its roots correspond to the statistics. Later on, other disciplines contributed to DM, such as: data bases, neural networks, evolutionary computation, and fuzzy logic [4].

One of the novel DM applications fields corresponds to EDM. It is a research area devoted to discover hidden patterns of student's attributes, behaviors and results. EDM produces mental models of student communities based on raw and conceptual data. Educational systems (e.g. WBES) are the main providers of information. Content, sequencing, SM, assessment, and evaluation repositories are a sample of the data source to be mined [5, 6].

In this approach, EDM is used to describe key attributes of a volunteers' sample. The subjects are members of the academic community. They participated in a Web-based experiment. The trial lasted six months and included several stages. One was devoted to outcome a SM to depict the volunteer. Another provided a course by means of a WBES. The assessment included pre and post measures about the course topic. At the beginning, 200 volunteers participated; but at the end, only eighteen successfully accomplished the whole experiment. Thereby, a critical question was made: Which are the attributes that distinguish students who abandon a course?

N.T. Nguyen et al. (Eds.): New Challenges for Intelligent Information, SCI 351, pp. 207–216.
springerlink.com   © Springer-Verlag Berlin Heidelberg 2011

The response to such a question is the object of the present work. Hence, the problem is tackled like a descriptive DM task. Moreover, the solution is thought of as the outcome of clusters. Thereby, a segmentation algorithm is used to scan the source information. It embraces personal data of the participants, a set of SM, and the assessment records. As a result, several groups of volunteers are produced. According to the attributes considered by the mining process, interesting highlights are found.

As well as this introduction, where a profile of DM, EDM, and the approach is stated, the article is organized as follows: In section two the related work is presented, whereas the acquisition of knowledge about the subjects is explained in section three. The description of the source data is given in section four, whilst in section five, the DM task, method, and algorithm are outlined. The results and the interpretation are shown in section six. Finally, some comments and the future work are given in the conclusion section.

2   Related Work

Research on the EDM arena is quite incipient. However, some WBES functionalities are the target of recent works. One of them corresponds to student modeling. There is one approach oriented to classify students with similar final marks [9]. It explores assessments and evaluations records. Some predictions about students' performance are carried out by statistical, decision, tree and rules induction DM methods.

Content is also another target of EDM. Rule mining is used to find out factors to improve the effectiveness of a course [2]. The approach generates advice to shape content. They are represented like *if-then* rules. As regards tutoring, dynamic learning processes are characterized by evolutionary algorithms [3]. The purpose is to deliver content that satisfies student requirements. Assessment logs represent a valuable data source for EDM. One approach examines events from online multiple choice tests [7]. It applies process discovery, conformance checking, and performance analysis.

3   Data Acquisition

This work is part of a research devoted to apply the student centered paradigm by the use of WBES. The aim is to enhance the apprenticeship of WBES users through the behavior adaptation of WBES to the users' needs. The usefulness of the SM was demonstrated by the execution of a trial [8]. Hence, the process devoted to tailor a set of SM is described in this section. In addition, personal data, assessment of the student's behavior, and evaluation measures taken from a universe of volunteers are also set.

First of all, a Web-based marketing campaign was launched to recruit participants for an experiment. The trial just informed that: A course about the "research scientific method" is available on-line. Furthermore, the definition of a psychological profile is included for the participants. What is more, all the activities are developed through Internet. In consequence, hundreds of individuals in Mexico submitted an application E-form. All candidate records were analyzed, but only two hundred were accepted.

Afterwards, four tests were adapted to be delivered on the Web with the purpose to set a SM of the participants. They were applied in the following order: learning preferences, personality traits, cognitive skills, and former knowledge about the course domain. During three months volunteers had the chance to respond the sequence of exams. Once a subject rightly answered a test, she/he was granted to access the next one. However, many volunteers gradually deserted. The assessment reveals a descending progression of participants that completed the corresponding exam of: 113, 102, 71, and 50. It means that only fifty subjects successfully completed the four tests!

Later on, the population of fifty volunteers was trained. They received introductory lessons about "science". Next, a pre-measure test was applied to the sample. The exam questioned participants about ten key concepts (e.g. law, theory). The Taxonomy of Educational Objectives (TEO) was used to estimate the mastering level about a key concept [1]. Thus, a value between 0 and 6 was attached to each key concept.

Once again, more participants abandoned the trial. Only eighteen subjects rightly finished the pre-measure. So they composed the final sample. It was randomly split into two comparative groups of nine members each. The WBES delivered random lectures to the control group. According to the advice provided by the SM, the WBES provided lectures to the experimental group. Volunteers took just one lecture about a specific key concept. Thus, a total of ten lectures were taught to the sample.

A post-measure was applied once the subject completed the lectures. So the same test was submitted to volunteers. The evaluation ranged from 0 to 60 for the participant. A range of [0, 540] corresponded to a group (i.e. 60 is the maximum value for a participant by nine members of a group). Thus, the apprenticeship is the difference between the post and the pre measures achieved by the participant.

Afterwards, descriptive and inference statistics were estimated. The hypothesis research, about the impact of the SM to the student's learning was demonstrated by: population mean, correlation, and linear regression. A confidence level of 0.95 and a significance level (α) of 0.05 were considered. A Pearson's coefficient of 0.828 and a probability of 0.006 was outcome for experimental group.

4   Data Representation

As a result of the earlier stated case study, a heterogeneous repository of factual, conceptual, and estimated data is outcome. Such information is the input of an EDM process. However, prior to mine the data, the identification of the sources is needed. The items representation is also accounted. Moreover, the media, format, and code used to store the information is relevant to be known as follows.

As regards the personal information, this is stored in a repository, whose records are encoded as extended markup language (XML) documents. They are normalized by XML schemas, which describe structure, elements, attributes and constraints. The documents represent the data given by volunteers as their request for participation.

The fields of the application E-form correspond to: id, gender, age group (e.g. 16-17, 18-19, 20-24, 25-34...), civil status (e.g. single, married, divorced...), occupation (e.g. name, type and number of occupations, whose instance value could respectively be: professor, academic, 2), highest scholastic degree (e.g. Bachelor, Master, Doctorate), profile of the academic level (i.e. for each academic degree: status, university type and prestige, area of study. An instance of possible values is respectively: candidate, public, good, engineering).

The SM holds several domains to depict the student. A domain reveals a set of attributes that characterize the volunteer. They are stated as concepts. The definition is given by a term, a meaning, and a state. The term labels the concept. The meaning is defined in an ontology. The state reveals the intensity level of the presence of such a concept in the subject. So a qualitative value is attached to the concepts' state. It is a kind of linguistic term that is well formalized by the fuzzy logic theory. The terms and the values of a domain are stored in a XML document, as the one presented bellow. The content and code of the domains are described next.

Extract of XML code to set SM repositories of four domains taken from [6]

```
01: <!- Learning domain, concept visual_spatial set from 02-07 lines -->
02: <domain id_term="learning_preferences">
03:   <instance id_concept="visual_spatial ">
04:     <level_linguistic_term>high</level_linguistic_term>
05:     <level_membership_degree>1.0</level_membership_degree>
::
10: <!- Personality domain, concept depression set from 11-14 lines -->
11: <domain id_term="personality">
12:   <instance id_concept="depression">
13:     <level_linguistic_term>low</level_linguistic_term>
14:     <level_membership_degree>1.0</level_membership_degree>
::
20: <!- Cognitive domain, concept verbal set from 21-24 lines -->
21: <domain id_term="cognitive">
22:   <instance id_concept="verbal">
23:     <level_linguistic_term>quite_low</level_linguistic_term>
24:     <level_membership_degree>1.0</level_membership_degree>
::
30: <!- Knowledge domain, concept theory set from 31-34 lines -->
31: <judge id_version="theory">
32:   <instance id_concept="knowledge_level">
33:     <level_linguistic_term>high</level_linguistic_term>
34:     <level_membership_degree>1.0</level_membership_degree>
::
40: <!- Assessment domain, log record set from 41-44 lines -->
41: <volunteer id_volunter="18">
42:   <session id_session="s_030">
43:     <access_date>20090915</access_date>
44:     <access_time>12.45:06</access_time>
```

The learning preference domain contains concepts about: nature disposed, musical-rhythmic, body-kinesthetic, interpersonal, visual-spatial, verbal-linguistic, logical-mathematical, and intrapersonal. The code lines 01 to 05 show a sample.

Regarding the personality domain, more that fifty traits were measured, such as: schizophrenia, paranoia, depression, social introversion, social responsibility, anger, cynicism, fears and impatience. Code lines 10 to 14 set a concept.

The cognitive domain estimates fourteen skills related to: verbal, memory, observation, performance, vocabulary, comprehension, causal reasoning, visual association and intelligence quotient. The verbal concept is stated in code lines 20 to 24.

The knowledge domain encompasses the measurement achieved for ten key concepts. It represents the identification of the key concept and the qualitative level outcome from the TEO. Likewise, the corresponding number of level is stored. Such a number is outcome as follows: ignorance: 0; knowledge: 1; comprehension: 2; application: 3; analysis: 4; synthesis: 5; evaluation: 6. Hence, volunteer's evaluation ranged between 0 and 60. The concepts are stored in a XML document. An instance is defined in code lines 30 to 34.

The assessment of the volunteers' behavior is also represented and constrained by XML documents and schemas. The logs reveal the occurrence of events, the number of trials, the lectures taken, the exercises performed, and the requests for help demanded by participants. A log record is given in code lines 40 to 44.

5 Educational Data Mining Approach

In order to respond the research question, the approach is tackled as a descriptive DM task. Thus, the task pursues to explain past events. In addition, the DM technique is the clustering. Hence, it produces groups of homogeneous instances. So the aim of this work is to find out peculiar characteristics between volunteers. The question is: which are the main differences between formal and informal people? According to the problem and the solution statement, a segmentation algorithm is used.

As regards the source data, the XML documents of the domains were transferred to a relational data base. It embraces 200 rows, where 82 correspond to women and 118 to men. The columns represent the concept's states values earlier stated.

The statement of a cluster model to be created needs three sorts of attributes, such as: *id* is the column that uniquely identifies each row; *input* is a set of one or more attributes whose values are used to generate clusters; *predictable* is an attribute whose values are used as references for further rows that lack such instance values.

A graphical sample of the clusters to be created is given in Fig. 1. The image corresponds to a cluster diagram. The window shows five groups of participants organized by their civil status. The density of the frequencies is revealed by the gray levels of the background. Likewise, there are relationships between pairs of clusters sketched by links. The aspect of the arc identifies its membership to a given group.

The design of the clusters accounts the *volunteer-id* data as the *id* attribute. The *trial-status* column corresponds to the predictable attribute. Several input attributes define the target of analysis carried out by the clustering model. Based on these parameters more than 20 clusters were produced by the segmentation algorithm.

Some of the concepts used as input attributes are: *age group*, *civil status*, *number of occupations*, *type of occupations* and *highest academic degree*. Moreover, a profile for each academic level is also accounted.

The cluster model is also drawn as a bar chart, like the one shown in Fig. 2. In the window, bars are organized into a matrix. The first row reveals the frequencies of the *subjects' civil status*. The second depicts the frequencies of volunteers who deserted and completed the trial. The first column identifies both groups. The second shows the *civil status* instances (e.g. married, single, divorced, free union, separated) and result (e.g. desertion, completion). The third column pictures the frequencies achieved by the instances in one bar. It corresponds to the universe. The remaining columns state the frequency of the cluster attached to each civil status instance.

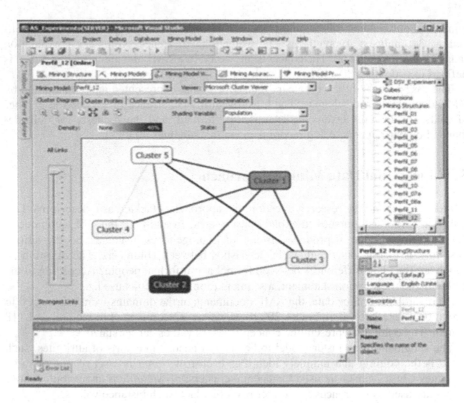

Fig. 1. A cluster diagram of five groups that organized by the civil status of the subjects

This view gives interesting correlations from the couples of clusters. An example is show in the sixth column. The top cluster identifies nine divorced subjects. The bottom cluster proclaims that all deserted! Also, the last column confirms such correlation, because the upper cluster points out that: two participants are separated. While, the lower cluster reveals that they also abandoned the trial.

These clustering models are carried out by a segmentation algorithm. It iteratively fixes up records from the data source into distinctive groups. The algorithm tries to cluster elements that hold specific instance values for several input attributes. These kinds of instance values reveal high frequencies of occurrence.

Wherefore, elements of a specific cluster contain common properties and establish meaningful relations. However, such properties are different from the instance values of the same set of main attributes held by elements of other groups. Moreover, the relationships between elements reveal several meanings for other clusters.

In addition, the segmentation algorithm measures how well the clusters depict groupings of the points. What is more, it aims at redefining the groupings to organize clusters that are more appropriate to set data. The process comes to an end when the algorithm is not able to improve the results any more by redefining the clusters.

As a complement to the clustering frequencies early sketched, Fig. 3 presents the percentage attached to each civil status instance. The former area devoted to draw the first three columns is hidden now by a frame. It holds three columns to identify the

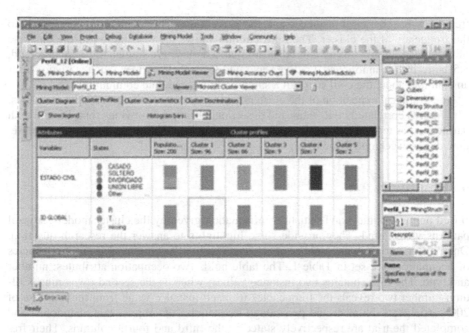

Fig. 2. Frequency of the correlation between civil status and accomplishment of the subjects

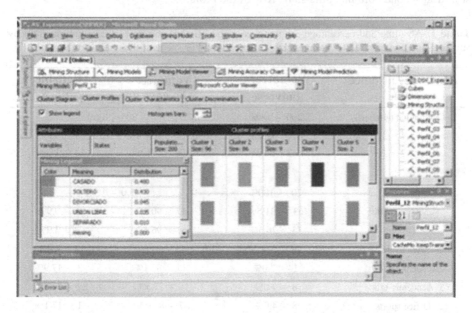

Fig. 3. Percentage of the correlation between civil status and accomplishment of subjects

frequency bar, the civil instances and the percentage. The length bar corresponds to the frequency of the cluster. The third column shows the distribution of the universe of volunteers. It gives away that: 48% are married, 45% are single, 4.5% are divorced, 3.5% are in free union, and 1% live separated from their partner. Such statistics offer an example about the civil status of a sample of the academic community.

The graphical exposition of the clusters facilities the association of input attributes. The figures correlated civil status and desertions. It appears obvious that: People with partner issues tend to face similar problems with other commitments too! In this case, they did not accomplish the whole trial.

6   Results

Based on the graphical and statistical information given by the cluster models, several patterns were found. They provide some highlights to answer the research question. One pattern concerns the number of occupations held by volunteers. The frequencies of occupations are set in Table 1. The table holds two occupation attributes: number and type. The first contains two instance values; whereas the second owns nine. Column number two reveals the frequencies for each instance value from the universe of 200 participants. Number and percentage frequencies of subjects that abandoned and completed the trial are respectively stated in the third and fourth columns. Their frequencies shown in any row are complementary. Therefore, the total is the frequency stored in second column at the same row of the table.

Table 1. Absolute and percentage frequencies of the number and type of occupations held by subjects of a universe of 200 volunteers, who abandoned or successfully achieved the trial

Attribute / instance value	Volunteers universe	Volunteers who gave up	Reliable subjects
Occupation number	200 - 100%	200 - 100%	
1	82 - 41.0%	78 - 95.1%	4 - 4.9%
2	118 - 59.0%	104 - 88.1%	14 - 11.9%
Occupation types	200 - 100%	200 - 100%	
Academic-office	63 - 31.5%	57 - 90.1%	6 - 9.9%
Office	53 - 26.5%	51 - 96.1%	2 - 3.9%
Academic	46 - 23.0%	43 - 93.6%	3 - 6.4%
Academic-sports	14 - 7.0%	11 - 82.1%	3 - 17.9%
Academic-arts	8 - 4.0%	6 - 72.5%	2 - 27.5%
Office-sports	8 - 4.0%	7 - 82.9%	1 - 17.1%
Office-arts	4 - 2.0%	4 - 100%	0 - 0%
Arts	3 - 1.5%	3 - 100%	0 - 0%
Sports	1 - 0.5%	1 - 100%	0 - 0%

The mining process generated several clustering models. The groups contain the following attributes: 1) number and type of occupations; 2) highest academic degree; 3) civil status; 4) Bachelor's status, study area, school prestige, and type. They represent different patterns. All of them were examined in order to respond the research question. The premise is to identify peculiar attributes or relationships.

As a consequence of the interpretation of the mining results, a set of correlational hypothesis are set. They pursue to describe subjects that successfully fulfilled the trial. The volunteers that deserted from the experiment are also depicted as follows:

A high level of trial accomplishment is outcome when the participant: 1) carries out more than one activity; 2) holds high academic degrees; 3) studies (or studied) her/his Bachelor degree in a university whose prestige is "excellent"; 4) actually studies at Bachelor's degree.

7 Conclusions

The findings produced by the EDM techniques promote the critique and reflection about the educational processes, components, and results. They provide empirical evidence that is useful to react before some events and prevent some sort of issues.

In this work, EDM reveals interesting patterns of subjects. They aim at extending the research to reinforce or refute the characteristics of academic groups. Whether they are confirmed, some measures could be stated to take care of such people. The purpose is to reduce the degree of academic desertion.

A future work considers including more student attributes to find out the psychological basis to explain different behaviors patterns. Moreover, it is necessary to apply other descriptive algorithms to discover new hidden knowledge.

Acknowledgments

The first author gives testimony of the strength given by his Father, Brother Jesus and Helper, as part of the research projects of World Outreach Light to the Nations Ministries (WOLNM). In addition, this research holds a partial support from grants given by: CONACYT-SNI-36453, CONACYT 118962, CONACYT 118862, SIP-EDI: DOPI/3189/08, SIP-20110398, COFAA-SIBE.

Referencces

1. Anderson, L.W., Krathwohl, D.R.: A Taxonomy for Learning, Teaching, and Assessing: A Revision of Blomm's Taxonomy. Longman, New York (2001)
2. García, E., Romero, C., Ventura, S., Castro, C.: An Architecture for Making Recommendations to Courseware Authors Using Association Rule Mining and Collaborative Filtering. Int. J. User Model and User Adapted Interaction 19, 99–132 (2009)
3. Guo, Q., Zhang, M.: Implement Web Learning Environment based on Data Mining. Int. J. Knowledge Based Systems 22(6), 439–442 (2006)
4. Hernández, J., Ranírez, M.J., Ferri, C.: Introduction to Data Mining. Pearson Prentice Hall, Madrid (2004)

5. Jung, J.J.: Social grid platform for collaborative online learning on blogosphere: a case study of eLearning@BlogGrid. Expert Systems with Applications 36(2), 2177–2186 (2009)
6. Cao, L.: In-depth behavior understanding and use: The behavior informatics approach. Inf. Sci. 180(17), 3067–3085 (2010)
7. Pechenizkiy, M., Vasilyeva, E., Aalst, W., De Bra, P.: Process Mining Online Assessment Data. In: Proceedings of the Second International Conference on Educational Data Mining, pp. 279–288 (2009)
8. Peña, A.: Student Model based on Cognitive Maps, Mexico City, Mexico: PhD Thesis, National Polytechnic Institute (2008)
9. Romero, C., Ventura, S., Espejo, P., Hervas, C.: Data Mining Algorithms to Classify Students. In: Proceedings of the First International Conference on Educational Data Mining, pp. 8–17 (2008)
10. Witten, I.H., Frank, E.: Data Mining: Practical Machine Learning Tools and Techniques, 2nd edn. Morgan Kaufmann, San Francisco (2005)

Part III

Service Composition and User-Centered Approach

Student Automatic Courses Scheduling

Janusz Sobecki and Damian Fijałkowski

Institute of Informatics, Wroclaw University of Technology
Wyb.Wyspianskiego 27, 50-370 Wrocław, Poland
{Janusz.Sobecki,Damian.Fijalkowski}@pwr.wroc.pl

Abstract. In the paper we present recommendation of student courses. We want to show that this problem has many aspects and a lot of parameters which can be specified by final user. We want to focus on recommendation aspect seen from user perspective and compare this expectations with possibilities of the system. We show the most important aspects of student courses recommendation.

Keywords: courses schedule, courses similarity, recommendation.

1 Introduction

According to [10] the main goal of the recommendation systems (RS) is to deliver customized information to a great variety of users of ever increasing web-based systems population. In this paper we focus on students and courses they can choose during studies. Many students have very special needs about courses. Some of those needs relate to time and day during a week, some relate do teachers and some to the place or building.

In this paper we concentrate on application for recommendation of student courses. In the following section we present the recommendation parameters and their priorities. In the section 3 the courses similarities and limitations problem are presented. The section 4 presents the result presentation problem. The final section presents conclusions and future work in the area of the user interface recommendation.

2 Parameters and Their Priorities

As we mentioned before there are a lot o parameters defining the best course for some student. We can divide them as below:

1. Functional (guidelines included in general studies schedule):
   a. necessary amount of points;
   b. necessary amount of hours;
   c. necessary courses included in study profile realized by student;
   d. type of the course (lecture, lab, exercise, project, …)
2. Nonfunctional:
   a. determinate:
      1) time of the course (day + hour);
      2) place of the course (building + room);
      3) probability of getting a good final result;

N.T. Nguyen et al. (Eds.): New Challenges for Intelligent Information, SCI 351, pp. 219–226.
springerlink.com   © Springer-Verlag Berlin Heidelberg 2011

   b.   indeterminate:
        1) lecturer reputation;
        2) previous contact with particular lecturer.

Some of the parameters presented above can be objectively measured but others describe subjective feelings and memories. Measuring each parameter is different and needs application of special techniques and algorithms. Parameters from 'administrative' group can be measured quite easy. We specify only numerical or Boolean condition – for example we specify that student has to collect 300 points during studies. This fact and information that studies takes 10 semesters let us predict that student should get about 30 points in each semester.

The personal parameters are even more difficult to measure. Student can for example specify that Monday at 3pm is a great time for him. In case when no course is available at that time we have a huge problem to say which other time is the best. Without other information we can't recommend anything. We can't assume that Monday at 5pm or 1pm is the best because that terms are nearest the specified one – they can be unacceptable by student.

The place of the course is very important thing for students, however it brings also some problems to measure differences between specified locations. In fact we don't focus on the course's place itself but rather on distance between places of two adjacent courses. That distance can be very important because in university courses schedule gaps between courses are set and have for example 15 minutes, which is the case of our university (Wroclaw University of Technology). Campuses of some universities are very big. Sometimes university has two or more campuses separated and placed in different parts of the city. In that case, time needed to move from one place to another can be longer than the gap between courses.

According to [10] probability of getting a positive mark from a particular course with a particular lecturer can be counted with Ant Colony Optimization (ACO). Also authors of [2] defines ACO as *'new natural computation algorithm from mimic the behaviors of ants colony, and proposed by Italian scholar M. Dorigo'*. Its etymology is presented in [3]: *'The original intention of ACO is to solve the complicated combination optimization problems, such as traveling salesman problem (TSP). ACO is based on observations about ant colonies and their mutual behavior. The behavior of a single ant is quite simple, but in the case of the whole ant colony it is more complicated and therefore ACO is able to complete complicated task'*.

Also the indeterminate parameters are very hard to measure. Lecturer reputation is based on subjective feelings of other students. It is possible to get numeral information (note) about each lecturer and based on this information calculate final note for a particular lecturer. We should remember that calculating the final note for a particular lecturer without including courses can give wrong results. It can happened when one lecturer has more than one course and opinions on some of his courses are significantly different opinions of rest of his courses. For example, the opinions of teachers from Wrocław University of Technology are collected in one of students forum known as "PolWro", there we can find there notes of many teachers from WUT. The notes are similar with students university notes (from 2.0 to 5.5) and are supplemented with additional opinion. We know that many students visit that forum before they decide to sign in a particular course.

Additionally each student has his own experience according to the previous courses that were taught by the particular teacher. Their opinions are mostly known only for them and can completely change their attitude to the course. This parameter is almost impossible to measure, without giving its value explicit by the particular student.

It is very important to realize that not all parameters are equally important for students. Unfortunately the importance of the parameters described above is different for each student. The importance is not only one difference. We have to know that each student has his own factors for each parameter. That makes recommendation process definitely harder to execute. Moreover algorithms creating three with results of recommendation work different with different order of parameters.

3 Courses Similarity and Limitations

Finding the course schedule that will fulfill all the defined criteria is very hard. It is extremely important to know how to measure distance between two courses. We can compare possibility of getting a positive final note but it is impossible to calculate that value for new courses, using only collaborative filtering methods. We have to define different areas where courses could be compared. We can compare type of the course – it is possible to check if some particular course is a lecture, laboratory, classes, seminar or project. That information can be used in recommendation process.

Each course in academic courses database has to be described in special format. That data can be converted to courses description in SOA paradigm. That transformation gives us a lot of possibilities for further calculations. Data in SOA approach are mostly presented in XML structure what is very useful in comparing process.

Each feature of a course can be transformed to numeric value using parameters specified by student and sum of all values of all parameters gives us a final value for a particular course. This is however hard to accomplish because people have a lot of problems with specification of parameters factors. This problem can be solved by other approach.

Instead of comparing courses calculating all of their features, what is very hard to do, we can compare students. It is much more easier to compare students than courses. Features of student can be compared with one set of factors specified in system and many of those factors are Boolean values. Each student in system database is described inter alia by the following parameters:

1. faculty;
2. specialization;
3. year of study;
4. previously realized courses.

This information is enough for the system to find similar students to the considered user/student. The system should search for students from the same year who have already made their choices for the particular semester or students that made these choices' in the previous semesters. Based on those students choices system can select the best courses for system user and even prepare proposition of schedule. We can distinguish the following most popular filtering methods [6]: Demographic Filtering

(DF), Collaborative Filtering (CF), Content-Based Filtering(CBF) and Hybrid Approach(HA), besides we can distinguish also case-based reasoning (CBR) and rule-based filtering (RBF).

We propose to apply a hybrid approach. Firstly we use demographic method to find proper students and collaborative to specify best courses. We also support courses filtering by content-based method in cases where collaborative method returns two or more equivalent answers and on the basis of user history in system, we chose one course from previously selected.

Calculating courses similarity isn't the only one problem. In many cases there could be many other limitations for student to attend a particular course. We can specify following kinds of limitations:

1. predefined limitations – list of courses which have to be realized before a particular course;
2. necessary payments – payment for a particular course:
    a. repeated course;
    b. course paid extra – for example some sport courses;
3. repeated courses – repeated course can't be replaced by other course without special permission.

4 Presentation of Results

Very important part of recommendation system is the form of the results presentation. Well presented answer from recommendation system can shorten time of student reaction for that answer. One obvious way of presentation of recommendation system answer is week courses schedule. Student can see all courses in one place and decide if that schedule meets his or her expectations. If the schedule doesn't meet specified expectations, student can reload whole schedule or try to change only one of presented courses. Reloading whole schedule is easier for system because system doesn't have to provide additional functionality.

In system which lets user to change one course in presented schedule very important is presentation of possible change options. Those options should be presented in a way consistent with the way in which student has specified parameters.

Basic types of recommendation results presentation are presented below:

1. Table view;
2. Tree view:
    a. horizontal;
    b. cascade;
3. Social networks view.

Table view is most common in many systems. It is easy to put data into table and order by one of columns. The biggest disadvantage of this approach is that person using recommendation system can't see connections between particular parts of the answer (courses) presented in rows. It is also very hard to group courses by more than one parameter. Table view is the best for simple questions to the system where answer is a list of results ordered by some counted value. It is possible to count that kind of value for recommended courses but there should be specified factor for each parameter and each parameter should be able to measure. As we mentioned in one of previous parts of this paper it is very hard to measure some parameters. Moreover specifying factors for each parameter is time-consuming and very difficult for people. People can order parameters from the most important to the least important but they have a lot of problems with assigning numeral values to those ordered parameters. The easiest way is to assume that factors are between 0 and 1 and are spread evenly but we know that this approach can't be proper in most cases.

Tree view seems to be better than table view because gives more complex results presentation for system user (student). When data is presented in tree student can very fast find paths in that tree which don't include the final leaf. Specific thing for courses recommendation is that there is a limited amount of courses, lecturer, times and places and (what is even more important) that not all combinations of those values are available in system. Generally speaking there is a very limited set of courses so some subtrees can include no courses and thanks to that student can very fast eliminate some of his needs and focus only on available options.

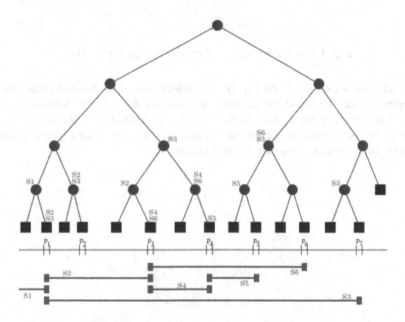

Fig. 1. Horizontal tree view with root, 4 levels, 13 nodes and 15 leafs [12].

Tree view can be presented in two ways (horizontal and cascade). Horizontal tree view (Fig.1) presents following levels of the tree from the root to leafs showing them on screen from up to down. This approach is good for trees with few nodes or trees with levels included few nodes. Limitations of this kind of tree is simple and based on human ability to read. People prefer to read longer sets of data or sentences from up to down. This can be seen in newspapers where text is divided into columns.

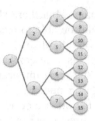

Fig. 2. Cascade tree view with root, 3 levels, 6 nodes and 8 leafs [13].

☐ 💻 My Computer
  ☐ 💾 Local Disk (C:)
    ⊞ 📁 DELL
    ⊞ 📁 Documents and Settings
    ⊞ 📁 I386
    ⊞ 📁 Inetpub
    ⊞ 📁 Intel
    ⊞ 📁 MSOCache
    ⊞ 📁 Program Files
    ⊞ 📁 Source
    ⊞ 📁 WINDOWS

Fig. 3. Cascade tree view – Windows files system tree [14].

Cascade tree view (Fig.2 and Fig.3) was created especially for computer systems. This approach can be found for example in Windows file system. Advantage of this kind of tree presentation is that mostly user in not limited with the size of screen. People using computers are accustomed to that presentation of tree so using that approach in recommendation system seems to natural.

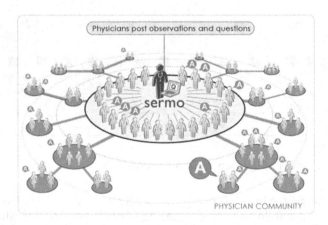

Fig. 4. Social network presenting connections between nodes [15].

Social network view is the latest from presented approaches. It was invented to present relations between people but can be also used in recommendation systems. Social network presentation is separate and huge branch of science and describing its mechanisms and algorithms would need another paper. From our point of view we assume that this approach works properly and could be used in student courses recommendation system.

Designing presentation module of any recommendation system we have to remember that different recommendation problems have to be solved in different periods of time. For some recommendation user can wait for minutes, hours and even days and it will be accepted as necessary time.

5 Summary

In this paper student automatic courses schedule system was described. We focused on most important issues and limitations. We also presented some solutions for mentioned problems and limitations. System was described mostly from student needs point of view. System has to meet students expectations so we think that this the best approach.

Summing up the paper we have to point that comparing course is very hard to do because of character of courses features and parameters. We presented that mostly it is easier to compare students and use their history in system for recommendation. Besides basing on students history system can use more nonfunctional courses features.

Acknowledgements

This work has been partially supported by the Polish Ministry of Science and Higher Education within the European Regional Development Fund, Grant No. POIG.01.03.01-00-008/08.

References

[1] Chen, Q., Aickelin, U.: Movie Recommendation Systems using an artificial immune system. Poster Proceedings of ACDM 2004, Engineers' House, Bristol, UK (2004)

[2] Colorni, A., Dorigo, M., Maoli, F., Maniezzo, V., Righini, G., Trubian, M.: Heuristics from Nature for Hard Combinatorial Optimization Problems. International Transactions in Operational Research 3(1), 1–21 (1996)

[3] Gao, W.: Study on Immunized Ant Colony Optimization. In: Third International Conference on Natural Computation (ICNC 2007), vol. 4, pp. 792–796 (2007)

[4] Herlocker, J.L., Konstan, J.A., Terveen, L.G., Riedl, J.: Evaluating Collaborative Filtering Recommender Systems. ACM Transactions on Information Systems 22, 5–53 (2004)

[5] Kobsa, A., Koenemann, J., Pohl, W.: Personalized Hypermedia Presentation Techniques for Improving Online Customer Relationships. Knowledge Eng. Rev. 16(2), 111–155 (2001)

[6] Montaner, M., Lopez, B., de la Rosa, J.P.: A Taxonomy of Recommender Agents on the Internet. Artificial Intelligence Review 19, 285–330 (2003)

[7] Nguyen, N.T., Sobecki, J.: Using Consensus Methods to Construct Adaptive Interfaces in Multimodal Web-based Systems. Universal Access in Inf. Society 2(4), 342–358 (2003)

[8] Sarwar, B., Konstan, J., Borchers, A., Herlocker, J., Miller, B., Riedl, J.: Using Filtering
     Agents to Improve Prediction Quality in the GroupLens Research Collaborative Filtering
     System. In: CSCW 1998, Seattle, Washington, USA, pp. 1–10 (1998)
[9] Sobecki, J.: Ant colony metaphor applied in user interface recommendation. New Gen-
     eration Computing 26(3), 277–293 (2008)
[10] Sobecki, J., Tomczak, J.: Student Courses Recommendation Using Ant Colony Optimiza-
     tion. In: ACIIDS 2010 (2010)
[11] van Setten, M.: Supporting People in Finding Information. Hybrid Recommender Sys-
     tems And Goal-Based Structuring, Enschede, The Netherlands, Telematica Instituut Fun-
     damental Research Series, No. 016 (2005)
[12] http://upload.wikimedia.org/wikipedia/en/e/e5/
     Segment_tree_instance.gif
[13] http://dvanderboom.files.wordpress.com/2008/03/
     binarytree.png
[14] http://dvanderboom.files.wordpress.com/2008/03/
     filesystemtreeview.png
[15] http://healthnex.typepad.com/photos/uncategorized/2007/06/
     08/aboutdiagram_03.gif

CAP Analysis of IEEE 802.15.3 MAC Protocol

Sana Ullah, Shahnaz Saleem, and Kyung Sup Kwak

UWB-ITRC Center, Inha University, 253 Yonghyun-dong, Nam-gu,
Incheon (402-751), South Korea
{sanajcs,roshnee13}@hotmail.com, kskwak@inha.ac.kr

Abstract. IEEE 802.15.3 is designed for high data rates multimedia applications in Wireless Personal Area Network (WPAN). The piconet is the basic topology of IEEE 802.15.3, which consists of a number of independent devices and a Piconet Coordinator (PNC). Medium Access Control (MAC) protocol plays a vital role in determining the energy consumption of all devices within a piconet. IEEE 802.15.3 MAC adapts a hybrid approach, i.e., a combination of CSMA/CA and TDMA, in order to support real-time communication while maintaining the desired Quality of Service (QoS). The channel is bounded by superframe structures where each superframe consists of a beacon, a Contention Access Period (CAP), and a Channel Time Allocation Period (CTAP). In this paper, we focus on the CAP part of the superframe. Numerical approximations are derived to analyze the performance of CSMA/CA used in CAP in terms of probability of successful transmission, average number of collisions, and normalized throughput. Numerical results are validated by extensive simulations using NS-2 simulator.

1 Introduction

The increasing demands of consumer devices allow the realization of sharing or streaming different types of multimedia files, such as digital images and high quality video/audio, between different devices. IEEE 802.15.3 standard is designed to accommodate these demands. It defines Physical (PHY) and Medium Access Control (MAC) specifications for high data rates multimedia applications at a greater distance while maintaining the desired Quality of Service (QoS) [1]. It supports different PHYs such as Narrowband (2.4 GHz) and Ultrawide Band (UWB). The IEEE 802.15.3 MAC adapts a combination of Time Division Multiple Access (TDMA) and Carrier Sensor Multiple Access/Collision Avoidance (CSMA/CA) mechanisms in order to support real-time applications in a reliable manner. It supports both isochronous and asynchronous data types. Contention free slots are allocated to the isochronous streams. In IEEE 802.15.3, when a number of independent devices want to communication with each other, they form a wireless adhoc network called piconet. The devices are the basic components of the piconet. At least one device in the piconet must be designated as a Piconet Coordinator (PNC). The PNC controls the entire operation of the piconet by sending beacons and control timing information. The beacons are always broadcasted in the beginning of the superframes.

N.T. Nguyen et al. (Eds.): New Challenges for Intelligent Information, SCI 351, pp. 227–236.
springerlink.com   © Springer-Verlag Berlin Heidelberg 2011

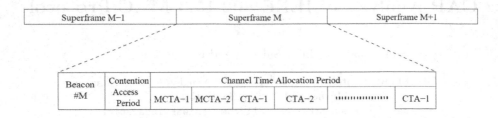

Fig. 1. The IEEE 802.15.3 Superframe Structure

In IEEE 802.15.3, the channel is bounded by superframe structures as given in Fig 1, where each superframe consists of three components: the beacon- which transmits control information to the piconet, Contention Access Period (CAP)-which uses CSMA/CA mechanism to send commands or asynchronous data to and from the PNC, and Channel Time Allocation Period (CTAP)- where devices are assigned specified time slots for isochronous streams. The piconet coordinator can sometimes use Management Channel Access Period (MCTA) instead of CAP unless restricted by the PHY. The MCTA uses slotted Aloha access scheme to send command frames to and from the PNC.

In this paper, we analyze the performance of the CAP for a number of contenting devices. Numerical approximations are used to model the behavior of these devices within a piconet under different conditions. We are interested to analyze the performance of the CAP in terms of probability of successful transmission, average number of collisions, and normalized throughput. This kind of numerical analysis is already considered by many authors (see Related Works) but most of them do not validate their analysis using well-know simulators such as Network Simulator (NS-2) [2]. We therefore simulate our model in NS-2 and observe the simulation results. The simulation results are closely compared with the numerical results. The analysis presented in this paper can help us determine the optimal duration of the CAP and the reasonable number of devices required in it.

The rest of the paper is organized into four sections. Section II presents the related works. Section III provides a brief description of IEEE 802.15.3 MAC. Section III and Section IV presents the numerical approximations and results. The final section concludes our work.

2   Related Works

Lots of research efforts are dedicated to analyze the performance of IEEE 802.15.3 MAC with respect to different parameters including retry limit and superframe variation [3]. In [4], the authors presented a simulation study of IEEE 802.15.3 MAC for a variety of scenarios using NS2. The capacity of p-persistent CSMA/CA for IEEE 802.15.3 is studied in [5]. In [6], the authors proposed a scheduling mechanism called Maximum Traffic (MT) scheduling for IEEE 802.15.3. This mechanisms guarantees the maximum data transmission within limited number of times lots. The throughput and delay analysis of IEEE 802.15.3 is presented in [7], where the authors considered Biachi's model [8] to derive closed form expressions of the average service time. In [9], the authors analyzed the performance of IEEE 802.15.3 MAC in terms of throughput, bandwidth utilization, and delay with various acknowledgement schemes. Other works on different acknowledgement schemes are presented in [10]. To handle the bursty traffic, the authors of [11] proposed an Enhanced Shortest Remaining Processing Time (ESRPT) algorithm, which decreases the job failure rates of delay sensitive applications and achieves better QoS. Kwan et.al presented the impact of IEEE 802.15.3 MAC on the Transmission Control Protocol (TCP) performance [12]. In [13], the authors analyzed the overhead associated with the IEEE 802.15.3 MAC when using UWB technology.

3   IEEE 802.15.3 Overview

3.1   General Description

The IEEE 802.15.3 MAC operates within a piconet as given in Fig. 2. The devices intending to communicate form an adhoc network where one device is designated as a PNC. The PNC always provide basic timing information of the piconet. In addition, it manages the QoS and energy consumption requirements of the piconet. Data can be communicated directly between two or more devices within the piconet after the PNC defines the superframe boundaries. The designation of the PNC depends on several important factors. A device willing to become the PNC first scans the channel to determine lowest interference from other networks. Once the channel is successfully found, the device assumes the role of the PNC.

As discussed earlier, the superframe structure of IEEE 802.15.3 MAC is a combination of the beacon, CAP, and CTAP periods. The beacon frame contains two types of fields, i.e., the piconet synchronization and the Information

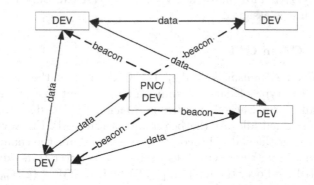

Fig. 2. IEEE 802.15.3 piconet elements

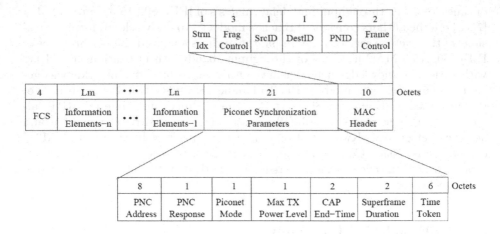

Fig. 3. IEEE 802.15.3 beacon format

Elements (IEs) as given in Fig. 3[1]. These two fields are used to define the duration of the entire superframe. The IEs provide additional information to all devices such as boundaries of the slots in the CTAP. The CAP is used to send requests to the PNC for the resource allocation and to send data to other devices in the piconet. The devices contend for the channel using the well-known CSMA/CA mechanism. Depending on the application, the PNC may use MCTA instead of CAP. The MCTA uses slotted Aloha protocol for resource allocation. There may or may not be one or more MCTAs in the superframe. The PNC is responsible to determine the appropriate number of MCTAs in the superframe. The CTAP is used to send isochronous data streams. It contains a number of CTAs (see Fig. 1) which have guaranteed start time and duration. The guard times are however required to protect the transmission in the adjacent CTAs from colliding. Additionally, a Shortest Interframe Spacing (SIFS) is used to ensure sufficient turnaround time between transmissions. The PNC is responsible to divide the CTAP into multiple CTAs. The CTA can only be used by the device to which it is assigned.

3.2   CSMA/CA in CAP

The CSMA/CA mechanism is used in the CAP part of the superframe. The PNC controls the type of data or command to be sent in the CAP. Initially, the device waits for a Backoff Interframe Spacing (BIFS) duration before the channel is determined idle. At the beginning of the CAP, the device waits for SIFS duration after the end of beacon transmission. After the channel is idle for BIFS/SIFS duraction, the device sets its backoff counter to a random integer uniformly distributed over the interval $[1, CW]$ where $CW \in (CW_{min}, CW_{max})$. The device starts decrementing the backoff counter only when the medium is

---

[1] Detailed description of each field is available in the standard.

idle for the duration of pBackoffSlot. The data is transmitted when the backoff counter reaches zero.

The CSMA/CA of IEEE 802.15.3 is the same as the traditional one, but there are some variations. The main differences, according to [6], are: 1) Retry_count: The retry count takes value in the range [0, 3], 2) Backoff_window (Retry_count): A table which has values [7, 15, 31, 63], 3) BW_random(Retry_count): A random integer which is selected from a uniform distribution over the interval [0, Backoff window (Retry count)].

4   Numerical Approximations of CAP

A finite number of stations, n, are considered in order to analyze the capacity of CAP. Initially, these devices send channel allocation requests to the PNC, but only one device can be served at a time and the remaining $n - 1$ devices will continue sending requests. We assume that each device has always a packet to send. Other assumptions are: 1) the channel is ideal, 2) Bit Error Rate (BER) is zero, and 3) No priorities are assigned to the devices. The following sections derive numerical approximations for the probability of successful transmission, average number of collisions, and normalized throughput.

4.1   Probability of Successful Transmission

Let p be the probability that the device transmits in a generic time slot. The value of p is given by the fraction of number of attempts by a device to the time the device takes in backoff, mathematically it can be represented by $p = E(A)/E(B)$ where $E(A)$ is the average number of attempts to transmit the data and $E(B)$ is the average backoff time.

Let p_c the conditional collision probability of any device. Now we have

$$p_c = \frac{1 - (1 - p)^n - np(1 - p)^{n-1}}{1 - (1 - p)^n} \qquad (1)$$

Once p is known, the probability p_t that in a slot there is at least one transmission is given by (according to [8]) the following equation.

$$p_t = 1 - (1 - p)^n \qquad (2)$$

Now the probability of successful transmission, p_{succ}, is given by

$$p_{succ} = \frac{np(1 - p)^{n-1}}{p_t} = \frac{np(1 - p)^{n-1}}{1 - (1 - p)^n} \qquad (3)$$

4.2   Average Number of Collisions

To calculate the average number of collisions $E(N_c)$, the $p(N_c)$ is given by

$$p(N_c = i) = p_c^i \times p_{succ} \qquad (4)$$

Now to find the expected value, consider a large number of collisions as given by

$$E(N_c = i) = \sum_{i=0}^{\infty} i \times p(N_c = i) = \sum_{i=0}^{\infty} i \times p_{succ}(1 - p_{succ})^i \qquad (5)$$

By considering $\sum_{i=0}^{\infty} i \times a^i = a/(1-a)^2$, Equation 5 will become

$$E(N_c) = \frac{1 - p_{succ}}{p_{succ}} \qquad (6)$$

Now substituting Equation 3 in Equation 6, we obtain the final equation for the average number of collisions

$$E(N_c) = \frac{1 - (1-p)^n}{np(1-p)^{n-1}} - 1 \qquad (7)$$

4.3   Normalized Throughput

The normalized throughput, say NT, is given by

$$NT = \frac{p_{succ}.p_t.E(D)}{(1 - p_t) + p_{succ}.p_t.\psi + p_t.(1 - psucc).\rho} \qquad (8)$$

where $E(D)$ is the average packet length, ψ is the average time the channel is sensed busy due to successful transmission, and ρ is the average time the channel is sensed busy during a collision. The equations for ψ and ρ are given by

$$\psi = BIFS + P_H + MAC_H + E(D) + 2\tau + SIFS + ACK \qquad (9)$$

$$\rho = BIFS + P_H + MAC_H + E(D) + \tau \qquad (10)$$

where P_H represents the PHY header and preamble. The MAC_H and τ respresents the MAC header and the propagation delay, respectively.

5   Results

We simulate our model in NS-2 simulator. Our simulation is based on the IEEE 802.15.3 MAC simulator developed by M. Demirhan [14]. The simulation parameters are listed in Table 1. The MAC and PHY headers are transmitted at a basic data rate of 22 Mbps. The original data rate is 44 Mbps.

Fig. 4 shows the probability of successful transmission, p_{succ}, for different devices within a piconet. It can be seen that the p_{succ} decreases with an increase in the number of devices. However this trend largely depends on the value of p (on the number of average attempts and the contention window). If we increase the value of p for all devices, the p_{succ} will decrease accordingly. Additionally, larger p would also affect the average number of collisions. Fig. 5 shows the average number of collissions as a function of network devices. As can be seen, when $n < 20$, there are few collisions. However, for $n > 20$, the $E(N_c)$ increases

Table 1. Parameters of IEEE 802.15.3

Symbol	Value	Symbol	Value
P_H	$7.997\mu s$	BIFS	$17.27\mu s$
Basic rate	22 Mbps	Data Rate	44 Mbps
MAC_H	$3.636\mu s$	SIFS	$10\mu s$
ACK	10 bytes	$E(D)$	60 bytes
Retry limit	2	τ	$1\mu s$

dramatically. Fig. 6 shows the normalized throughput for a number of devices. As we increase the number of devices, the normalized throughput increases to a maximum point and then it decreases. One of the main reasons is that the overall bandwidth is occupied at a certain stage, after which the throughput cannot be increased even if we increase the number of devices. Again this depends on p. If the bandwidth is not fully occupied, larger p results in higher throughput. As the network becomes saturated, a larger p will decrease the throughput as a function of network devices. Its worth noting that the simulation results are very close to analytical results.

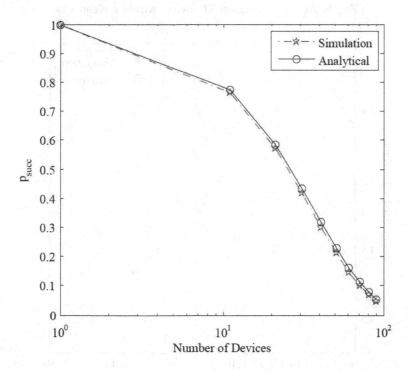

Fig. 4. p_{succ} vs. Number of devices within a piconet

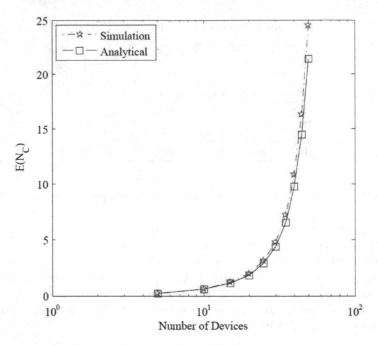

Fig. 5. $E(N_c)$ vs. Number of devices within a piconet

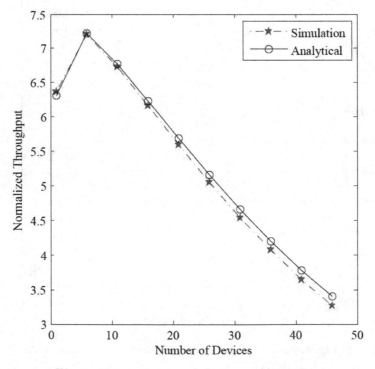

Fig. 6. NT vs. Number of devices within a piconet

The above results can be used to optimize different parameters such as the value of p and can determine the upper bounds for the reasonable number of devices required within a piconet. They can also define the required CAP duration for different devices and traffic loads.

6 Conclusions

In this paper, we studied the CAP period of IEEE 802.15.3 MAC protocol using numerical approximations and NS-2 simulations. We analyzed the performance of the CSMA/CA protocol, which is the basic access mechanism in the CAP, in terms of probability of successful transmission, average number of collisions, and normalized throughput. It was observed that the overall performance of the piconet depends on p and the number of devices. The obtained results can be used to optimize the CAP size and the number of required devices within a piconet.

Acknowledgement

This research was supported by the The Ministry of Knowledge Economy (MKE), Korea, under the Information Technology Research Center (ITRC) support program supervised by the Institute for Information Technology Advancement (IITA) (IITA-2009-C1090-0902-0019).

References

1. IEEE Draft Std 802.15.3: Wireless Medium Access Control (MAC) and Physical Layer (PHY) Specifications for High Rate Wireless Personal Area Networks, Inst. Elec. Electron. Eng., New York, USA (June 2002)
2. Network Simulator, http://www.isi.edu/nsnam/ns/ (Date visited: October 2010)
3. Cai, L.X., et al.: Performance Analysis of Hybrid Medium Access Protocol in IEEE 802.15.3 WPAN. In: Proc. 4th IEEE Consumer Communications and Networking Conference, Las Vegas, USA, January 2007, pp. 445–449 (2007)
4. Chin, K.W., Lowe, D.: Asim ulation study of the ieee 802.15.3 MAC. In: Australian Telecommunications and Network Application Conference (ATNAC), December 2004, Sydney, Australia (2004)
5. Cali, F., Conti, M., Gregori, E.: Dynamic Tuning of the IEEE 802.11 Protocol to Achieve a Theoretical Throughput Limit. IEEE/ACM Trans on Networking 8, 785–799 (2000)
6. Tseng, Y.H., Wu, H.K., Chen, G.H.: Maximum Traffic Scheduling and Capacity Analysis for IEEE 802.15.3 High Data Rate MAC Protocol. In: The Proc. of VTC 2003, October 2003, vol. 3, pp. 1678–1682 (2003)
7. Liu, X., Zeng, W.: Throughput and delay analysis of the IEEE 802.15.3 CSMA/CA mechanism considering the suspending events in unsaturated traffic conditions. In: IEEE 6th International Conference on Mobile Adhoc and Sensor Systems, October 12-15, pp. 947–952 (2009)

8. Bianchi, G.: Performance analysis of the IEEE 802.11 distributed coordination function. IEEE J. of Select. Areas of Commun. 18(3) (March 2000)
9. Mehta, S., Kwak, K.S.: Performance Analysis of mmWave WPAN MAC Protocol. In: Third International Conference on Convergence and Hybrid Information Technology, November 11-13, vol. 2, pp. 309–316 (2008)
10. Chen, H., Guo, Z., Yao, R., Li, Y.: Improved Performance with Adaptive Dly-ACK for IEEE 802.15.3 WPAN over UWB PHY. IEICE TRANS. Fundamentals E88-A(9) (September 2005)
11. Liu, X., Dai, Q., Wu, Q.: An Improved Resource Reservation Algorithm for IEEE 802.15.3. In: 2006 IEEE International Conference on Multimedia and Expo, July 9-12, pp. 589–592 (2006)
12. Chin, K.W., Lowe, D.: A Simulation Study of TCP over the IEEE 802.15.3 MAC. In: The IEEE Conference on Local Computer Networks, November 17, pp. 476–477 (2005)
13. Goratti, L., Haapola, J., Oppermann, I.: Energy consumption of the IEEE Std 802.15.3 MAC protocol in communication link set-up over UWB radio technology. Wireless Personal Communications 40(3), 371–386 (2007)
14. http://hrwpan-ns2.blogspot.com/

Indoor Positioning System Designed for User Adaptive Systems

Peter Brida, Frantisek Gaborik, Jan Duha, and Juraj Machaj

University of Zilina, Faculty of Electrical Engineering, Department of Telecommunications
and Multimedia, Univerzitna 8215/1, 010 26 Zilina, Slovakia
{Peter.Brida,Jan.Duha,Juraj.Machaj}@fel.uniza.sk

Abstract. The paper deals with the indoor positioning solution based on utilization of IEEE 802.11 network and fingerprinting method. This solution was
designed as a part of user adaptive integrated positioning system that provides
location-based services in indoor environment. The integrated positioning system is engineered for ubiquitous positioning, i.e. anywhere and anytime. The
system for indoor environment is called WifiLOC and it is implemented as a
mobile-assisted positioning system. The architecture and fundamental principles of the WifiLOC system are presented in this paper. The positioning system
is based on the fingerprinting method, which utilizes signal strength information
for position estimation.

Keywords: Fingerprinting localization, indoor positioning system, architecture,
location based service, user adaptive system.

1 Introduction

The idea of User Adaptive System (UAS) lies in interaction between user and system
through his mobile device. Such interaction can be linked with user's requests. These
requests would include for example user current position or tracking information. The
UAS also may take advantages of location based systems oriented to providing information support based on current user's position. Modern sophisticated services
also deliver content to the user according its actual position [1] – [3]. On the other
hand, success of the services consists also in performance of localization. Very attractive group of localization methods applicable in various wireless network platforms is
called database correlation methods or fingerprinting. The implementation of the
appropriate database plays very important role from effective function of the UAS.
This paper designs a suitable indoor positioning solution for systems based on database correlation method. Various positioning systems based on wireless platforms
were proposed in [4], [5]. But the more existing systems the more problems arise.
Creation of new technologies is almost impossible because of many existing standards
and limitations as well as exhausted resources e.g. frequency spectrum for radio
waves. Also, entire design of new localization system would be very costly or otherwise complicated as well. Many producers therefore only attach existing systems (e.g.
GPS - Global Positioning System) and do not attempt to improve them.

N.T. Nguyen et al. (Eds.): New Challenges for Intelligent Information, SCI 351, pp. 237–245.
springerlink.com   © Springer-Verlag Berlin Heidelberg 2011

The most popular algorithms used in indoor environment are based on IEEE 802.11 standard [6], [7]. Majority of these algorithms is based on RSS (Received Signal Strength) and fingerprinting algorithm.

Many researchers have paid an attention on the issue of mobile positioning using the received signal strength information [8] - [16]. This interest could be based on the fact that signal strength information can be simply measured by MS (Mobile Station) and does not need an expensive additional implementation costs compared with other methods.

The fingerprinting method seems to be more suitable for Non-Line-of-Sight (NLoS) environment compared to trilateration way and is also more immune against multipath. The main advantage is that it does not need to create new infrastructure, but it only needs to utilize information about existing surrounding networks.

We decided to base our solution on the IEEE 802.11 platform, because it is widespread. The basic idea results from the utilization of the platform apart from its main purpose, which is mainly to cover user data communication. Our approach adds a value to the IEEE 802.11 communication platform by providing user positioning. Furthermore, MS must not be modified and only localization server has to be added to the network. Only the software application for MS has to be implemented as a communication interface for a user and the localization server. Therefore, we decided to use Fingerprinting method based on received signal strength information in this work.

The rest of the paper is structured as follows. Section 2 introduces theoretical principles of fingerprinting method. In Section 3, implementation of the positioning system is presented. Section 4 concludes the paper and suggests some future studies.

2   Fingerprinting

Fingerprinting algorithm can be implemented in various ways from mathematical point of view. They can be divided into deterministic and probabilistic algorithms. In our work we deal with fingerprinting based on deterministic algorithm - Nearest Neighbor method (NN), which can achieve sufficient accuracy as well as more complicated method, when density of radio map is high enough [15].

Accuracy of this method in radio networks is according to [11] determined by two factors. Firstly, signal properties vary much at relatively small area. For instance, in few meters range, signal from an AP (Access Point) can get attenuated, get lost or be replaced with a stronger one. Secondly, these signals are relatively constant in time. It allows data gathering and their use in future.

A disadvantage of this method is sensitivity for environment changes such as object movement in the building (e.g. people, furniture), which altogether affect signal properties. It is necessary to update the map, but basically, walls and furniture affect the signal most of all and therefore update is not required so often.

Fingerprinting method consists of two phases. At first, it is creation of the radio map for an area where planned localization service is desired (see Fig. 1). Radio map is basically a database of spots with known position (coordinates) coupled with various radio signal properties, e.g. RSS, signal angles or propagation time. This phase is called off-line phase. Generally, the off-line phase can be performed by either measurements in a real environment or by prediction as described in [17]. In first case, it is very time consuming, but there are precise and real RSS information used in calculations. On the other hand, prediction of RSS is more comfortable, but the data is highly dependent on a quality of map model of given environment. There is a compromise between the demanding effort and accuracy in [12].

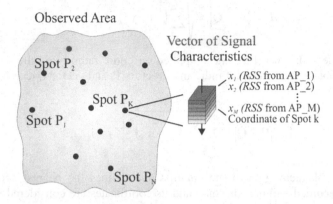

Fig. 1. Radio map for fingerprinting using RSS.

After the radio map is created, second phase can take place. MS measures signal properties at unknown spot. Then the radio map is searched to find a best match from existing spots. The match is actually the nearest point from database and is considered MS position (for NN method). This step is called on-line phase. Next sections deal with fingerprinting in more details.

2.1 Localization Algorithm

The most common method how to find the best match and actually perform localization is Euclidean algorithm. Let's assume a fingerprint in radio map, which is characterized by vector **P**:

$$\mathbf{P} = [x_j] = [x_1, \ldots, x_M], \tag{1}$$

where x_j characterizes the spot, i.e. values of the signal properties (e.g. RSS). M represents number of signal properties used for radio map creation. In general, let's consider the radio map contains fingerprints of N spots.

$$\mathbf{P}_i = [x_{ij}] = [x_{i1}, \ldots, x_{iM}], \ i = 1, \ldots, N . \tag{2}$$

Unique identifiers of neighbor AP are stored in radio map as well as spot coordinates and they are coupled with x_i, but are not shown here for model simplicity. The whole radio map contains all fingerprints $\mathbf{P}_i$ and creates the set **S** written as

$$\mathbf{S} = \{\mathbf{P_i} : i = 1 \ldots N\}. \tag{3}$$

In case of MS localization, the signal properties are measured at unknown spot, a new fingerprint **Q** is obtained

$$\mathbf{Q} = [y_j], \quad j = 1, \ldots, M . \tag{4}$$

The Euclidean distance d_k between vectors $\mathbf{P_k}$ and $\mathbf{Q}$ is defined as

$$d_k = \left| \mathbf{P_k} - \mathbf{Q} \right| = \sqrt{\sum_{j=1}^{M} (x_{kj} - y_j)^2} . \tag{5}$$

When Euclidean distance formula is applied on entire radio map, the vector of distances $\mathbf{D}$ is obtained between all radio map vectors $\mathbf{P_i}$ and vector $\mathbf{Q}$ can be calculated as

$$\mathbf{D} = [d_i] = \left[\left| \mathbf{P_i} - \mathbf{Q} \right| \right] = \left[\sqrt{\sum_{j=1}^{M} (x_{ij} - y_j)^2} \right], \quad i = 1, \ldots, N . \tag{6}$$

The element of vector D with minimum value defines the nearest spot to Q. Its position is recorded within radio map and its coordinates are considered the location estimate.

3 WifiLOC Positioning System

The basic properties of the WifiLOC are described in this part. The system utilizes signal information for positioning from surrounding Wifi networks. The system is based on fingerprinting positioning method and signal strength information as mentioned above.

3.1 WifiLOC Architecture

We decided to design WifiLOC as the mobile-assisted positioning concept. The mobile-assisted positioning means that the necessary measurements are done in a localized mobile station and measured results are forwarded to the network part LCS (localization server). The position is estimated (calculated) at server side. The architecture of the system is depicted in Fig. 2.

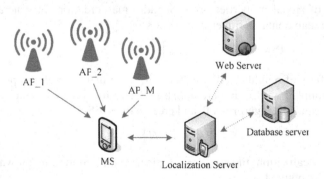

Fig. 2. Radio map for fingerprinting using RxLev.

The system is based on client-server architecture. From different point of view the entire architecture could be divided into three almost independent parts:

1. localization server,
2. network of access points,
3. mobile station - user, client.

The division was purposely performed, because of function of particular parts.

Localization Server

The core component of the system is the localization server. It is main brain of the system and provides more functions: communication with clients, position estimation, radio map database... Therefore, it consists of more functional entities: database server, web server and communication platform. LCS is built on Ubuntu operating system [18].

Radio map is saved on database server based on MySQL platform, which is free to use [19].

Communication between client-server is implemented via Simple Object Access Protocol (SOAP) web services. Web services are deployed at web server. They are used to upload measurement data to database server as well as exchange data when localization occurs. Web services technology offers reliable data transport and cross-platform object exchange.

Communication between client and LCS could be implemented by various standard communication links depending on availability. Obviously, Wifi is used by default, but Bluetooth, UMTS (Universal Mobile Telecommunications System) or GSM (Global System for Mobile Communications) were successfully tested.

Network of Access Points

Network of access points can consist of various network provider APs. It is one of the biggest advantages of the WifiLOC, because in this case it is not necessary to build new network for this purpose. Hence, initial costs are minimal.

The signals from all fixed AP are passively scanned and utilized for positioning. The system relies on fact that transmission power is not changed.

Mobile Station

Mobile station can be any mobile phone, personal device, tag or laptop equipped with IEEE 802.11 chipset. Localization application is developed in Java language SDK (Standard Development Kit) due to its easy implementation and cross-platform compatibility.

3.2 Positioning Data Visualization

Positioning data visualization is the same at the client application and LCS. Obviously, it depends on hardware equipment of the MS. Given positioning data is shown by means of web page, which is generated by web server, arbitrary web browser could be used to display them. Generally, the positioning result would be depicted in three levels:

1. Local - the web page displays local map (ground plan) and estimated position (spot) in local space. It is exact information about corridor or room where client is situated. This information is important for better orientation of the client in current area.
2. Sub-Global - in this level the actual building is identified where client is situated. This additional information could be used by client for better orientation in case of the next navigation in the building complex.
3. Global - the web page displays map and estimated position in global space. Maps are created using Google Maps™ API. This information is important for better global orientation of the client.

This information could be supplemented by exact coordinates of the place or picture which can be seen by user. It depends on implementation of the WifiLOC system, e.g. there could be information about the nearest physician surgery in hospital etc.

In Fig. 3 Client application can be seen on the mobile terminal. The application displays in text form that the user "Smolyk" is currently located in room number ND321. The user can decide to show different way for position data visualization via icons in application (see Fig. 3). There are shown various types of positioning data visualization in Fig. 4. The visualization could be seen at client and server.

Fig. 3. WifiLOC client application environment.

Design of "extensible" positioning system was the main idea during development phase. "Extensible" means that system should be able to implement in all relevant Wifi networks. The system is extensible from service extension point of view. The service portfolio could be extended without necessary modifications in mobile stations. The system does not have any limits from a number of users' point of view.

Next system advantage consists of the possibility to fill up data to database by a user in case that its terminal is able to use local coordinates, because WifiLOC is based on local coordinates system. This option can be used in case that a user needs further radio map (higher density of RSS information). These data can be easily converted to GPS coordinates as reference coordinates. Hence, advantage of the system consists of this fact, because results of mobile positioning can be also presented in WGS-84 (World Geodetic System) and are compatible with GPS and maps, which are

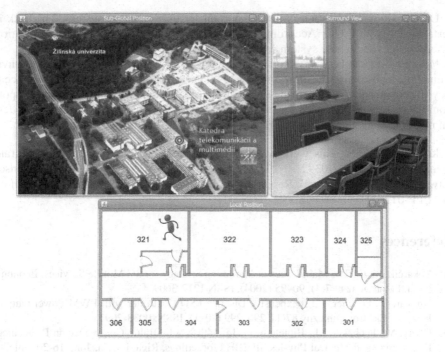

Fig. 4. Positioning data visualization.

based on WGS-84. The process of map creation is started and finished from the device which performs measurements to radio map. Positioning data (RSS and coordinates) is measured during this process. When all desired data is measured, they are sent to the server. In that case, reference points are marked manually in the map. The other way lies in association of reference points with fixed points, e.g. doors, offices and others. Finally, the most important advantage is that the WifiLOC system is independent on surrounding Wifi network operators. The system only utilizes information about signal strength from these networks. Arbitrary network can be used for communication between user and localization server.

4   Conclusion and Future Work

We designed the positioning solution based on fingerprinting method and network based on IEEE 802.11 standard. It is called WifiLOC. This system seems to be good appropriate solution for user adaptive systems which need to know user position in indoor environment. The solution uses received signal strength information for its function. The solution is implemented as mobile-assisted positioning. The mobile terminal independently collects the necessary data from surrounding access points. The measured data is sent from mobile terminal to the localization server for position estimation. The server estimates position and the information about position is sent back to the terminal when it is requested. Position information can be displayed on the map of terminal screen as well as at the server.

User position information is very important for user adaptive system. This task is performed by WifiLOC. According user's position can be offered various location based services and whole system can be adapted.

Future works can be focused on investigation of influence of various kinds of environments on positioning accuracy or how to increase of positioning accuracy. An accuracy improvement could be made with current localization algorithm, but more spots from database would be used for position estimation. It could be also based on different map matching algorithm or on the use of other extra data for position estimation.

Acknowledgements. This work was partially supported by the Slovak VEGA grant agency, Project No. 1/0392/10 "The research of mobile nodes in wireless sensor networks" and by the Slovak Research and Development Agency under the contract No. LPP-0126-09.

References

1. Tokarčíková, E.: The Main Issues of Marketing Strategy for 3G Mobile Services. Economics and Management 7(4), 90–95 (2004), ISSN 1212-3609
2. Krejcar, O., Frischer, R.: Inexpensive Discrete PSD Controller with PWM Power Output. Przegląd Elektrotechniczny 87(1), 296–299 (2011), ISSN: 0033-2097
3. Cerny, M., Penhaker, M.: Biotelemetry. In: 14th Nordic-Baltic Conference an Biomedical Engineering and Medical Physics, IFMBE Proceedings, Riga, Latvia, June 16-20, vol. 20, pp. 405–408 (2008)
4. Chen, Y., Kobayashi, H.: Signal Strength Based Indoor Geolocation. Communications 1, 436–439 (2002)
5. Wallbaum, M.: WhereMoPS: An Indoor Geolocation System. In: The 13th IEEE International Symposium on Personal, Indoor and Mobile Radio Communications, vol. 4, pp. 1967–1971 (2002)
6. IEEE Standard for Information Technology-telecommunications and Information Exchange Between Systems-local and Metropolitan Area Networks-specific Requirements - part 11: Wireless LAN medium access control (MAC) and physical layer (PHY) Specifications, IEEE Standard 802.11-2007 (2007), ISBN: 978-0-7381-5656-9
7. Doboš, Ľ., Duha, J., Marchevsky, S., Wieser, V.: Mobilné rádiové siete (Mobile Radio Networks), 1st edn. EDIS, Žilina (2002) (in Slovak)
8. Brida, P., Cepel, P., Duha, J.: A Novel Adaptive Algorithm for RSS Positioning in GSM Networks. In: CSNDSP 2006 Proceedings, Patras, Greece, July 19-21, pp. 748–751 (2006)
9. Krejcar, O.: Problem Solving of Low Data Throughput on Mobile Devices by Artefacts Prebuffering. EURASIP Journal on Wireless Communications and Networking, Article ID 802523, 8 pages (2009), doi:10.1155/2009/802523, ISSN: 1687-1499
10. Hata, M., Nagatsu, T.: Mobile Location Using Signal Strength Measurements in a Cellular System. IEEE Trans. on Vehicular Technology VT-29, 245–252 (1980)
11. Otsason, V.: Accurate Indoor Localization Using Wide GSM Fingerprinting. Master's thesis, Tartu, Estonia (2005)
12. Anne, K.R., Kyamakya, K., Erbas, F., Takenga, C., Chedjou, J.C.: GSM RSSI-based Positioning Using Extended Kalman Filter for Training ANN. In: proceedings of VTC 2004, Los Angeles (2004)

13. Takenga, C., Kyamakya, K., Quan, W.: On the Accuracy Improvement Issues in GSM Location Fingerprinting. In: Vehicular Technology Conference, VTC-2006 Fall, pp. 1–5 (2006)
14. Li, B., Wang, Y., Lee, H.K., Dempster, A.G., Rizos, C.: A New Method for Yielding a Database of Location Fingerprints in WLAN. IEE Proc. Communications 152(5), 580–586 (2005)
15. Tsung-Nan, L., Po-Chiang, L.: Performance Comparison of Indoor Positioning Techniques Based on Location Fingerprinting in Wireless Networks. In: International Conference on Wireless Networks, Communications and Mobile Computing, vol. 2 (2005)
16. Brida, P., Cepel, P., Duha, J.: Mobile Positioning in Next Generation Networks (ch. XI). In: Kotsopoulos, S., Ioannou, K. (eds.) Handbook of Research on Heterogeneous Next Generation Networking: Innovations and Platforms, pp. 223–252. IGI Global (Information Science Reference), New York (2008), ISBN 978-1-60566-108-7 (hardcover)
17. Chrysikos, T., Georgopoulos, G., Birkos, K., Kotsopoulos, S.: Wireless Channel Characterization: On the Validation Issues of Indoor RF Models at 2.4 GHz. In: First Panhellenic Conference on Electronics and Telecommunications (PACET), Patras, Greece, March 20-22 (2009)
18. Canonical Ltd. Ubuntu Homepage, http://www.ubuntu.com
19. MySql Homepage, http://www.mysql.com

Localized Approximation Method Using Inertial Compensation in WSNs

ChangWoo Song[1], KyungYong Chung[2], Jason J. Jung[3],
KeeWook Rim[4], and JungHyun Lee[1]

[1] Dept. of Information Engineering, INHA University
[2] Dept. of Computer Information Engineering, Sangji University
[3] Dept. of Computer Engineering, Yeunnam University
[4] Dept. of Computer and Information Science, Sunmoon University
{up3125,majun9821}@hotmail.com, kyjung@sangji.ac.kr,
j2jung@gmail.com, rim@sunmoon.ac.kr, jhlee@inha.ac.kr

Abstract. Sensor nodes in a wireless sensor network establish a network based on location information, set a communication path to the sink for data collection, and have the characteristic of limited hardware resources such as battery, data processing capacity, and memory. The method of estimating location information using GPS is convenient, but it is relatively inefficient because additional costs accrue depending on the size of space. In the past, several approaches including range-based and range-free have been proposed to calculate positions for randomly deployed sensor nodes. Most of them use some special nodes called anchor nodes, which are assumed to know their own locations. Other sensors compute their locations based on the information provided by these anchor nodes. This paper uses a single mobile anchor node to move in the sensing field and broadcast its current position periodically. We provide a weighted centroid localization algorithm that uses coefficients, which are decided by the influence of mobile anchor node to unknown nodes, to prompt localization accuracy. In addition, this study lowered the error rate resulting from difference in response time by adding reliability for calculating and compensating detailed location information using inertia.

Keywords: Wireless Sensor Networks, Inertial, Localization Accuracy.

1 Introduction

Wireless sensor network(WSN) is composed of a large number of sensor nodes that are densely deployed in a field. Each sensor performs a sensing task for detecting specific events. The sink, which is a particular node, is responsible for collecting sensing data reported from all the sensors, and finally transmits the data to a task manager. If the sensors can't directly communicate with the sink, some intermediate sensors have to forward the data[1].

There are several essential issues (e.g., localization, deployment, and coverage) in wireless sensor networks. Localization is one of the most important subjects for

N.T. Nguyen et al. (Eds.): New Challenges for Intelligent Information, SCI 351, pp. 247–255.
springerlink.com   © Springer-Verlag Berlin Heidelberg 2011

wireless sensor networks since many applications such as environment monitoring, vehicle tracking and mapping depend on knowing the locations of the sensor nodes [2].

To solve the localization problem, it is natural to consider placing sensors manually or equipping each sensor with a GPS receiver. However, due to the large scale nature of sensor networks, those two methods become either inefficient or costly, so researchers propose to use a variety of localization approaches for sensor network localization. And when GPS (Global Positioning System) is used, it is hard to get accurate location information because of the reception rate varying according to place. These approaches can be classified as range-based and range-free. Firstly, the range-based approach uses an absolute node-to-node distance or angle between neighboring sensors to estimate locations. Common techniques for distance or angle estimation include received signal strength indicator (RSSI), time of arrival (TOA), time difference of arrival (TDOA), and angle of arrival (AOA). The approaches typically have higher location accuracy but require additional hardware to measure distances or angles. Secondly, the range-free approach does not need the distance or angle information for localization, and depends only on connectivity of the network and the contents of received messages. For example, Centroid method, APIT method , DV-HOP method, Convex hull, Bounding box, and Amorphous algorithm have been proposed. Although the range-free approach cannot accomplish as high precision as the range-based, they provide an economic approach. Due to the inherent characteristics (low power and cost) of wireless sensor networks, the range-free mechanism could be a better choice to localize a sensor's position, so we pay more attention to range-free approach in this paper[3-4][11-12].

Recently research is being made on the strap-down inertial navigation positioning system using inertia. However, this method has the shortcoming that sensor errors accumulate continuously when the system cannot be initialized for a long period. To cope with this shortcoming, it was suggested in the past to increase location initialization using a larger number of RFIDs, but early investment costs increase along with the expansion of space required for positioning. In order to solve this problem, this study introduces a method that enables the compensation of node location in the middle of travelling by calculating the distance between nodes when a node has reached another node knowing its location.

2   Related Work

2.1   Issues in Wireless Sensor Networks

Wireless sensor networks provide the means to link the physical world to the digital world. The mass production of integrated, low-cost sensor nodes will allow the technology to cross over into a myriad of domains. In the future, applications of wireless sensor networks will appear in areas we never dreamed. Listed below are just a few places where sensor networks can and will be deployed.

• Earthquake monitoring
• Environmental monitoring

- Factory automation
- Home and office controls
- Inventory monitoring
- Medicine
- Security

Although still in its infancy, wireless sensor network applications are beginning to emerge. A recent study on Great Duck Island in Maine used sensor networks to perform monitoring tasks without the intrusive presence of humans.

When monitoring plants and animals in the field, researchers have become more concerned about the effects of human presence. In the Smart Kindergarten project, using pre-school and kindergarten classrooms as the setting, the plan is to embed networked sensor devices unobtrusively into familiar physical objects, such as toys. The environment will be able to monitor the interactions of children and teachers in order to promote the development of skills. Researchers at Wayne State University believe implanted biomedical devices called smart sensors have the potential to revolutionize medicine. Proposed applications include an artificial retina, glucose level monitors, organ monitors, cancer detectors, and general health monitors.

Realization of sensor networks needs to satisfy the constraints introduced by factors such as fault tolerance, scalability, cost, hardware, topology change, environment, and power consumption. Because these constraints are highly stringent and specific for sensor networks, new wireless ad hoc networking techniques are required.

Many localization techniques say that a node can either be localized or it can't. We present a location discovery algorithm that provides, for every node in the network, a position estimate, as well as an associated error bound and confidence level. We provide a versatile framework that allows users to perform localization queries based on the required accuracy and certainty[5-10].

2.2 Weighted Centroid Localization Algorithm

In the algorithm, mobile anchor node confronts the right to decide the location of the centroid through weighted factor to reflect. The use of weighted factor reflected the intrinsic relationship between them[3].

We embody this relationship through the formula of the weighted factor:

$$X = (X1/d1 + X2/d2 + X3/d3)/(1/d1 + 1/d2 + 1/d3)$$
$$Y = (Y1/d1 + Y2/d2 + Y3/d3)/(1/d1 + 1/d2 + 1/d3). \tag{1}$$

Figure 1 illustrates Known 3 mobile anchor nodes coordinate (X1, Y1), (X2, Y2), (X3, Y3), unknown node to anchor nodes distance d1, d2, d3. According to the formula can be calculated unknown node coordinates (X, Y). Compared to ordinary centroid algorithm, 1/d1, 1/d2, 1/d3 is weighted factor. The factor 1/d1, 1/d2, 1/d3 indicates that mobile anchor node with a shorter distance to unknown nodes has a larger infect its coordinates. We can improve the localization accuracy from these inner relations.

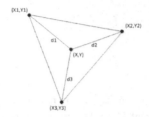

Fig. 1. Scheme of the adaptive weighted centroid localization algorithm

Weighted Centroid Localization Algorithm process:

① The mobile anchor node periodically sends its own information.

② Unknown node received information, only records the same location of the mobile anchor node average RSSI.

③ Unknown node received over threshold m in the position information then RSSI value in accordance with the smallest sort of mobile anchor node location. And to establish the mapping between RSSI value and the distance from unknown node to the mobile anchor node. The establishment of three sets:

Mobile anchor node_set = {a1, a2, ... , am};
Distance_set = {d1, d2, ... , dm};
Mobile anchor node position_set = {(X1, Y1), (X2, Y2), ... , (Xm, Ym)};

④ RSSI value with the first few large location of mobile anchor node of the calculation:

Based on the preceding analysis, In the mobile anchor node_set Select RSSI value of large node location then the composition of the triangle set. This is very important. Triangle_set = {(a1, a2, a3), (a1, a2, a4), ... (a1, a3, a4), (a1, a3, a5) ... };

⑤ n location of mobile anchor nodes can be composed of C_n^3 triangles. The use formula ① calculates C_n^3 coordinate.

⑥ Calculates the mean value(X,Y) of C_n^3 coordinate. The (X, Y) is Unknown node coordinate.

3   Localized Approximation Method

3.1 Inertia Information

Inertia information indicates how long a moving object has moved from its current position. For actual utilization of the information, accordingly, it needs to be transformed to the navigation coordinate system.

• Inertial coordinate system
In this coordinate system, the origin is the center of earth, and axis X and Y are positioned on the equator regardless of the earth's rotation, and axis Z coincides with the earth's rotation axis.

• Global coordinate system
In this coordinate system, the origin is the center of earth, axis X is the point where the longitude intersects the equator, axis Z is the axis running toward the North Pole, and axis Z is the direction turning 90 counterclockwise from axis X.

• Navigation coordinate system
In this coordinate system, the origin is the center of the moving object, and axis X, Y and Z are north, east and down, respectively.

Inertia information is largely composed of accelerometer and gyroscope data. An accelerometer produces data on what direction the sensor will move in from its current position based on axis X, Y and Z where the sensor is located, and a gyroscope produces data on how much the sensor will rotate from its current position. Because of the characteristic of inertia information, it cannot know its absolute coordinates. For this reason, initial alignment is very important. In the general initial state of inertia information, body acceleration is as in (2)[13].

$$f^b = [f_x f_y f_z] = c_n^b = \begin{bmatrix} \sin\theta g^n \\ -\sin\phi\cos\theta g^n \\ -\cos\phi\cos\theta g^n \end{bmatrix} \tag{2}$$

Because the initial speed in this state is 0, acceleration for axis X and axis Y is also 0. Accordingly, acceleration at a standstill state can be expressed as $f^b = [0,0,-g^n]^2$. This is acceleration at a standstill state for axis X, Y and Z, and the initial coordinate transformation matrix obtained using the attitude angles of the sensor, which are φ (roll), θ (pitch), and ψ (yaw), is as in (3)[13].

$$C_b^n = \begin{bmatrix} c(\theta)c(\psi) & s(\phi)s(\theta)c(\psi)-c(\phi)s(\psi) & c(\phi)s(\theta)c(\psi)+s(\phi)s(\psi) \\ c(\theta)s(\psi) & s(\phi)s(\theta)s(\psi)+c(\phi)c(\psi) & c(\phi)s(\theta)s(\psi)+s(\phi)c(\psi) \\ -s(\theta) & s(\phi)c(\theta) & c(\phi)c(\theta) \end{bmatrix} \tag{3}$$

The linear acceleration of the body coordinate system measured in this way is transformed to the acceleration of the navigation coordinate system using the coordinate transformation matrix, and then position and speed are obtained by calculating the navigation equation. The equation for calculating speed is as in (4)[13].

$$v^n = c_b^n f^b - (2w_{ie}^n + w_{en}^n) \times v^n + g^n \tag{4}$$

The result of (4) $v^n = [v_N\ v_E\ v_D]^2$ is the element of speed expressed in the navigation coordinate system and indicates the angular velocity of the navigation coordinate system for the inertial coordinate system, and w_{en}^n indicates the angular velocity of the earth fixed coordinate system for the inertial coordinate system. These can be expressed numerically as in (5), (6) and (7)[13].

$$w_{ie}^n = [\Omega \cos L, 0, -\Omega \sin L] \tag{5}$$

$$w_{en}^n = [\rho_N, \rho_E, \rho_D] = [l \cos L, -L, l \sin L]$$
$$= \left[\frac{v_E}{(R_t + h)}, -\frac{v_N}{(R_m + h)}, \frac{v_E \tan L}{(R_t + h)}\right] \tag{6}$$

$$R_m = \frac{R_0(1 - e^2)}{(1 - e^2 \sin e^2 L)^{3/2}}$$
$$R_t = \frac{R_0}{(1 - e^2 \sin^2 L)^{1/2}} \tag{7}$$

L is the value of axis X in the coordinate system, l of axis Y, and h of axis Z. In addition, R_c is the equator radius of the earth ellipsoid, e earth eccentricity, and Ω the angular velocity of the earth's rotation. When the numerical expression is expressed using inertia information obtained from a sensor, it is as in (8)[13].

$$L = \int_0^t \frac{v_N}{(R_m + h)} d\tau + L(0) = \frac{V_N}{R_m + h}$$

$$l = \int_0^t \frac{V_E}{(R_t + \cos L)} d\tau + l(0) = \frac{V_E}{(R_m + h)\cos L} \tag{8}$$

$$h = \int_0^t (-v_D) d\tau + h(0) = -V_L$$

(8) is applied to the coordinate system of a navigation object for location calculation through compensation. A navigation object cannot know its location from coordinates, but it can estimate the current location based on the distance to an object whose location is known. Accordingly, error can be reduced by expressing speed with the direction of a specific sensor and its distance for a specific period of time. If there is a node near the sensor, a better result of positioning can be obtained by giving a high weight to the position measured by the sensor. On the contrary, if the sensor is distant from nodes its accuracy becomes low and therefore a high weight should be given to the position measured using inertia information. Accordingly, for location compensation between sensor nodes, reliability is added as a new criterion. If D is the distance to a known sensor node, it can be expressed as (9).

$$D = c \times TOF \times d_0$$
$$c = 331.5 + 0.60714 \times Temperature_{Celsius} \tag{9}$$

If D_0, which is obtained from (9), is the distance to the previous sensor, and D_i is the distance to the currently measured sensor, speed y moving toward an arbitrary vector can be defined as (10).

$$V^{ult} = \frac{D_i^2 - D_0^2}{t_i} \tag{20}$$

In order to increase the accuracy of v^{ult} indicating speed, reliability variable k is added. Variable k is expressed differently according to the type and characteristic of sensor but, in general, data are applied based on the error variation according to the distance to the closest sensor. Figure 2 shows error variation between the mean value and values measured in real time using data received from a sensor.

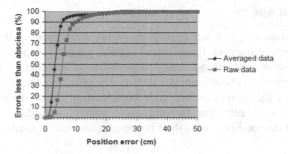

Fig. 2. Error variation between the mean value and values measured in real time

4 Experiments

4.1 Simulation Environment

The key metric for evaluating a localization technique is the accuracy of the location estimates versus the communication and deployment cost. To evaluate this proposed method we use UNIX, programs with the C language. We have carried on the computer simulation to the above algorithm. Simulation condition: The mobile anchor node reference MICA2 mote; Uses outdoor launches the radius 200 to 300m; Deployment area is 200*200. The unknown node arranges stochastically; the unknown node is 220. The mobile anchor node has 6 kinds of situations: 9, 12, 16, 20, 25, and 30 positions.

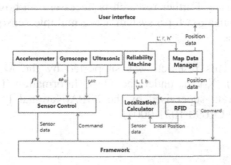

Fig. 3. System overview

Figure 3 outlines the operation of the system. The sensor controller sends acceleration and angular velocity information measured with an accelerometer and gyroscope as well as ultrasonic information. The location calculator calculates the current location of the moving node using information received from each node. If there is no location initialization, the location is ignored. The calculated location coordinates L, l and h, and speed information from sensors are input into reliability device, and compensated information L', l' and h' are calculated. The calculated location information of the moving node is compared with known map data by the map manager, and the actual location is estimated and sent to the user interface and the location calculator.

4.2 Results and Analysis

The simulation uses adaptive weighted centroid localization algorithm and maximum likelihood estimation method. Localization accuracy mainly depends on the numbers of the mobile anchor node broadcasting its positions or the anchor density. It is very easy for our method to change anchor density by adjusting the interval time or the moving length of the mobile anchor node broadcasting its positions or by changing the moving interval of spiral line. In comparison with other methods, this is one of the advantages with our method, and it does not require additional hardware.

5   Conclusion and Future Works

In order to resolve the shortcoming of the general positioning method using the weighted centroid localization algorithm, this study designed a method that can calculate detailed location information using inertia and reduce the error rate resulting from difference in response time by adding reliability as a new variable for compensation. Through an experiment, we found that the new method has higher reliability against long-term errors compared to the existing method. Furthermore, it can reduce the long-term error rate at a low cost.

If there are many opportunities available for sensor initialization, however, the new method may increase errors resulting from the synchronization of each sensor and the reliability level tends to depend on the type of sensor. Furthermore, for high reliability, reference objects need to be entered into map data in advance. Despite these shortcomings, however, if the reliability level is set based on empirical or experimental data, we can build a positioning system of high reliability at a relatively low cost compared to network size. Further research is necessary on how to reduce the error rate of sensor synchronization for a system using multiple sensors and on the design of a filter for sensor bias.

Acknowledgments. "This research was supported by the MKE(The Ministry of Knowledge Economy), Korea, under the ITRC(Information Technology Research Center) Support program supervised by the NIPA(National IT industry Promotion Agency)" (NIPA-2010-C1090-1031-0004).

References

1. Ssu, K.F., Ou, C.H., Jiau, H.C.: Localization with mobile anchor points in wireless sensor networks. IEEE Trans. on Vehicular Technology 54(3), 1187–1197 (2005)
2. Hu, L., Evans, D.: Localization for mobile sensor networks. In: Proc. of ACM MobiCom (2004)
3. Song, C.W., Ma, J.L., Lee, J.H., Chung, K.Y., Rim, K.W.: Localization Accuracy Improved Methods Based on Adaptive Weighted Centroid Localization Algorithm in Wireless Sensor Networks. International Journal of Computer Science and Information Security 8(8), 284–288 (2010)

4. Laurendeau, C., Barbeau, M.: Centroid localization of uncooperative nodes in wireless networks using a relative span weighting method. EURASIP J. Wirel. Commun. Netw. (2010)
5. Nagpal, R., Shrobe, H.E., Bachrach, J.: Organizing a global coordinate system from local information on an ad hoc sensor network. In: Zhao, F., Guibas, L.J. (eds.) IPSN 2003. LNCS, vol. 2634, pp. 333–348. Springer, Heidelberg (2003)
6. Niculescu, D., Nath, B.: DV based positioning in ad hoc networks. Journal of Telecommunication Systems 22(4), 267–280 (2003)
7. Kim, Y.C., Kim, Y.J., Chang, J.W.: Distributed Grid Scheme using S-GRID for Location Information Management of a Large Number of Moving Objects. Journal of Korea Spatial Information System Society 10(4), 11–19 (2008)
8. Lee, Y.K., Jung, Y.J., Ryu, K.H.: Design and Implementation of a System for Environmental Monitoring Sensor Network. In: Proc. Conf. APWeb/WAIM Workshop on DataBase Management and Application over Networks, pp. 223–228 (2008)
9. Hammad, M.A., Aref, W.G., Elmagarmid, A.K.: Stream window join: Tracking moving objects in sensor network databases. In: SSDBM (2003)
10. Niculescu, D., Nath, B.: Position and orientation in ad hoc networks. Ad hoc Networks 2(2), 133–151 (2002)
11. Chen, Y., Pan, Q., Liang, Y., Hu, Z.: AWCL: Adaptive weighted centroid target localization algorithm based on RSSI in WSN. Proc. IEEE ICCSIT 9, 9–11 (2010)
12. Wang, J., Urriza, P., Han, Y., Cabrić, D.: Performance analysis of weighted centroid algorithm for primary user localization in cognitive radio networks. In: Proc. ACSSC, Pacific Grove, pp. 7–10 (2010)
13. Titterton, D.H., Weston, J.L.: Strapdown Inertial Navigation Technology. Peter Pegerinus (1997)

Performance Testing of Web Map Services

Jiří Horák, Jiří Ardielli, and Jan Růžička

VSB Technical University of Ostrava, Institute of Geoinformatics, 17. listopadu 15,
70833 Ostrava Poruba, Czech Republic
{jiri.horak,jiri.ardielli,jan.ruzicka}@vsb.cz

Abstract. Network services represent a key part of INSPIRE. They shall be available to the public and accessible via the Internet. The advent of prototypes using OGC WMS on PDA's requires to test performance of selected servers and its suitability for using such service on PDA platform. The paper proposes a way how to test and measure the availability of network view services (WMS) for end users. Map servers of two main providers of Czech spatial network services according INSPIRE (CENIA, Czech Environmental Information Agency, and COSMC, Czech Office for Surveying, Mapping and Cadastre) has been tested using performance testing. The testing helps to understand how the system will handle the load caused by many concurrent users. Any provider is able to test a latency of application, but for end users it is important to measure the overall latency (the response time) and also other characteristics like error occurrence, availability and performance. The response time, error occurrence, availability and performance were measured during two months of repeated access using same scenarios and different number of concurrent virtual users. Better performance and benchmarking were obtained testing of COSMC map services, where the average response time is bellow 1 s.

Keywords: WMS, PDA, INSPIRE, performance, testing.

Introduction

Network services represent a key part of INSPIRE. They should be available to the public and accessible via the Internet. The regulation of network services establishes the obligatory quality criteria.

Utilization of OGC WMS on mobile devices is in the last time supported by proposals of new architecture solutions based on middleware equipped with different Data Access Plug-ins [1].

The PDA users are still limited by many technical, software and interface factors. Using map services on a PDA depends mainly on a communication speed and suitability of map rendering on the given display conditions (illumination issues etc.). Nevertheless the basic condition of using such services is a satisfactory quality of service, which is measured by various characteristics.

The end user's requests and its ITC conditions may strongly vary. Also the satisfaction of an end user is a subjective term.

N.T. Nguyen et al. (Eds.): New Challenges for Intelligent Information, SCI 351, pp. 257–266.
springerlink.com   © Springer-Verlag Berlin Heidelberg 2011

The complex evaluation of end user's conditions and prediction of its satisfaction should utilize both content driven and function capability measurements as well as quantitative measurements which control the service accessibility and performance.

These IT aspects of services are vital for end users. They can be measured by the overall latency (a measurement of the time processing a request) and also by other characteristics like concurrency (measurement of the influence when more the one user operates web-application), availability (a measurement of the time a web-enabled application is available to take request) and performance (average of the amount of time that passes between failures).

Quality of Service

The evaluation of the Spatial Data Infrastructure (SDI) and its components, namely geoportals, usually starts considering the cost savings/cost reduction objective, but in recently the evaluation and comparison has been based on control-driven performance measurement systems focusing on inputs and outputs [12].

Quantitative measurements used in SW testing offer easy quantification, better standardization (for many aspects some industry or official standards of IT can be applied), repeatability, possibility of automated and long-term processing and evaluation, understandability [16].

Software testing methods are traditionally divided into black-box (functional) testing and white-box (glass-box, structural) testing [9].

Concerning analysis of web client-server systems it is needed to distinguish server-side and client-side analysis. Server-side testing can utilize click stream analysis, web server log files, de-spidering the web log file and exploratory data analysis [14]. Recommended metrics include Number of Visit Actions, Session Duration, Relationship between Visit Actions and Session Duration, Average Time per Page, Duration for Individual Pages. It is recommended to measure results for individual pages and then aggregated results, otherwise metrics will be distorted due to the processing of mixture of both navigation pages and content pages.

The server side testing can also be utilized to explore a dependency between complexity of the picture and the velocity of image rendering and delivery. Such testing may lead to improvement of the service using suitable preprocessing (e.g. FastCGI for University Minnesota Map Server [2]), tiling/caching [13] or load balancing [18].

Hicks et al. [8] stresses potential benefits of client-side automated testing and distinguishes following 5 basic types: Test precondition data, Functional testing, Stress test functionality, Capacity testing and Load and performance testing.

The most frequent form of testing is load testing (alternative terms are performance testing, reliability testing or volume testing). Load testing generally refers to the practice of modeling the expected usage of a software program by simulating multiple users accessing the program concurrently [19].

When the load placed on the system is raised beyond normal usage patterns, in order to test the system's response at unusually high or peak loads, it is known as stress testing. The load is usually so great that error conditions are expected results, although no clear boundary exists when an activity ceases to be a load test and becomes a stress test.

Any good load testing has to be based on realistic usage patterns. In the case of web site testing, the load generator has to mimic browser behavior. It usually means to submit a request to the Web site, wait for a period of time after the site sends a reply to the request, and then submits a new request [15]. The load generator can emulate thousands of concurrent users to test web site scalability. Each emulated browser is called a virtual user. The virtual users' behavior must imitate the behavior of actual real users (follow similar patterns, use realistic think times, and react like frustrated users, abandoning a web session if response time is excessive).

Following metrics are usually applied:

Time delay - time delay can be described by latency or by response time.

Latency is a time interval between the instant at which an instruction control unit issues a call for data and the instant at which the transfer of data is started [9].

Instead of latency a response time is usually used. The response time can be specified as the time taken for the application to complete any given request from the end user. However, to be complete, any such metric must account for the number of users that are likely to be interacting with the application at any given time [3].

Error occurrence - Error occurrence can be described using simple average of the amount of time that passes between failures. Another possibility is to evaluate percentage of errors occurred during the session or some iteration of requests.

Availability - Availability measures the percentage of time customers can access a Web-based application [15]. Availability goals typically vary according to the application type. Critical applications require the best availability conditions. Geoportals as an important part of SDI should be a reliable source of data and services and that is why the requested availability is set up to 99% of the time [10].

Inspire Network Services

Directive 2007/2/EC of the Council and the European Parliament establishes the legal framework for setting up and operating an Infrastructure for Spatial Information in Europe (INSPIRE) based on infrastructures for spatial information established and operated by the member states [5].

To assure the availability and accessibility of data a Regulation for Network Services has been adopted by the Commission of the European Communities [10]. This Regulation sets out the Requirements for Network Services (Article 3) and the Access to the Network Services (Article 4).

According to Article 3 of this Regulation the Network Services shall be in conformity with the requirements concerning the quality of services set out in Annex I of this Regulation. According to Annex I the following Quality of Service criteria relating to performance, capacity and availability shall be ensured [10].

Performance - Performance means the minimal level by which an objective is considered to be attained representing the fact how fast a request can be completed within an INSPIRE Network Service.

The response time for sending the initial response to a Discovery service request shall be maximum 3 seconds in normal situation. For a 470 Kilobytes image (e.g. 800x600 pixels with a color depth of 8 bits), the response time for sending the initial response to a Get Map Request to a view service shall be maximum 5 seconds in normal situation.

Normal situation represents periods out of peak load. It is set at 90% of the time.

Response time means the time measured at the Member State Service location, in which the service operation returned the first byte of the result.

Because the response time is measured at the service location, the criteria can be specified as server-side measurements

Capacity - Capacity means limit of the number of simultaneous service requests provided with guaranteed performance.

The minimum number of served simultaneous requests to a discovery service according to the performance Quality of Service shall be 30 per second.

The minimum number of served simultaneous service requests to a view service according to the performance quality of service shall be 20 per second.

Availability - Availability means probability that the Network Service is available. The probability of a Network Service to be available shall be 99% of the time.

The transposition of the INSPIRE Directive in Czech Republic is coordinated by the Ministry of the Environment and the Ministry of Interior in collaboration with the Czech Office for Surveying, mapping and the Cadastre (COSMC). Moreover, the Czech Environmental Information Agency (CENIA) has been responsible for IN-SPIRE implementation since 2006. The transposition into national law is established by the law 380/2009.

CENIA has created and maintained a central geoportal (http://geoportal.cenia.cz) for the INSPIRE relevant data themes as a component of the Portal of the Public Administration, which in turn is under the responsibility of the Ministry of Interiors [5]. CENIA has developed several web mapping services with basic administrative, topographic and environmental data to provide all the necessary data for authorities. The implementation emphasizes on operational capability of the server using e.g. two parallel running WMSs. The geoportal offers an access to 4 terabyte of data through 90 local and 15 remote map services. Currently a new national geoportal is in under development.

The Czech Office for Surveying, mapping and the Cadastre (COSMC) is an autonomous supreme body of the state administration of surveying, mapping and cadastre in the Czech Republic (http://www.cuzk.cz/). COSMC have been playing a major role in data production, because it is responsible for 5 Annex I, 2 Annex II, and 2 Annex III INSPIRE data themes (geographic and cadastral reference data, DTM and ortho-imagery, buildings and land use themes). In 2005 the COSMC launched a new geoportal in Czech and English languages which includes a business module, web map services, the GPS stations network, and geodetic control points [6]. The newer version of the server has been operating since mid 2010. COSMC metadata are currently compliant with the national and cadastral standards.

Methodology of Testing

The goal of the testing is to verify usability (suitable conditions) of selected network service providers for end users. This accent on end user approach requires selection of the client-side black-box testing. The results of testing will be used for evaluation of suitability of service for PDA users.

For the client-side testing we have selected three important aspects (metrics) influencing the end-user acceptance and utilization of network service: performance, error occurrence and response time.

These metrics are measured during a long-term performance testing which repeats many requests with approximately the same load and monitor the behavior of the service using indicated metrics for a long time.

The testing of network services was undertaken for CENIA and COSMC (see previous chapter) using appropriate URLs for WMS services. The WMS standard compliance has not been tested.

The next step was to define a realistic usage patterns to mimic real user's behavior. The usage experiences and demands were reflected in the construction of various scenarios with defined sequences of visited web pages (GetMap requests) and time planning. Basic layers were selected from following data themes - Administrative units, Orthophoto, Topographic maps, Natural and Landscape protected areas, Cadastral maps, Definition points of parcels and buildings. It represents a mixture of data themes from Annex I and II. The scenario starts with simple overview maps (i.e. digital cadastral maps both in raster and vector models) and scanned topographical maps. Later requests combine more layers and create complex maps. It represents the final stage of a map composition prepared for printing.

Scenarios for the long-term performance testing utilize iterations of requests which are repeatedly replayed every 60 minutes. Each block contains increased number of accesses to individual selected pages. In the case of CENIA portal the scenario contains 19 selected web pages, for COSMC geoportal the number of visited web pages is 10. Main settings for the long-term performance testing:

Duration:	2 months
Number of iterations:	1 to 24
Number of virtual users:	tests with 1, 3, 5 and 10 users
Interval between runs:	60 minutes
Delays between users:	10 ms

The nature of load tests facilitates an automation and repeatability of the testing. A wide range of different web Site Test Tools and Site Management Tools (more than 420 tools) is documented on http://www.softwareqatest.com/qatweb1.html, where the special category Load and Performance Test Tools includes overview of some 40 tools. Short recommendations for tools suitable for performance, load and stress testing can be found in [7]. A good comparison of 8 selected load test tools can be found in [4] including description of scenario preparation, execution, form of results and monitoring capability.

The direct testing of WMS service on PDA was not performed due to the limitation of suitable SW. For the indicated purpose we have selected a WAPT software v.3.0.

The client-side testing is negatively influenced by unknown network and client status. To improve the evaluation of the network services it is necessary to exclude time when network is overloaded or unavailable or the client registers some problems. For this purpose testing of selected high reliable servers is performed for continual monitoring of network status. The core evaluation includes identification of time when more than half of monitored servers (services) are unavailable or badly available (long latency). These time slots are excluded from processing of WMS service testing results.

The testing results of CENIA services are also influenced due to a usage of middleware software for generating WMS requests. Current testing of the given predefined map extents does not address different conditions of geographical layers, namely various densities of map features. Thus the next improvement of testing will consist in generating fully or partly random walk across the map.

Testing Results

The limit for acceptable response time is set to 10 seconds, because to this time the user's attention is focused on the dialogue of application [11]. The advice to split the response time to 0.1 s ("fluent reaction"), to 1 s (a noticed delay, but no action is needed) and to 10 s (anything slower than 10 s needs a percent-done indicator as well as a clearly signposted way for the user to interrupt the operation [17]) is used for many years. Concerning the availability aspect we can apply the limit of 99% [10].

The Long-Term Performance Testing of CENIA Geoportal

In most cases (iterations) the average web transaction (the average response time) is under 2 seconds that is a satisfactory. Such indicator does not provide us with information how often end user faces a bad situation with long response time when he can decide to abandon the session. More suitable is the evaluation of the maximum request processing time.

The maximum web transaction (the maximum response time) shows a large variability of results. Even if we exclude the network error anomalies (issued from network error occurrences), still we record cases when the response time exceeds 10 s. In some cases also the response time around 50 s has occurred. These long time intervals have been recorded for pages including ortho-imagery and raster topographical background.

The overall performance of tested application can be evaluated as a stable one. During increased number of iterations the system performance was approximately constant (usually between 10 and 40 pages per second), for more virtual users even slightly decreasing (fig. 1). It indicates a probable insufficient cache system on the server side, because an efficient cache system for these repeatedly visited pages will provide the increased performance.

Concerning error occurrences the most frequent is network error. It appears every day, sometimes repeatedly. WAPT has no features of transient network analyzers, thus it is impossible to identify the network element where the error occurs. The second type of error, the HTTP error, was rarely registered.

Small amount of errors (around 1%) was registered for most of the iterations. Unfortunately, the variability of error occurrence is quite large and in some test runs we obtained about 30% of errors. These results cannot be a satisfactory for end users.

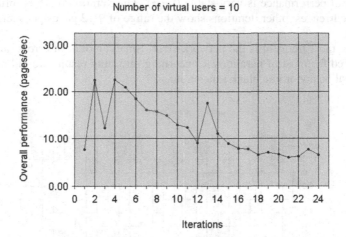

Fig. 1. Overall performance.

In some cases the service returns a blank image. The reason of this behavior is not known; according to the provider it may occur during an internal refresh of services.

The Long-Term Performance Testing of COSMC Geoportal

The average response time is usually bellow 1 s, which is excellent for end users. Only 1 page (n. 8) with requested raster layers and 2 sets of definition points indicates average access time around 3 s. The maximum response time do not exceed 10 s (except of abnormal iteration n. 16). Most of pages show the maximum time bellow 4 s, more demanding page n. 8 below 6 s (fig. 2).

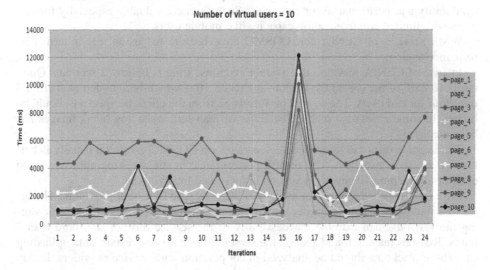

Fig. 2. Maximum response time.

The overall performance is very stable (fig.3). The iteration n. 15 is influenced by the error occurrences; other iterations show the range of 9-12 pages per second, which is quite low.

Again the most frequent is the network error. Small amount of errors (around 1%) was registered for most of iterations. Concerning returning results we did not register any abnormal behavior like blank images etc.

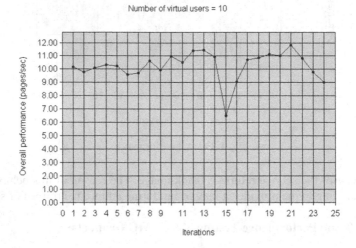

Fig. 3. Overall performance.

Conclusions

The client-side testing of web map services provides a tool how to evaluate their availability and performance for end users. Such tests are valuable especially for end users with limited communication capacity like mobile users.

WMS services of CENIA and COSMC have been tested using a long-term performance testing.

In case of CENIA services, the average response time is bellow 2 seconds. Quite rarely (0.25% of the cases) the responses exceed 10 s, which is regarded as a critical threshold for end users. The overall performance from the client perspective should be higher. In some cases also specific behavior of map server sending blank images has been recorded.

The average response time for selected pages on COSMC geoportal usually is bellow 1s, which is excellent. The overall performance in stress testing approves good conditions reaching 40 pages per second in increased load volume.

Errors of services (service not available) were recorded for both providers. The error distribution is unequal, usually cumulative. The occurrence may reach 70 % during the session. Such extreme inaccessibility of the service did not last more than 2 hours. Reasons may be found also on the server side - a service restart, data uploading etc. These blackouts should be analyzed in cooperation with service providers. It may substantially influence the service availability and consecutively, the user confidence and their willingness to exploit the service.

References

1. Brisaboa, N.R., Luaces, M.R., Parama, J.R., Viqueira, J.R.: Managing a Geographic Database from Mobile Devices Through OGC Web Services. In: Chang, K.C.-C., Wang, W., Chen, L., Ellis, C.A., Hsu, C.-H., Tsoi, A.C., Wang, H. (eds.) APWeb/WAIM 2007. LNCS, vol. 4537, pp. 174–179. Springer, Heidelberg (2007), ISBN 978-3-540-72908-2
2. Brock, A.: A Comparison of ArcIMS to MapServer. In: Proceedings Open Source Geospatial 2005 MUM/EOGEO 2005, Minneapolis, USA, June 16-18 (2005)
3. Brown, S., Dalton, S., Jepp, D., Johnson, D., Li, S., Raible, M.: Pro JSP 2. A-Press (2005), ISBN 978-1-59059-513-8
4. Buret, J., Droze, N.: An Overview of Load Test Tools (2003),
   http://clif.objectweb.org/load_tools_overview.pdf
   (accessed February 25, 2009)
5. Craglia, M., Campagna, M. (eds.): Advanced Regional Spatial Data Infrastructures in Europe, JRC Scientific and technical report, Joint Research Centre, Institute for Environment and Sustainability, February 10, p. 132 (2009), ISBN: 978-92-79-11281-2,
   http://sdi.jrc.ec.europa.eu/ws/Advanced_Regional_SDIs/
   arsdi_report.pdf (accessed February 25, 2009)
6. Czech Office for Surveying, Mapping and Cadastre: Annual report 2007. ČUZK, Prague (2008), ISBN 978-80-86918-54-9
7. Gheorghiu, G.: Performance vs. load vs. stress testing (2005),
   http://agiletesting.blogspot.com/2005/02/
   performance-vs-load-vs-stress-testing.html (accessed February 25, 2009)
8. Hicks, G., South, J., Oshisanwo, A.O.: Automated testing as an aid to systems integration. BT Technology Journal 15(3), 26–36 (1997)
9. Institute of Electrical and Electronics Engineers: IEEE Standard 610.12-1990: IEEE Standard Glossary of Software Engineering Terminology (1990), ISBN 1-55937-067-X
10. INSPIRE Commission Regulation (EC) No 976/2009 of 19 October 2009 implementing Directive 2007/2/EC of the European Parliament and of the Council as regards the Network Services (2009), http://eur-lex.europa.eu/LexUriServ/
    LexUriServ.do?uri=OJ:L:2009:274:0009:0018:EN:PDF
    (accessed December 27, 2010)
11. Krejcar, O., Cernohorsky, J.: New Possibilities of Intelligent Crisis Management by Large Multimedia Artifacts Prebuffering. In: Proceedings I.T. Revolutions 2008, Venice, Italy, December 17-19. LNICST. Springer, Heidelberg (2008), ISBN 978-963-9799-38-7
12. Lance, K.T., Georgiadou, Y., Bregt, A.: Understanding how and why practitioners evaluate SDI performance. International Journal of Spatial Data Infrastructures Research 1, 65–104 (2006)
13. Liu, Z., Devadasan, N., Pierce, M.: Federating, tiling and Caching Web map servers (2006), http://www.crisisgrid.org/cgwiki/images/4/49/
    GIC_Oct27_2006_Final.ppt (accessed February 25, 2009)
14. Markov, Z., Larose, D.: Data Mining the Web - Uncovering Patterns in Web Content, Structure, and Usage, 213 p. Wiley, New York (2007), ISBN 978-0-471-66655-4
15. Menascé, D.A.: Load Testing of Web Sites. IEEE Internet Computing 6(4), 70–74 (2002)

16. Mendes, E., Mosley, N. (eds.): Web Engineering, p. 438. Springer, New York (2006), ISBN-10 3-540-28196-7
17. Nielsen, J.: Usability Engineering. Morgan Kaufmann, San Francisco (1994), ISBN 0-12-518406-9
18. Pancheng, W., Chongjun, Y., Zhanfu, Y., Yingchao, R.: A Load Balance Algorithm for WMS. In: IEEE International Proceedings of Geoscience and Remote Sensing Symposium, Toulouse, France, September 20-24, vol. 5, pp. 2920–2922 (2004)
19. Wang, H., Li, Y., Li, X.: A Mobile Agent based General Model for Web Testing. In: Proceedings of Second International Symposium on Computational Intelligence and Design, Changsha, China, December 12-14, vol. 1, pp. 158–161 (2009), ISBN 978-0-7695-3865-5

Sending Social Impression via a Mobile Phone

KangWoo Lee[1], Kwan Min Lee[1], and Hyunseung Choo[2]

[1] Department of Interaction Science, Sungkyunkwan University
53 Myeongnyun-dong, 3-ga, Jongno-gu, Seoul, 110-745, South Korea
[2] School of Information and Communication Engineering, Sungkyunkwan University
300 Chon-Chon Dong, Jang-Ahn Ku, Suwon, 440-746, South Korea
kangwooster@gmail.com

Abstract. Multimedia caller identification (MCID) is a new application for mobile phones that displays video contents. Its roles in impression formation were examined in automobile telemarketing context. The expertise and kindness of a car dealer are presented in both visual MCID and verbal conversation contents. Experimental results show that the MCID content elicits higher ratings of kindness, whereas conversation content elicits higher ratings of expertise. Information related to car sales can be effectively delivered by how the visual and verbal contents are combined. These results also suggest that showing visually kind and verbally intelligent would be the best strategy for achieving more positive evaluation from clients of telemarketing.

Keywords: MCID, Social Impression, Attractiveness, Types of Cues.

1 Introduction

Mobile phones open up new opportunities to improve facilitation of communication and to encourage various mobile commerce activities. They have become powerful devices for linking people to people and establishing a new relationship between salesperson and clients. Technology has made the devices multi-faceted which not only enables users to communicate and message but it also enables users to enjoy the new feature like multimedia caller identification [1]. This new technology allows users to send their own created contents to others as well as to receive them during ringing. The contents can be various depending on a sender's expressive purpose – sending emotional states, displaying commerce advertisement, or making an impression of the sender. A receiver may identify who the sender is, what might make him/her a call, and so on even before a conversation proceeds.

The technology of MCID has not been investigated its potential in social communication context. It may be required that a strategic way be displayed user created contents (UCC) that matches the following phone conversation. An important question, then, is how UCC in MCID displays to cope with the following phone conversation in a strategic way, and what role the technology plays in forming social impression that users intend to deliver.

Generally, social impression shaped by computer technology has focused on types of information available such as verbal vs. non-verbal behaviors, different communication

N.T. Nguyen et al. (Eds.): New Challenges for Intelligent Information, SCI 351, pp. 267–276.
springerlink.com

modes such as text based vs. visual based communication mode[2] [3] [4] [5]. That is, researches on computer based communication (CMC) have mainly put their interest in how the impression is altered by adding or removing certain communicative cues [6]. In contrast to the current trends of technology based communication researches, few studies have examined the impression formed by the combination of communicative contents relative to different types of communicative medium.

This paper examines the following issues in telemarketing situation in which a dealer explains an automobile product and service on a mobile phone. In this telemarketing situation, we consider the impression formation built by two different types of communicative medium – MCID and verbal conversation – that convey the expertise and kindness of the dealer. We also consider the how the formed impression influences on costumer's rate about the product as well as service. The sales situation considered in our study would happen in natural sales setting where a salesperson's personal attributes are associated with certain aspects of product. An important consideration here is that the MCID features the following phone conversation. A possible functionality of MCID is to send multimedia contents containing user's attributes and to imprint what kind of person he is to others; as a result, what he says in the phone conversation can be more credible or more persuasive.

2 Related Works

2.1 Multimedia Caller Identification

Caller identification (CID) is a telephone service that allows a receiver to identify a caller before he or she answers his/her telephone. The service can also provide a name associated with the calling telephone number. A caller's number is displayed on the called party's phone during the ringing signal. CID is the most common telephone (including both wire and wireless phone) service across worldwide. Beside of identifying the called party, it can be used for evasive purpose such as avoiding telemarketers or other unwanted calls. Calling Name Presentation (CNAP) can be considered as the extent of CID in a way that supports text messages. So, it allows a caller to express himself/herself using simple text message, and to present it the called subscriber. Even though the message can be composed by caller in CNAP, it still is limited of using alphanumeric characters.

We developed a new kind of caller identification technology so-called 'Multimedia Caller Identification' (MCID) that is more advanced and enjoyable form of CID. MCID has appeared in richer soil of information technology in which people express themselves in the form of user created contents (UCC) with various media such as text, music, image and video. That is, photos, music, video clips generated by users or commercial multimedia contents purchased from contents providers can be displayed on the other party's phone. Figure 1 conceptually illustrates the MCID service. First, a user creates his/her own multimedia contents or purchases commercial contents that are transformed into an advanced multimedia format (AMF). Second, a caller's telephone number and is registered in a MCID server. Third, the registered contents are retrieved from the server if a user makes a call. Finally, the contents are shown on the other party's phone's display during ringing.

This service is not limited simply to show a caller's identification, but can be used in various purposes. For instance, this service can be used as a name card containing a caller's business information or as a company and product advertisement. Because various applicable potentials of this service, it takes a growing interest of Korean. The MCID service has been recently commercialized in South Korea. The prominent tele-communication companies such as Korea Telecom, SK telecom, LG telecom provide the MCID services. Now, more than 1,000,000 users have signed up on the service.

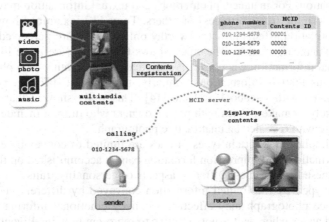

Fig. 1. The conceptual illustration of MCID service

2.2 Thin Slices

A multimedia content of MCID is displayed during phone ringing and finished as soon as one receives a phone. This means that the content is shown up within a very limited time slice that may last from couple of seconds to several seconds, and a very instantaneous impression is established in this short duration. In impression literature, this is called as a 'thin slice' that is a brief and dynamic sample of a person's behavior less than 5 minutes and is often video clip taken from a long video recoding [7]. Re-searches on thin slices have shown that the ability of observers to assess people's per-sonalities and associated consequences with relatively high accuracy from brief and impoverished information. For instance, people are surprisingly good at predicting a teacher's effectiveness [8], judging salesperson [9], rating personalities [10]. For in-stance, Hari et al. [9] examined the influence of different sales people and different gestures on the sales outcome - intention to buy. They demonstrated that the thin slice judgment on a sales situation is very accurate in predicting the intention to buy a product.

Studies on first impressions of the people's face provide further insights on how social impression is shaped. The judgments of personal traits as attractiveness, like-ability, trustworthiness, aggressiveness, and competence are made after as little as 100 ms [11]. In fact, the judgments of first impressions are formed surprisingly quickly. People are able to extract the cues required for impressions formation from static pho-tographs less than a second [12]. Especially, survival related feature such as angry faces are judged more quickly and even unconsciously. In contrast, non-survival related traits such as intelligence is not extracted quickly comparing to survival related feature.

2.3 Visual vs. Verbal Cues in Impression Formation

First impressions of other persons are often constructed on various cues including verbal and visual information. However, these cues do not equally contribute to impression formation. For instance, photographs and textual information may differently contribute in forming trustworthiness of others. Toma [13] examined how the nature of online information affects how trustworthy online daters are perceived using both visual and textual information. The study shows that people are more likely to trust online personal information when it is given in textual form rather than photograph or photograph plus textual information. Similarly, the roles of different types of cue were investigated with Facebook website [14]. This study showed that participants were more likely to make friends with profile owners who did not include visual cues than with those who revealed an unattractive photograph.

This may implies that certain types of cues are limited to conveying emotional or personal information, and impression formation can be accomplished on the inference driven from these limited cues. Different aspects of personality traits can be exaggerated or stereotyped by the limited information conveyed by different cues [15]. For instance, a face photograph may effectively convey emotional information such as happiness and anger whereas it is not effective to represent how intelligent he/she is.

2.4 Personal Attractiveness in Sales

Numerous researchers have explored the relationship between perceived attractiveness of salesperson and salesperson's performance. The attractiveness is a general concept that encompasses various aspects including expertise, personality properties, physical appearance etc. These aspects have been widely investigated in both experimental and simulated situations. For instance, customers perceive attractive salespersons more favorably and their purchase intention is more strongly influenced by explanation of attractive salesperson [16]. The impression formed by service providers' kindness can influence on customers' satisfaction [17]. Also, expertise of salesperson effectively persuades customers to buy a product. In general, first impressions of salesperson affect sales encounter's attitude toward the salesperson's proposal as well as perceived quality of a product. This may result from the customer referral information commonly available to salespeople or from physical appearance of salespersons. These studies suggest that a salesperson's impressions can be strategically used to meet customer's need by presenting his/her personal attractiveness.

3   Experiment

In the experiment we investigate impression formed from two different types of communication medium – MICD and telephone – in automobile sales situation. Also, we investigate how the formed impression affects customer's evaluation on what the seller has presented.

3.1 Participants

48 undergraduate and graduate students participated in the study examining the effects of MCI on impression formation. Participants received 5000 Korean won (about 4.5 US dollar) in a coupon at the completion of the questionnaire after experiments. Participants were assigned randomly to the 4 conditions of a 2 (script contents: expertise vs. kindness) by 2 (MCID contents: expertise vs. kindness) design. A male experiment cooperator was recruited using university website ads. He was paid 10000 Korean (about 9 US dollar) won per an hour. The cooperator took a role to explain a car and to persuade participants to buy it on a mobile phone.

3.2 Apparatus

The Mtelo application ran on a Nokia N76 mobile phone, with a 240 x 320 pixels resolution and 2.4 inches screen dimension, displaying an AMF file. It is a folder type phone. The folder is needed to open up so as to talk on the phone.

3.3 Stimulus

Two video clips were used to express expertise and kindness of the caller. For the kind MCI condition, the video clip consisted of a sequence of image in which a male person in full dress was smiling pleasantly and then bowing politely. On the other hand, for the expertise MCI condition, the video clip of consisted of a sequence of image in which the same person wearing glasses in a white shirt appeared, and various car interior parts – steering wheel, control panels, electronic modules, etc were displayed (see Fig. 2). The brownish colored background was employed for the kind MCI condition, whereas bluish colored background was employed for the expert MCI condition. Images in the both conditions contained catch-phases in the bottom of the

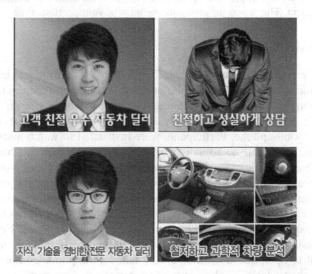

Fig. 2. The contents of MCID used in the experiments. The upper content emphasizes the kindness of the caller, whereas the lower content emphasizes the expertise of the caller.

images. For instance, the catch-phrase 'the kindest car dealer appointed by clients' was used to express the caller's kindness, whereas the catch-phrase 'the car dealer with expert knowledge and technique' was used to express the caller's expertise. Each video clip has a 128 by 160 pixel resolution and lasted for 23 seconds.

For verbal conversation, two different scripts were used to converse with clients on the phone. One script emphasizes the dealer's kindness and car services, whereas the other script emphasizes the dealer's expertise and quality of a car. The latter script is presented in Table 1.

Table 1. A script presents the expertise of the car dealer and introduce more technical aspects of a car.

[Greeting] Good morning, Sir/Madam.
[Asking availability] Are you available for couple of minutes?
(Participants were instructed to answer "yes")
[Introducing a caller] This is Mr. Kim Junmyoung (the caller's name). I have 10 year career in analyzing and assessing automobiles. Based on systemic analysis and evaluation of car performance, price, management, repair, we make an offer the best car for you.
[Introducing a car] I would like to introduce New Pride (car name) that utilizes various new technologies. Design of New Pride is highly evaluated by virtue of its dynamic impression as well as its friendly image with smooth curvature. A new VGT engine equipped in New Pride makes 100 horse power, and curtails the fuel consumption and noise. A more stable body cell of New Pride provides better protection. In our car shop we provide various services to treat you with our best. We offer free expendable supplies and emergency car recue service to enhance costumer's satisfaction and convenience. Your satisfaction is our priority. We always keep it in our mind.
[Catalog] If you tell me your home address, we are going to send a catalog containing systemic and analytical information about New Pride.
(Yes or no response for receiving the catalog)
[Appreciation] We again promise to introduce the best car with the scientific analysis and careful assessment. Thank you for giving me minutes of your precious time

3.4 Experiment Procedure

A participant reported to an experimental room for participation and introduced a seat by an experimenter who was one of the authors. At another experimental room opposite to the room there was the experimental cooperator. So, they never had a chance to see each other. Participants were briefly explained the basic purpose of the study and how to use a mobile phone. Participants were instructed to watch a screen of the mobile phone and to identify who called when it rang. The experimenter informed the participants that the caller would ask favors twice. First, the caller would ask whether participants were available to converse on the phone. Second, the caller would ask whether participants wanted catalogs containing detailed information of a car explained by the caller. For the first request participants were instructed to say 'yes' because the experiment could not be proceeded if they said 'no'. However, participants were instructed to answer freely for the second request. Participants were also instructed not to open questionnaires until a talk on the phone completed. When participants clearly understood the experiment procedure and were ready, the experimenter moved into the room where the experimental cooperator was and asked to make a call. In a short time later, the phone call from the experiment cooperator was given to a participant. The conversation would start when participants received the phone. The conversation usually lasted for 2-3 minutes.

3.5 Assessments

After the conversation completed, participants was asked to fill a questionnaire. The questionnaire consisted of 18 items that could be classified into 6 categories of participants' responses including perceived expertise, perceived kindness, perceived quality of automobile, perceived quality of car service, further consulting intention, likability. Each of these items was rated on a 10-point scale ranging from very disagreeable to very agreeable. Cronbach's alpha reliabilities of the items were measured: perceived expertise (Cronbach α = .898), perceived kindness (Cronbach α =.746), perceived quality of automobile (Cronbach α =.802), perceived quality of car service (Cronbach α =.795), further consulting intention (Cronbach α =.714) and likability (Cronbach α =.817).

4 Results

An ANOVA was performed on the 6 categories of participants' responses. The perceived expertise of the car dealer was significantly affected by the expert script (F(1, 48) = 3.853, p = .056), whereas the perceived kindness of the car dealer was significantly affected by the MCID kindness contents (F(1, 48) = 5.375, p < .05). No interaction effects between MCID and conversation contents on the perceived impression of the car dealer were found. The results indicate that the perceived impressions of the car dealer seem to be influenced by the different types of cues. Fig. 3 illustrates the impressions of the car dealer shaped by the conversation and MCID contents. Similarly, the perceived quality of the introduced car was significantly affected by the expert script (F(1, 48) = 4.682, p < .05), whereas the perceived kindness of the car

Table 2. Means and standard deviations for the 6 categories of the responses

	Expert multimedia content		Kind multimedia content	
	Expert script	Kindness script	Expert script	Kindness script
Perceived expertise	5.7778 (1.7885)	4.8403 (1.7431)	6.0278 (1.7722)	5.0278 (1.5206)
Perceived kindness	7.1563 (1.2399)	7.5729 (1.2206)	7.9167 (1.2807)	8.1510 (.7018)
Quality of Automobile	5.9205 (1.1898)	5.5590 (1.1635)	6.4953 (1.3321)	5.4316 (.8141)
Quality of Service	6.6771 (1.5004)	6.8073 (1.1449)	7.6146 (1.3069)	7.6458 (1.2403)
Intention of further consultation	6.5417 (2.3008)	5.7500 (1.8278)	7.1250 (2.5417)	6.8750 (1.7854)
Likability	5.5000 (1.7838)	5.1667 (1.5275)	6.5000 (1.6330)	6.5833 (2.0671)

dealer was significantly affected by the kind MCID ($F(1, 48) = 6.282$, $p < .05$). No statistically significant effects were found in the intention of further consultation, while the likability of the dealer was strongly influenced by the kind MCID ($F(1, 48) = 9.631$, $p < .01$). We did not find any interaction effects between the scripts and MCID contents throughout the assessed categories.

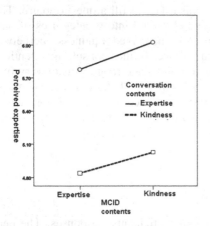 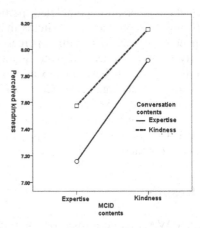

Fig. 3. Impressions of the car dealer formed by the MCID and conversation contents. The left and right figure illustrates the perceived expertise and kindness, respectively.

5 Discussion

One objective of the present study was to demonstrate possible roles of MCID in forming impression and its contribution to telemarketing activities. We found some interesting aspects from the experiment that are worth to discuss here.

In general, the perceived expertise and quality of automobile were mainly influenced by the expert script, while the perceived kindness and quality of service were mainly influenced by the kind MCID. These results indicate that some attributes of person perception can be easily associated with certain properties of sales items. For instance, kindness can be easily associated with the perceived quality of service, rather than the perceived quality of the automobile. This may imply that the desirable impression of a salesperson relative to sales items can be strategically chosen in order to persuade his/her costumers.

There seem to be effective medium to present the desirable personal impression. The expertise of a salesperson cannot be effectively conveyed by the visual MCID contents, rather can be effectively conveyed by verbal conversation on mobile phone. In contrast, the kindness of a salesperson can be easily perceived by the visual MCID contents. These differences may result from the expressivities of different types of contents. Since the expertise of a salesperson is related to the knowledge of sales products, this personal attribute cannot be effectively presented in the MCID contents that are displayed in the very limited time duration of ringing. For the kindness of a salesperson, on the other hand, a smiling face has strong effect on forming the impression with the visual MCID contents.

It is also interesting that the participants tended positively to rate on the 6 categories when the kindness is visually presented and the expertise is verbally presented. This may imply that there is a certain combination of cues that effectively appeals to the participants' perception. Participants perceived that the salesperson was not only kind but also intelligent.

6 Conclusion

The results of the present study suggest that the usage of different types of cues impact how impression is shaped and how effectively sale items are presented. The results have practical implication for designing MCID contents as well as the strategic presentation of information. Furthermore, since no previous research has examined the role of MCID in impression formation, this research can serve as a milestone for more future studies.

Acknowledgments. This research was supported in part by MKE and MEST, Korean government, under ITRC NIPA-2010-(C1090-1021-0008), FTDP(2010-0020727) and PRCP(2010-0020210) through NRF, respectively.

References

1. Website of the company that develops MCID technology, http://www.mtelo.com/
2. Washburn, P.V., Hakel, M.D.: Visual and verbal content as influences on impression formed after simulated employment interview. Journal of Applied Psychology 58(1), 137–141 (1973)
3. Berry, D.S.: Accuracy in social perception: Contributions of facial and vocal information. Journal of Personality and Social Psychology 61, 298–307 (1991)
4. Naylor, R.W.: Nonverbal cues-based first impressions: Impression formation through exposure to static images. Marketing Letters 18, 165–179 (2007)
5. Rashotte, L.S.: Written versus visual stimuli in the study of impression formation. Social Science Research 32, 278–293 (2003)
6. Carlson, J.R., George, J.F., Burgoon, J.K., Adkins, M., White, C.H.: Deception in computer mediated communication. Academy of Management Journal 13, 5–28 (2004)
7. Ambady, N., Krabbenhoft, M.A., Hogan, D.: The 30-Sec. Sale: Using Thin Slices to Evaluate Sales Effectiveness. Journal of Consumer Psychology 16(1), 4–13 (2006)
8. Ambady, N., Rosenthal, R.: Half a minute: Predicting teacher evaluations from thin slices of nonverbal behavior and physical attractiveness. Journal of Personality and Social Psychology 64, 431–441 (1993)
9. Hari, J., Stros, M., Marriott, J.: The 30-Second-Sale: Snap impressions of a retail sales person influence consumers decision making. Anspruchsgruppenorientierte Kommunikation 1, 53–66 (2008)
10. Borkenau, P., Mauer, N., Riemann, R., Spinath, F.M., Angleitner, A.: Thin Slices of Behavior as Cues of Personality and Intelligence. Journal of Personality and Social Psychology 86, 599–614 (2004)
11. Willis, J., Todorov, A.: First impressions: Making up your mind after 100 ms exposure to a face. Psychological Science 17, 592–598 (2006)

12. Berry, D.S.: Accuracy in social perception: Contributions of facial and vocal information. Journal of Personality and Social Psychology 61, 298–307 (1991)
13. Toma, C.L.: Perceptions of trustworthiness online: the role of visual and textual information. In: 2010 ACM Conference on Computer Supported Cooperative Work, pp. 13–22. ACM Press, New York (2010)
14. Wang, S.S., Moon, S.-I., Kwon, K.H., Evans, C.A., Stefanone, M.A.: Face off: Implications of visual cues on initiating friendship on Facebook. Computers in Human Behavior 26(2), 226–234 (2010)
15. Hancock, J.T., Dunham, P.J.: Impression formation in computermediated communication revisited: An analysis of the breadth and intensity of impressions. Communication Research 28, 325–347 (2001)
16. DeShields Jr., O.W., Kara, A., Kaynak, E.: Source effects in purchase decisions: The impact of physical attractiveness and accent of salesperson. International Journal of Research in Marketing 13(1), 89 (1996)
17. Grandey, A.A., Fisk, G.M., Mattila, A.S., Jansen, K.J., Sideman, L.A.: Is "service with a smile" enough? Authenticity of positive displays during service encounters. Organizational Behavior and Human Decision Processes 96, 38–55 (2005)

Part IV

Intelligent Management and e-Business

Application of Fuzzy Feedback Control for Warranty Claim

Sang-Hyun Lee[1], Sang-Joon Lee[2], and Kyung-Il Moon[3]

[1] Department of Computer Engineering, Mokpo National University, Korea
[2] School of Business Administration, Chonnam National University,
Yongbong-dong, Buk-gu Gwangju, Korea
s-lee@chonnam.ac.kr
[3] Department of Computer Engineering, Honam University, Korea

Abstract. Many companies analyze field data to enhance the quality and reliability of their products and service. In many cases, it would be too costly and difficult to control their actions. The purpose of this paper is to propose a model that captures fuzzy events, to determine the optimal warning/detection of warranty claims data. The model considers fuzzy proportional-integral-derivative (PID) control actions in the warranty time series. This paper transforms the reliability of a traditional warranty data set to a fuzzy reliability set that models a problem. The optimality of the model is explored using classical optimal theory; also, a numerical example is presented to describe how to find an optimal warranty policy. This paper proves that the fuzzy feedback control for warranty claim can be used to determine a reasonable warning/detection degree in the warranty claims system. The model is useful for companies in deciding what the maintenance strategy and warranty period should be for a large warranty database.

Keywords: Fuzzy reasoning; Fuzzy Control; PID tuning; Warranty claims data; Time series.

1 Introduction

In marketing durable products, a warranty is an important instrument used to signal higher product quality and greater assurance to customers. A warranty is defined as a contractual obligation of a manufacturer in selling a product, to ensure that the product functions properly during the warranty period. Manufacturers analyze field reliability data to enhance the quality and reliability of their products and to improve customer satisfaction. In many cases, it would be too costly or otherwise unfeasible to continue an experiment until a reasonably significant number of products have failed, but there are many sources from which to collect reliability-related data.

Suzuki et al. (2001) note that the main purposes and uses of warranty claim data are early warning/detection, understanding relationships within warranty claims data, determining warranty degree, comparing reliability among products, constructing a warranty claims database, and predicting future warranty claims and costs[13]. The important aspects of a warranty and the concepts and problems inherent in warranty

N.T. Nguyen et al. (Eds.): New Challenges for Intelligent Information, SCI 351, pp. 279–288.
springerlink.com   © Springer-Verlag Berlin Heidelberg 2011

analysis are discussed[4,6,10,11]. In exploring the optimality of a warranty policy, many issues must be considered, such as the renewed period of warranty[4], the cost structure for replacement[2], repair[14], periodic replacement of shock[3], optimal time for repair[12], optimal time of repair or replacement[7], minor and catastrophic failure[3], optimal maintenance[1] and associations among price, quality, and warranty period[9]. Some of these studies assume that the product failure process follows a non-homogeneous procession[8].

Since randomness is not merely an aspect of uncertainty in many fields of application, the fuzziness of the environment cannot be neglected in modeling an observed process. It is difficult to capture warranty claims data vis-à-vis reliability in polluted and imprecise situations—especially for new and durable products, non-mass products, and products with short development times. Usually, there is no comparative reliability information available, and warranty claims data tend to be based on subjective evaluation or rough estimate. To mitigate such problems, the modeling of reliability distribution must be based on the fuzziness of warranty claims data.

It is desirable to incorporate fuzzy theory into warranty claims data distribution for the warranty model. In this paper, we propose a model that captures fuzzy events, to determine the optimal warning/detection of warranty claims data. The model considers fuzzy proportional-integral-derivative (PID) control actions in the warranty time series.

2 PID Tuning and Fuzzy Reasoning

In general, outlier detection is based on the UCL(Upper Control Limit) by using a Shewart control chart. However, the control chart has some probabilistic restrictions, such as unexpectedly out-of-bounds points, even though it is in a steady state. In particular, most of these problems are bound up with instant errors. Instant errors with respect to warranty claims data can be obtained by using a weighted moving average, which is defined as some differences between the standard nonconforming fraction and types of UCL.

A PID controller attempts to correct the error between a measured process variable and a desired set point by performing calculations and then outputting a corrective action by which one can adjust the process as needed. The PID controller algorithm involves three separate parameters: the Proportional, Integral, and Derivative values. The Proportional value determines the reaction to the current error, the Integral value determines the reaction based on the sum of recent errors, and the Derivative value determines the reaction to the rate at which the error has been changing. By "tuning" the three constants in the PID controller algorithm, the PID controller can provide a control action designed for specific process requirements. Now, we suggest a PID control method that uses instant errors.

There are several methods for tuning a PID loop. The most effective methods generally involve the development of some form of process model when choosing Proportional, Integral, and Derivative values based on the dynamic model parameters. Manual "tune by feel" methods can be inefficient. The choice of method will depend largely on whether or not the loop can be taken "offline" for tuning, and the response time of the system. If the system can be taken offline, the best tuning method often involves subjecting the system to a step-change in input, measuring the output as a function of time, and using this response to determine the control parameters. In this section, PID tuning is used to obtain some parameters K_p, K_i, and K_d, which determine the reaction thresholds corresponding to the Proportional, Integral, and Derivative values, respectively. First, a step-response method can be used as the basic PID tuning principle (see Table 1, Fig. 1). Draw a tangent line to the rising curve, and find L, T, and K, which corresponds to 63% of the steady-state value. Usually, the PID controls are obtained from these values as follows.

Table 1. PID control

	K_p	K_i	K_d
P control	0.3~0.7 T/KL		
PI control	0.37~0.6 T/KL	0.3~0.6 T/KL	
PID control	0.6~0.95 T/KL	0.6~0.7 T/KL	0.3~0.45 T/KL

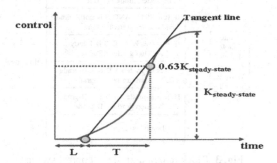

Fig. 1. Derivation of PID parameters

Rules-based logic has been used to capture human expertise in classification, assessment, diagnostic, and planning tasks, and probability has traditionally been used to capture decision-making under uncertain conditions. A number of methods have been used to propagate probabilities during rules-based inference. The weakness of such techniques is that they do not reflect the way human experts reason under uncertainty. Probabilistic reasoning is concerned with uncertain reasoning concerning well-defined events or concepts. On the other hand, fuzzy logic is concerned with reasoning vis-à-vis "fuzzy" events or concepts. Problems above the threshold or with respect to alert grade identification in our warranty system are related to fuzzy concepts. When humans reason with terms such as "alert grade," they do not normally have a fixed threshold in mind, but they do have a smooth fuzzy definition. Humans can reason very effectively with such fuzzy definitions; therefore, in order to capture human fuzzy reasoning, we need fuzzy logic.

We can produce the warning degrees of the warranty time series by using a fuzzy reasoning method (see Fig. 2).

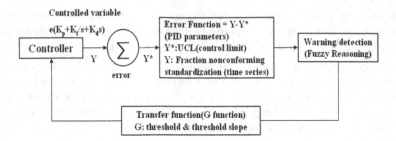

Fig. 2. Fuzzy feedback control of warranty claim time series

This fuzzy reasoning can be interpreted as an application of the fuzzy logic matrix concerning the probability that comprises the self-threshold and the relative threshold. Now, consider the design of a fuzzy controller for the threshold terms; the block diagram of this control system appears in Fig. 3.

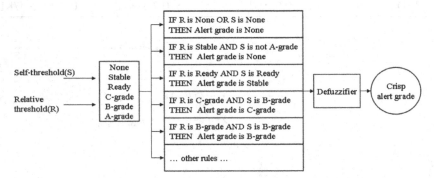

Fig. 3. Block diagram of fuzzy control system

There are two input variables—self-threshold and relative threshold—and a single output variable, the fuzzy warning degree. The fuzzy set mappings are shown in Fig. 4.

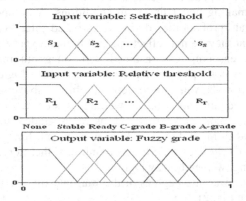

Fig. 4. Fuzzy set mappings

The most common shape of membership functions is triangular—although trape-zoids and bell curves are also used—though the shape is generally less important than the number of curves and their placement. Three to seven curves are generally appropriate for covering the required range of an input value—or the "universe of discourse," in fuzzy jargon. The truth values of the input variables are obtained from the probability through the normalization of the self-threshold and relative threshold. The descriptive statistics of the threshold terms are needed to obtain the normal probability distribution. These statistics can be computed with a P-bar control chart. If the rule specifies an AND relationship between the mappings of the two input variables, the product of the two is used as the combined truth value. The fuzzy warning degrees can be defined as None, Stable, Ready, C-grade, B-grade, and A-grade. Then, the rule set is provided in Table 2.

Table 2. Fuzzy rules

R \ S	S0	S1	S2	S3	S4	S5
R0	None	None	None	None	None	None
R1	None	None	None	None	None	C-grade
R2	None	None	Stable	Ready	C-grade	C-grade
R3	None	Stable	Ready	C-grade	C-grade	B-grade
R4	None	Ready	C-grade	C-grade	B-grade	A-grade
R5	None	C-grade	C-grade	B-grade	A-grade	A-grade

In practice, the controller accepts the inputs and maps them into their membership functions and truth values. These mappings are then fed into the rules. The appropriate output state is selected and assigned a membership value at the truth level of the premise. The truth values are then defuzzified. The results of all the rules that have been fired are defuzzified to a crisp value by one of several methods. There are dozens in theory, with each having various advantages and drawbacks. The "centroid" method is very popular, in which the "center of mass" of the result provides the crisp value; another approach is the "height" method, which takes the value of the largest contributor. The centroid method favors the rule with the output of the greatest area, while the height method obviously favors the rule with the greatest output value.

3 Application of Automobile Warranty Claims

This section introduces practical cases of applying the proposed fuzzy feedback control to the warranty claims data from the automobile industry. To determine the warning/detection degree of warranty claims data, this study uses warranty claim data arranged for reporting to a general company.

3.1 Warranty Claims Time-Series Analysis

Fig. 5 denotes the warranty time-series analysis; the ratio of the warranty claims is calculated as a percentage of the warranty claim counts, divided by the production-sale counts and then multiplied by the cumulative hazard function. In this analysis, it can be seen that C per 100 rises rapidly around April 26, 2004. Thus, a warning time would be about April 26, and early warning/detection of the warranty claims can be attained for more than one month, compared to C per 100.

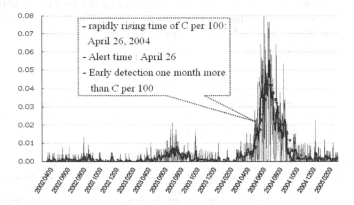

Fig. 5. Warranty time-series analysis of A-Car part Generator

3.2 Shape Parameter Estimation of Operation Group

The shape parameter can be obtained through an eight-operation code group per car type (see Fig. 6). It is compared to the individual operation code by estimating the shape parameter value per group. In short, it is difficult to warranty a time-series identification, if it is adjusted by estimating the shape parameter of an individual operation code. It is the same with some reverse corrections.

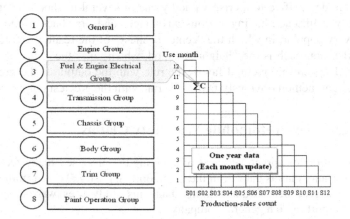

Fig. 6. Operation group category and data matrix

The warranty claims are accumulated through the use months by multiplying production-sales count by the corresponding count, thus bringing together the corresponding count of the operation code within each group. Since the error occurs in a case where the total claims in one month of use is zero, two months of use is considered the starting point. Table 3 represents the statistical analysis table for the shape parameter estimation in connection with Fig. 6.

Table 3. Reliability analysis table

Use month	Production-sales counts	Warranty claims	Failure rate	Cumulative Failure rate	R(t)	X=ln(t)	Y=ln(ln(1/Rt))
1	11,358,600	341	0.00600426	0.00600426	0.99399574	0	-5.112276212
2	10,030,200	599	0.01194393	0.01794819	0.98205181	0.693147181	-4.011224042
3	9,161,200	801	0.01748679	0.03543498	0.96456502	1.098612289	-3.322070954
4	7,776,400	757	0.01946916	0.05490415	0.94509585	1.386294361	-2.874064819
5	6,739,000	746	0.02213978	0.07704393	0.92295607	1.609437912	-2.523560505
6	5,762,200	891	0.03092569	0.10796962	0.89203038	1.791759469	-2.169321739
7	4,925,000	1,067	0.04332995	0.15129957	0.84870043	1.945910149	-1.807590109
8	4,111,400	1,407	0.06844384	0.21974341	0.78025659	2.079441542	-1.393792619
9	3,574,600	1,086	0.06076204	0.28050545	0.71949455	2.197224577	-1.111070595
10	2,921,000	1,054	0.07216707	0.35267251	0.64732749	2.302585093	-0.832632369
11	1,901,400	134	0.01409488	0.36676739	0.63323261	2.397895273	-0.783252527
12	1,044,600	787	0.15067969	0.51744708	0.48255292	2.484906650	-0.316541623

3.3 PID Tuning and Fuzzy Reasoning

An instant error can be produced by the weighted moving average over seven days, in terms of the standardized fraction nonconforming and the difference between the control type UCL and the analyzing type UCL. A PID tuning function of the instant error can be obtained by approximating, as a third-polynomial equation, with seven days of data. The Gauss-Jordan method is used for curve-fitting; each parameter value is managed according to part significance per each car type, and overshooting and lacks of response are updated through feedback from the system manager. Now, consider the fuzzy rules in Table 2, where R and S are associated with the fuzzy values None, Stable, Ready, C-grade, B-grade and A-grade, with the triangular membership functions shown in Fig. 7.

For given inputs x1 and x2, the non-fuzzy output u is computed using the centroid method. The input surface generated by the self-threshold and the relative threshold is shown in Fig. 8. Crisp alert grades based on finer partitions of the output u (such as 12 grades) can be considered under the assumption of exponential distribution as the following.

If u < 0.4, then warranty time series grade = None
Else if u > 0.4 and u < 0.45, then warranty time series grade = Stable
Else if u >= 0.45 and u < 0.5, then warranty time series grade = Ready_C
Else if u >= 0.5 and u < 0.55, then warranty time series grade = Ready_B
Else if u >= 0.55 and u < 0.6, then warranty time series grade = Ready_A
Else if u >= 0.6 and u < 0.7, then warranty time series grade = C

Else if u >= 0.7 and u < 0.8, then warranty time series grade = CC
Else if u >= 0.8 and u < 0.85, then warranty time series grade = B
Else if u >= 0.85 and u < 0.9, then warranty time series grade = BB
Else if u >= 0.9 and u < 0.95, then warranty time series grade = A
Else if u >= 0.95 and u < 0.99, then warranty time series grade = AA
Else if u > 0.99, then warranty time series grade = S

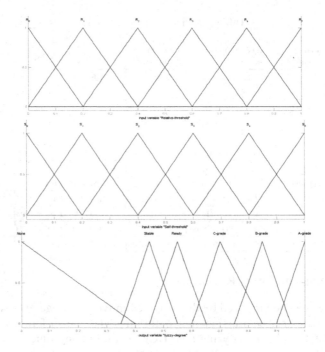

Fig. 7. Membership functions

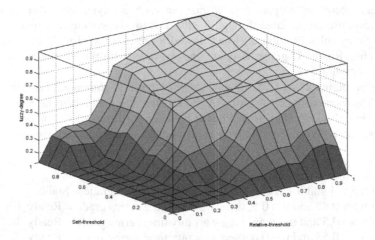

Fig. 8. Surface based on 36 rules

The threshold and alert grade module (Fig. 9) can be used to obtain the history values of warranty time-series detection. In the "Probability Method Selection" menu, the selection of "Regression" is related to finding a probability distribution through the mean, and standard estimation by regression method. "ΣCount" and "First Detection Date" denote the detection count per alert degree and the first detection date in connection with each alert degree, respectively. "Current Grade & Risk Information" denotes some current analysis values of time-series detection. Alert degree is nil until now, and risk is 32.16%. Also, the self-threshold is 0.08, and the relative threshold is 0.32. In Fig. 18, the yellow-green graph denotes the standardized fraction nonconforming, the yellow graph the Bollinger curve, the red graph the control type chart, and the pink graph the analyzing type chart.

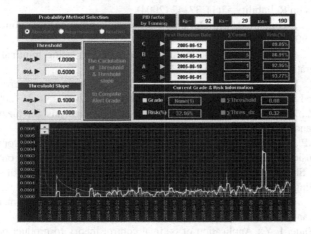

Fig. 9. Threshold and alert grade module

4 Conclusion

Today, a warranty system is very important to enhancing a product's sales. An appropriate warranty period and the cost of this warranty should be determined, based on the precise identification of warranty claims data. This study proposed a methodology for creating a flexible formula for warning degrees, as determined in a fuzzy environment. Although the fuzzy control model adopted here is more complex than the classical model, the evidence shows that the classical model is inadequate in a fiercely competitive environment. There are several reasons for this. In reliability analysis, one must collect a relatively large amount of lifetime data, because the classical reliability estimation is typically based on precise lifetime data; however, with new industrial technologies, demanding customers, and shorter product development cycles, the lifetimes of products have become ambiguous. It is time-consuming, expensive, and sometimes impossible to obtain enough exact observations to suit the lifetime distribution—and with few available data points, it is difficult to estimate the lifetime distribution parameters using conventional reliability analysis methods. Hence, to enhance the success of marketing a new product, fuzzy theory or other theories should be implemented, by capturing experience, subject judgment, and available lifetime data to fit the reliability distribution in a faster way. However, in order to formulate a warranty model, companies still need to collect relevant data for all parameters of the model. The cost and time needed to collect such data must be estimated before considering warranty models. This study is thus useful to companies in deciding what their maintenance strategies and warranty periods should be; it also allows for an extended warranty price to be derived, if the cost elasticity function is available.

References

1. Bai, J., Pham, H.: Discounted warranty cost of minimally repaired series systems. IEEE Transactions on Reliability 53(1), 37–42 (2004)
2. Barlow, R.E., Hunter, L.C.: Optimum preventive maintenance policies. Operation Research 8, 90–100 (1960)
3. Boland, P.J., Proschan, F.: Periodic replacement with increasing minimal repair costs at failure. Operations Research 30(6), 1183–1189 (1982)
4. Chukova, S., Hayakawa, Y.: Warranty cost analysis: Non-renewing warranty with repair time. Applied Stochastic Models in Business and Industry 20(1), 59–71 (2004)
5. Huang, H.Z., Zuo, M.J., Sun, Z.Q.: Bayesian reliability analysis for fuzzy lifetime data. Fuzzy Sets and Systems 157(12), 1674–1686 (2006)
6. Iskandar, B.P., Murthy, D.N.P., Jack, N.: A new repair-replace strategy for items sold with a two-dimensional warranty. Computers and Operations Research 32(3), 669–682 (2005)
7. Jhang, J.P., Sheu, S.H.: Opportunity-based age replacement policy with minimal repair. Reliability Engineering and System Safety 64(3), 339–344 (1995)
8. Juang, M.G., Anderson, G.: A Bayesian method on adaptive preventive maintenance problem. European Journal of Operational Research 155(2), 455–473 (2004)
9. Lin, P.C., Shue, L.Y.: Application of optimal control theory to product pricing and warranty with free replacement under the influence of basic lifetime distributions. Computers and Industrial Engineering 48(1), 69–82 (2005)
10. Murthy, D.N.P., Blischke, W.R.: Warranty and reliability. In: Balakrishnan, N., Rao, C.R. (eds.) Handbook of Statistics: Advances in Reliability, pp. 541–583. Elsevier Science, Amsterdam (2001)
11. Murthy, D.N.P., Djamaludin, I.: New product warranty: A literature review. International Journal of Production Economics 79(3), 231–260 (2002)
12. Sheu, S.H.: A generalized model for determining optimal number of minimal repairs before replacement. European Journal of Operational Research 69(1), 38–49 (1993)
13. Suzuki, M.R., Karim, K., Wang, L.: Statistical analysis of reliability warranty data. In: Balakrishnan, N., Rao, C.R. (eds.) Handbook of Statistics: Advances in Reliability, pp. 585–609. Elsevier Science, Amsterdam (2001)
14. Tilquin, C., Cléroux, R.: Periodic replacement with minimal repair at failure and general cost function. Journal of Statistical Computation Simulation 4(1), 63–67 (1975)

Exploring Organizational Citizenship Behavior in Virtual Communities

Sung Yul Ryoo[1] and Soo Hwan Kwak[2,*]

[1] Graduate School of Business, Sogang University, Seoul, Korea
syryoo@sogang.ac.kr
[2] Department of Business Administration, Silla University, Pusan, Korea
shkwak@silla.ac.kr

Abstract. Virtual communities have been drastically emerging with the development of internet technology. However, although organizational citizenship behavior is regarded as a key factor of the sustainability of virtual communities, little is known about the antecedents of OCB in the context of virtual communities. Thus, this study investigates the relationships between user evaluations and user involvement. Also, this study examines the relationships between user involvement and organizational citizenship behavior. The findings indicate user values such as utilitarian, hedonic, and social value affect member's virtual community involvement, and then virtual community involvement has an effect on members' OCB in virtual communities.

Keywords: virtual communities, organizational citizenship behavior, user evaluations, user involvement.

1 Introduction

Virtual communities are considered as great opportunities for businesses to earn economic returns [1]. A virtual community means "*a group of people with similar interests who interact with one another using the internet*" [2]. Although the predominant purposes of virtual communities are information and knowledge exchange [3], virtual communities can be classified as transaction and other business, purpose or interest, relations or practices, fantasy, and social networks [2].

According to a review of the literature on virtual communities, much research focused on the factors facilitating people join a virtual community and the factors motivating members' active participation, in particular, such as the knowledge contribution behavior of users [3-5].

However, virtual communities are unlikely to be sustainable unless properly stimulated [6]. The success of virtual communities can be achieved when many members are willing to stay and behave spontaneously regarding virtual communities [7]. In the context of virtual communities, members' spontaneity refers to organizational

---

* Corresponding author.

N.T. Nguyen et al. (Eds.): New Challenges for Intelligent Information, SCI 351, pp. 289–298.
springerlink.com   © Springer-Verlag Berlin Heidelberg 2011

citizenship behavior (OCB) [8]. OCB is regarded as *"behavior that contributes to the maintenance and enhancement of the social and psychological context that supports task performance"* [9].

Managing virtual communities effectively is important in gaining competitive advantage and sustainability of virtual communities. Thus, realizing this promise depends on firm's ability to develop and maintain communities in which individuals can have both the opportunity and the motivation to participate and contribute. [6]. OCB may be critical for sustainable virtual communities. This factor strengthens the relational ties among community members.

Although much previous research provides a few insights into sustainability of virtual communities, few research on OCB has been done to study the consequences of OCB in IS adoption context [10-11]. However, little is known about the antecedents of OCB in the context of virtual communities.

This study aims to explore the factors that stimulate OCB in virtual communities. For this purpose, this study incorporates OCB theory with IS literature. The research model of this study was developed to empirically examine the relationships between user evaluations about using virtual communities and virtual community involvement and the links between virtual community involvement and OCB.

The following section of this paper reviews the antecedents of user involvement and its impacts on OCB and then develops research model and hypotheses, based on relevant theories of virtual communities and OCB theory. The third section presents research methodology. The fourth section describes research method and empirical results. We conclude with a discussion of the results and their implications.

2 Theoretical Background and Hypotheses

2.1 The Values of Virtual Communities

Virtual communities have been drastically emerging with the development of internet technology. Much research identified several reasons why users are willing to use virtual communities. Based on Cheung and Lee [4], these factors are classified into five types: purposive value, self-discovery, entertainment value, social enhancement, and maintaining interpersonal interconnectivity.

Meanwhile, in recent, Kim et al. [12] suggests that user's evaluation criteria to IT-related service use has three dimensions: utilitarian value, hedonic value, and social value. Focusing on automatic use of IT services that are used on a daily basis such as virtual communities, they explain that these three values have impacts on IT usage through usage intention. As most users of virtual communities use virtual communities on a daily basis, we adopt this classification of values of using virtual communities.

Based on social identity theory [13], people begin to explore several alternatives and compare them until involvement can be developed. After establishing personal and social identity, people continue to communicate their identity and reinforce the current identity. In this process, if groups which people participate in fulfill the needs of their members well, the members may have a high involvement [4]. For example, Lin [14] provide an empirical support that social value of virtual communities have a positive impact on members' involvement. Also, Cheung and Lee [4] provide that hedonic value have a significant effect on members' involvement.

Hence, we hypothesize:

H1: *The utilitarian value of virtual communities will positively influence user involvement.*

H2: *The hedonic value of virtual communities will positively influence user involvement.*

H3: *The social value of virtual communities will positively influence user involvement.*

2.2 Organizational Citizenship Behavior in Virtual Communities

Armstrong and Hagel Ⅲ [1] suggest that the potential economic value of the virtual communities stem from the number of participants, frequent use, interactions among users, and user's willingness to engage in virtual communities. However, these elements need not only the support of service firm but also user's voluntary behavior to enlarge his communities and improve the quality of his communities.

These voluntary behaviors can be explained in terms of organizational citizenship behavior (OCB). In the context of virtual communities, OCB, based on Barnard's definition [15], can be defined as *"willingness of participants to contribute efforts to the cooperative virtual communities."*

Although several IS researcher adopted organizational citizenship behavior theory to enhance IS success model [10-11], few studies have been devoted to an empirical work in the context of virtual communities. Yu and Chu [16] examined the antecedents of organizational citizenship behavior in online gaming communities. Also, Shin and Kim [8] investigated the relationship between organizational citizenship behavior and knowledge contribution in virtual communities such as discussion forum.

Meanwhile, user involvement in virtual communities is regarded as the key factor of motivating members' participation [17]. And then, member's involvement will have a favorable impact on OCB in virtual communities [18].

Therefore, we hypothesize:

H4: *User involvement will positively influence user's organizational citizenship behavior in virtual communities.*

A conceptual research model has been developed as shown in Figure 1. The model and several hypotheses are developed from prior research.

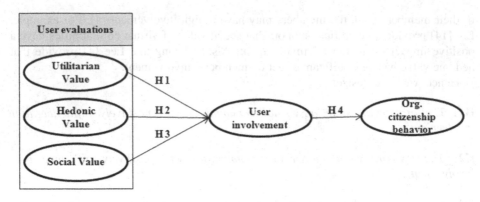

Fig. 1. Research Model

3 Research Methodology

3.1 Data Collection

A total of 567 students responded to the survey between June 1st and 30th, 2010. Responses with missing data were excluded. The final sample to test the proposed hypothesis consists of 534 data. Table 1 shows the demographical characteristics of the sample. About 49.8% of the respondents were males and the rests were females. Most of respondents (51.7%) spend 1 to 8 hours per week in virtual communities.

Table 1. The characteristics of sample

Items	Categories	Frequency	Ratio (%)
Gender	Female	268	50.2
	Male	266	49.8
	Total	534	100.0
Age	20-29	497	93.1
	30-39	23	4.3
	40-49	7	1.3
	over 50	7	1.3
Usage time/week	below 1 hour	94	17.6
	1-8	276	51.7
	8-16	100	18.7
	16-24	35	6.6
	24-32	13	2.4
	32-40	10	1.9
	over 40	6	1.1
	Total	534	100.0

3.2 Measures

To assure a survey questionnaire, we adopted existing and previously validated scales and modified those to the research context of this study. Table 2 delineates the operational definitions of the research constructs and their sources. All items have a seven-point Likert scale, from 1=strongly disagree to 7=strongly agree.

Table 2. The operational definitions and sources of measurement

Variables	Operational definitions	Sources
Utilitarian value	Utilitarian value refers to the effectiveness and efficiency that result from the use of virtual communities.	[12]
Hedonic value	Hedonic value refers to the fun or pleasure derived from the use of virtual communities.	[12]
Social value	Social value refers to enhancement of a user's social image by the use of virtual communities.	[12]
VC involvement	VC involvement refers to the level of members' attachment and sense of belonging to the particular virtual communities.	[3]
Organizational citizenship behavior	Organizational citizenship behavior refers to the level of spontaneity of members' behaviors that contribute to the particular virtual communities.	[8]

4 Data Analysis and Results

We used structural equation modeling called Partial Least Squares (PLS) to analysis the research model. PLS is based on an iterative combination of principal components analyses and regression, and it aims to explain the variance of the constructs in the model [19]. Specifically, SmartPLS 2.0 was used for the analysis.

4.1 Measurement Model

We first assessed the measurement model for reliability, convergent validity, and discriminant validity. Table 3 shows that the reliability of all the measurement scales is good. All composite reliabilities (CR) are greater than 0.9, which is very higher than the recommended cut-off (0.7). As the results in Table 3 and 4 show, convergent validity was evaluated by using the rule of thumb of accepting items with individual item loadings of 0.50 or above, composite reliability (CR) of 0.70 or above, and average variance extracted (AVE) of 0.50 or above [20]. Based on the results in Table 3 and 4, all constructs met the 0.70 CR and 0.50 AVE criterions, supporting convergent validity.

Table 3. The result of convergent validity

Construct	number of items	composite reliability	AVE
Utilitarian value	3	0.90	0.75
Hedonic value	4	0.95	0.84
Social value	3	0.91	0.77
VC involvement	4	0.93	0.76
Organizational citizenship behavior	6	0.93	0.70

Table 4. Factor structure matrix of loadings and cross-loadings

	Util	Hed	Soc	VC	OCB
Util1	0.77	0.47	0.38	0.42	0.30
Util2	0.91	0.59	0.47	0.59	0.36
Util3	0.91	0.61	0.48	0.58	0.38
Hed1	0.62	0.89	0.45	0.61	0.37
Hed2	0.59	0.91	0.53	0.70	0.38
Hed3	0.60	0.94	0.51	0.69	0.39
Hed4	0.58	0.92	0.52	0.68	0.41
Soc1	0.45	0.46	0.85	0.65	0.45
Soc2	0.46	0.51	0.91	0.66	0.47
Soc3	0.46	0.48	0.86	0.62	0.40
VC1	0.59	0.71	0.67	0.89	0.48
VC2	0.43	0.50	0.59	0.80	0.44
VC3	0.57	0.73	0.63	0.92	0.51
VC4	0.57	0.59	0.66	0.87	0.51
OCB1	0.29	0.29	0.40	0.43	0.78
OCB2	0.36	0.35	0.48	0.50	0.85
OCB3	0.31	0.35	0.45	0.49	0.89
OCB4	0.37	0.41	0.39	0.49	0.79
OCB5	0.37	0.38	0.39	0.47	0.88
OCB6	0.32	0.34	0.41	0.44	0.86

Discriminant validity was assessed in two ways. First, we examined the cross loadings showed that no item loads more highly on another construct than its own construct (see Table 4). Second, we compared the square root of AVEs from each

construct with its correlations with the other constructs as a test of discriminant validity (shown on the Table 5 diagonal). Based on the results in Table 4 and 5, all constructs passes both discriminant validity tests [20].

Table 5. The correlation of latent variables and the square roots of AVEs

Construct	Utilitar-ian value	Hedonic value	Social value	VC involvement	Organiza-tional citizenship behavior
Utilitarian value	**0.86**				
Hedonic value	0.65	**0.91**			
Social value	0.52	0.55	**0.88**		
VC involvement	0.62	0.73	0.73	**0.87**	
Organizational citizenship behavior	0.40	0.42	0.50	0.56	**0.84**

4.2 Structural Model

This study tests research hypotheses by examining the significance of structural paths
in the PLS analysis.

Table 6 and Figure 2 summarize the results of the PLS analyses. The results show
that the path from utilitarian value to virtual community involvement (H1) ($t = 1.48$;
$p<0.1$) was positive and significant. We also found a positive effect of hedonic value
on virtual community involvement (H2) ($t = 3.86$; $p<0.00$), and a positive influence of
social value on virtual community involvement (H3) ($t = 5.12$; $p<0.00$). As for Hypothesis 4, we found that, as predicted, organizational citizenship behavior in virtual
community was positively associated with virtual community involvement ($t = 5.71$;
$p<0.00$). Thus, all hypotheses one through four are supported.

Table 6. Path Analysis Results

		Original Sample (O)	T Statistics (|O/STERR|)	p-value	Supported
H1	Util -> VC	0.13*	1.48	0.069	yes
H2	Hed -> VC	0.41***	3.86	0.000	yes
H3	Soc -> VC	0.44***	5.12	0.000	yes
H4	VC -> OCB	0.56***	5.71	0.000	yes

(df=533), (1-tailed test), ***$p<0.01$, ** $p <0.05$, * $p <0.1$.

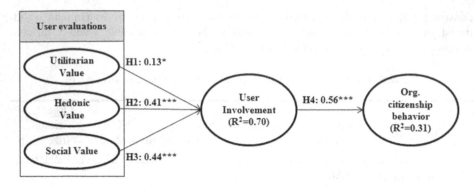

Fig. 2. Result of PLS Analysis

Further, following the approach suggested by Baron and Kenny [21], we examined the mediating effect of virtual community involvement on the relationships between user evaluations and organizational citizenship behavior.

Baron and Kenny recommended three conditions as follows: (1) the significant effects of the independent variable on the mediating variable, (2) the significant influence of the mediating variable on the dependent variable, and (3) when condition 1 and 2 are controlled, no significant relationships between the independent variable and the dependent variable (full mediation) or reduction in strength of significance (partial mediation).

We first tested the direct effect of user evaluations on organizational citizenship behavior. As a result, the social value had direct positive effect on organizational citizenship behavior ($t = 2.76$; $p<0.00$). However, utilitarian value and hedonic value ($t = 0.88$, *not significant*; $t = 1.23$, *not significant*) had not a significant link with organizational citizenship behavior. And then, in the mediation model, the positive effect of social value on organizational citizenship behavior shrinks (0.36 to 0.19). Taken together, these results suggest that the effect of social value on organizational citizenship behavior is partially mediated by virtual community involvement, while the effects of utilitarian and hedonic value on organizational citizenship behavior are fully mediated by virtual community involvement.

5   Discussion and Conclusion

5.1   Implications for Theory and Practice

The aim of this study is to develop a research model to provide an explanation of factors influencing organizational citizenship behavior in virtual communities. The results from this study support that the importance of user evaluations such as utilitarian, hedonic, and social value. Also, the findings suggest that virtual community involvement have important roles in explaining the relationship between user evaluations and organizational citizenship behavior in virtual communities. The presence of partially mediated effects also suggests that omission of virtual community involvement from the full research model could lead to an underestimation of the total effects of user evaluations on organizational citizenship behavior in virtual communities.

This study makes substantial contributions to theory and has some implications for practice. From the theoretical perspective, the results of this study support that the values of virtual communities such as utilitarian, hedonic, and social value affect members' virtual community involvement from automatic perspective. The findings of this study also highlight the predominant roles of virtual community involvement as the antecedents of OCB in virtual communities.

From the practical perspective, this study suggest that members' virtual community involvement have a positive significant mediating effect on the relationship between values of virtual communities and OCB in virtual communities. Therefore, VC managers need to increase the members' commitment or involvement, which leads to members' spontaneous OCB in virtual communities. To do this, VC managers need to provide various tools compatible with each member's purpose to maximize the values of virtual communities.

5.2 Limitations and Future Research

Although this study presents empirical evidence regarding the effects of user evaluations on organizational citizenship behavior in virtual communities, the findings should be interpreted in the context of its several potential limitations.

First, most respondents are skewed toward 20's and 30's person. Hence, the interpretation of our results is subject to the constraints of characteristics of various age groups. Second, there could be other variables being motivated by organizational citizenship behavior in virtual communities. While this study selected the antecedents of organizational citizenship behavior in virtual communities, there are various consequences such as knowledge contribution in virtual communities or members' continuance intentions.

References

1. Armstrong, A., Hagel III, J.: The Real Value of ON-LINE Communities. Harvard Business Review 74, 134–141 (1996)
2. Turban, E., King, D., Lang, J.: Introduction to Electronic Commerce. Pearson Education, Upper Saddle River (2009)
3. Zheng, J., Kim, H.-w.: Investigating Knowledge Contribution from the Online Identity Perspective. In: Twenty Eighth International Conference on Information Systems (ICIS), Montreal, Canada (2007)
4. Cheung, C.M.K., Lee, M.K.O.: Understanding the sustainability of a virtual community: model development and empirical test. Journal of Information Science 35, 279–298 (2009)
5. Chen, I.Y.L.: The factors influencing members' continuance intentions in professional virtual communities - a longitudinal study. Journal of Information Science 33, 451–467 (2007)
6. Koh, J., Kim, Y.G., Butler, B., Bock, G.-W.: Encouraging Participation in Virtual Communities. Communcations of the ACM 50, 69–73 (2007)
7. Bourhis, A., Dubé, L.: 'Structuring spontaneity': investigating the impact of management practices on the success of virtual communities of practice. Journal of Information Science 36, 175–193 (2010)

8. Shin, H.K., Kim, K.K.: Examining identity and organizational citizenship behaviour in computer-mediated communication. Journal of Information Science 36, 114–126 (2010)
9. Organ, D.W.: Organizational citizenship behavior: It's construct clean-up time. Human Performance 10, 85–97 (1997)
10. Yoon, C.: The effects of organizational citizenship behaviors on ERP system success. Computers in Human Behavior 25, 421–428 (2009)
11. Yen, H.R., Li, E.Y., Niehoff, B.P.: Do organizational citizenship behaviors lead to information system success?: Testing the mediation effects of integration climate and project management. Information & Management 45, 394–402 (2008)
12. Kim, S.S., Malhotra, N.K., Narasimhan, S.: Research Note–Two Competing Perspectives on Automatic Use: A Theoretical and Empirical Comparison. Information Systems Research 16, 418–432 (2005)
13. Hogg, M.A., Abrams, D.: Social Identifications: A Social Psychology of Intergroup Relations and Group Processes. Routledge, London (1988)
14. Lin, H.-F.: Determinants of successful virtual communities: Contributions from system characteristics and social factors. Information & Management 45, 522–527 (2008)
15. Barnard, C.I.: The functions of the executive. Harvard University Press, Cambridge (1938)
16. Yu, C.-P., Chu, T.-H.: Exploring knowledge contribution from an OCB perspective. Information & Management 44, 321–331 (2007)
17. Kang, I., Lee, K.C., Lee, S., Choi, J.: Investigation of online community voluntary behavior using cognitive map. Computers in Human Behavior 23, 111–126 (2007)
18. Cohen, A., Keren, D.: Individual Values and Social Exchange Variables. Group & Organization Management 33, 425–452 (2008)
19. Chin, W.W., Marcolin, B.L., Newsted, P.R.: A partial least squares latent variable modeling approach for measuring interaction effects: Results from a Monte Carlo simulation study and an electronic-mail emotion/adoption study. Information Systems Research 14, 189–217 (2003)
20. Gefen, D., Straub, D.W., Boudreau, M.-C.: Structural equation modelling and regression: Guidelines for research practice. Communications of the AIS 4, 1–79 (2000)
21. Baron, R.M., Kenny, D.A.: THE Moderator Mediator Variable Distinction in Social Psychological Research - Conceptual, Strategic, and Statistical Considerations. Journal of Personality and Social Psychology 51, 1173–1182 (1986)

Integration of Causal Map and Monte Carlo Simulation to Predict the Performance of the Korea e-Procurement System

Kun Chang Lee[1] and Namho Chung[2,*]

[1] Professor of MIS at SKK Business School
WCU Professor of Creativity Science at Department of Interaction Science
Sungkyunkwan University
Seoul 110-745, Republic of Korea
kunchanglee@gmail.com
[2] Associate Professor
College of Hotel & Tourism Management
Kyung Hee University
Seoul 130-701, Republic of Korea
Tel.: +82 2 9612353; Fax: +82 2 9642537
nhchung@khu.ac.kr

Abstract. Among e-Government systems, the Korea ON-line E-Procurement System (KONEPS) is unique in that it requires the automated completion of a wide variety of jobs across the entire procurement process: from exchanging documents, opening bids, contracting, and shopping and paying electronically to sharing information on goods, services, and participants. KONEPS is also a rare example of a successful government-administered e-procurement system that has saved a great deal of money and time in the Government-to-Business procurement process. Most e-procurement system studies focus on a rhetorical analysis of case studies and do not employ rigorous methods, making them unsuitable for assessment by potential customer countries. This study proposes a generalized approach to estimating the performance of e-procurement systems by integrating the Monte Carlo-Assisted Causal Mapping Simulation (MOCA-CAMS). Simulations in which MOCA-CAMS is applied to KONEPS indicate a very promising performance for KONEPS under uncertain conditions in customer countries. This study will provide a foundation for further discussion of this increasingly important area of e-procurement research and practice.

Keywords: e-Government; e-Procurement systems; Korea ON-line E-Procurement System (KONEPS); Monte Carlo simulation; Causal Mapping.

1 Introduction

Government functions have undergone a rapid transformation since the latter part of the 1990s, largely due to the impact of technologies that have enabled delivery of

---

* Corresponding author.

N.T. Nguyen et al. (Eds.): New Challenges for Intelligent Information, SCI 351, pp. 299–308.
springerlink.com   © Springer-Verlag Berlin Heidelberg 2011

services over the Internet. The private sector has taken great strides in utilizing these technologies; new service industries, better delivery of services and faster, cheaper communication are just a few of the by-products of this technological revolution. Perhaps inevitably, these private sector transformations have created a similar expectation for improved delivery of government services, but most governments have been very slow to effect such improvements.

This administrative focus has gradually shifted to a customer focus, however, providing citizens and trading partners alike with direct access to services, information, and transactions via the Internet. Implementation of "electronic government" (e-government) [1] systems has not been a simple task, however: many governments lack the fundamental infrastructure, organizational culture, understanding and resources necessary to make the transition to e-government, and those that do not lack them may have different expectations. Some countries may see the transformation to e-government as an excuse to establish basic infrastructure, while others see it as an opportunity to deliver innovative services to their citizens. The Republic of Korea is one of the latter.

Korea is committed to adopting new technologies and re-engineering processes that make government run more efficiently, and its national IT policies have created an environment in which a wide spectrum of e-government, e-procurement, and government-to-business (G2B) services thrive. This study measures the performance of one of these services, the Korean ON-line E-Procurement System (KONEPS).

In order to do this, a number of factors relevant to the system and the parties involved in the G2B transaction must be considered. In this case, a causal mapping method [2] has been adopted to systematically incorporate all of the relevant factors. In our approach, called MOCA-CAMS (Monte Carlo-Assisted Causal Mapping Simulation), when all of the relevant factors have been explicitly sorted and a set of possible causal relationships and causal coefficients (causalities) between the factors induced, uncertainties are dealt with using the Monte Carlo simulation [3, 4]. A wide spectrum of application situations are hypothetically generated versus competitors' systems, and statistically significant results that predict the performance of KONEPS in various uncertain situations are derived. This paper will present a review of emerging research on e-governments in Section 2, focusing on e-procurement systems, causal mapping, and Monte Carlo simulation. Section 3 describes the proposed MOCA-CAMS methodology as it is applied to KONEPS, and presents our results. Concluding remarks appear in Section 4.

2 Literature Review

2.1 e-Government

The rapid technological changes that have transformed the world economy and society have begun to transform governments, as well [1]. The impact of the Internet on governments has been tremendous: from static web sites to the direct provision of services to citizens, the transformation from government to e-government has demanded a change of focus from administration to service.

Discussions in the public administration literature range from micro issues of IT managers in the public sector [5] to macro issues of national IT infrastructure [6]. The government depository literature follows up with policy issues in specific sectors, and the early literature on e-government draws from these rich discussions; however, the emerging picture of e-governments is more complex than that of "IT in the public sector." Early discussions addressed how e-government relates to organizational issues, technology issues, business models, etc. Perhaps most e-government studies are confined to analyzing broad and generic case studies because e-government applications dictate the boundaries they use. To our knowledge, this study is the first to attempt to analyze the performance of KONEPS in various simulated situations, and to predict its performance after adoption by other countries.

2.2 e-Procurement and KONEPS

e-procurement systems are utilized for such tasks as exchanging documents, opening bids, contracting, shopping and paying electronically, and sharing information on goods, services, and participants. While business-to-business e-commerce has prospered, e-procurement by government, or "government-to-business" (G2B), has not progressed as much. A number of U.S. States have already reported successful experiences with e-procurement, such as Virginia's EVA system. Still, it is very hard to find good e-procurement models at the national or federal level. A study on Chilean e-procurement systems, ChileCompra, reported disappointing results in systems development and agency usage in recent years [7].

e-procurement systems in Korea have had significantly more success. The Public Procurement Service of Korea (PPS, http://www.pps.go.kr) launched KONEPS (http://www.g2b.go.kr) on September 30, 2002 in an effort to "establish a nationwide web-based procurement system, dealing with the whole procurement process, including acquisition of all information on the national procurement projects, procurement requests, bids, contracting, and payment for 27,000 public organizations and 90,000 private firms" [8]. Using KONEPS, most public organizations in Korea, from central and local government agencies to public enterprises, can purchase and contract with little more than a personal computer and an Internet connection. KONEPS has attracted favorable attention from many international and non-government organizations. PPS's effort to reform and develop e-procurement systems has been recognized with awards from other Korean government agencies, as well. The most recent report of the Organization for Economic Co-operation and Development [9] stated that KONEPS has "a strong pull-through effect on information and communications technology use in the private sector," with "no further action required."

2.3 Causal Mapping

This study proposes a new approach to predicting the performance of KONEPS when it is adopted for use in another country. We suggest using causal mapping to formulate the complicated causal relationships among relevant factors in KONEPS and decision makers' judgment(s), and applying the Monte Carlo simulation to derive robust inference results. Causal mapping helps to design a causal map (CM) for the target problem in which a set of relevant variables are organized in a diagram.

CMs have proven particularly useful for solving problems in which a number of decision variables and uncontrollable variables are causally interrelated. Since their introduction by Tolman [10], CMs have been extensively developed and modified to analyze social and political situations [11]. Recently, CMs have been used in way-finding, business redesign, knowledge management, the Bosporus crossing problem, and the design of electronic commerce web sites.

In this study, the CM is used as a simulation vehicle to predict the performance of KONEPS when it is adopted by other countries. Since the CM represents the causal relationships among the elements of a given object and/or problem, it is composed of (1) concept nodes that represent the factors describing a target problem; (2) arrows that indicate causal relationships between two concept nodes; and (3) causality coefficients on each arrow indicating a positive (or negative) strength with which a node affects other nodes. The CM allows a set of identified causality coefficients to be organized in an adjacency matrix, yielding a simulation based on it. Such simulations enable designers to identify the most relevant design factors for enhancing outcome variables, such as customer satisfaction. A more objective method is therefore required to quantify the causality coefficients and to help organize an adjacency matrix to perform a CM simulation. In this study, the CM describing the characteristics of KONEPS and its usage environment was derived from several rounds of interviews with KONEPS experts and researchers.

There are two primary advantages to CM simulation. First, activation of specific nodes can result in a chain of effects on the other nodes through positive or negative causal relationships defined in the CM until equilibrium is attained. Second, a variety of what-if sensitivity simulations can be performed effectively, according to the decision-maker's intent. Through these simulations, decision-makers can identify the relevant decision variables and their acceptable values, ensuring they get the intended results.

The CM allows experts to freely draw causal pictures of their problems. We view the CM as a dynamic system which settles down to a specific stable state after several iterations. The causal dynamic system represented by CM responds to external stimuli, and we interpret its equilibrium-seeking behavior as a CM-based simulation or inference. Similarly, we can apply this CM-based simulation approach to predicting the performance of KONEPS.

2.4 Monte Carlo Simulation

The Monte Carlo simulation is very effective when there is no empirical data about the target problem and a certain level of uncertainty exists for some decision variables. By generating random numbers for the uncertain variables in line with the appropriate probability distribution, the Monte Carlo simulation provides approximate solutions to a variety of mathematical problems. Since KONEPS is going to be adopted in countries other than Korea, customers may be concerned how effectively it will work, or they might worry about its quality and user-friendliness compared to competitors' systems. Moreover, customers from other countries have their own cultures and ethnic groups that are quite different from those in Korea. They may need to feel certain that KONEPS will work before they decide to adopt it as their primary e-procurement system. Therefore, the Monte Carlo simulation can and should be applied to all of the uncertainties embedded in the decision-making process for adopting KONEPS.

The Monte Carlo method was invented in the 1940s by scientists working on the atomic bomb, who named it for the city in Monaco famed for its casinos and games of chance. Its core idea is to use random samples of parameters or inputs to explore the behavior of a complex system or process to be solved. The scientists faced physics problems, such as models of neutron diffusion, which were too complex for an analytical solution, so they had to be evaluated numerically. They had access to one of the earliest computers (MANIAC), but their models involved so many dimensions that exhaustive numerical evaluation was prohibitively slow. The Monte Carlo simulation proved to be surprisingly effective at finding solutions to these problems. Since that time, Monte Carlo methods have been applied to an incredibly diverse range of problems in science, engineering, finance, and business applications in virtually every industry.

In order for the Monte Carlo simulation to successfully predict the performance of KONEPS, uncertainties should be clarified so that variables describing KONEPS contain a certain level of uncertainty. MOCA-CAMS provides a series of simulation results based on (1) causal relationships among the variables regarding KONEPS; (2) uncertainties about some variables; and (3) probability distribution functions like uniform, beta, and normal distribution.

3 Methodology

3.1 MOCA-CAMS

MOCA-CAMS predicts the expected performance of an e-government system like KONEPS when it is adopted for use by another country. It consists of two phases: in the first, the causal relationships hidden in the target problem are defined. In this study, since KONEPS is itself a target system, the first phase of MOCA-CAMS is devoted to defining the causal relationships of KONEPS. In the second phase, the Monte Carlo simulation is incorporated to deal with the uncertainties of some highly volatile variables in KONEPS.

MOCA-CAMS begins with the statistical assumption that a 95% confidence level is maintained. With the confidence level at 95%, 5,000 runs of random number generation were performed to secure the statistical validity of the MOCA-CAMS simulation results. All the factors considered in this simulation are assumed to have a 5-point Likert scale according to each condition, where 1 denotes Very Bad, 2 Bad, 3 Neutral, 4 Good, and 5 Very Good. To ensure systematic and consistent calculation of the simulation results, each Likert scale is transformed into a specific value between interval [-1, +1]. In this simulation, the 5-scale Likert value is transformed as follows: 1 into -0.5, 2 into -0.2, 3 into 0.2, 4 into 0.6, and 5 into 1.0. This scale transformation is dependent upon the decision-maker's judgment; however, due to the law of large numbers, interpretation of the simulation results will remain consistent irrespective of the transformation rule. On the basis of the transformation rule, output values over 0.6 and 1.0 indicate that the outputs are "Good" and "Very Good" respectively, on the 5-point Likert scale.

3.2 Derivation of Causal Map

Use of the MOCA-CAMS simulation mechanism first requires definition of the nodes representing the components or factors that consist of the target problem. In our case, the target problem is predicting the performance of KONEPS when it is adopted for use by another country. Factors of the target problem were determined through three rounds of interviews with two procurement experts and two professors working in e-procurement-related fields in Korea. The nodes selected for the MOCA-CAMS simulation are described in Figure 1.

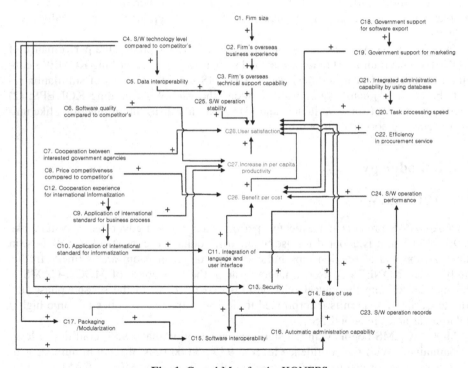

Fig. 1. Causal Map for the KONEPS

Three nodes (C26, C27, and C28) represent the output variables, and five nodes (C3, C4, C6, C8, and C24) represent the characteristics of KONEPS that a customer country would want to know about versus those of a competitor. The five nodes should therefore be dealt with such that they could vary sharply according to the customer country's judgment and assessments about KONEPS. Henceforth, the MOCA-CAMS simulation process should represent the high volatility of the five nodes using the Monte Carlo simulation mechanism. We consider three kinds of probability distribution functions to represent the uncertainties embedded in nodes C3, C4, C6, C8, and C24: uniform distribution, beta distribution, and normal distribution. Methods of determining causal relationships among variables consisting of the target problem or KONEPS, and their corresponding causality coefficients, include using the statements of decision-makers [11, 12], questionnaires prepared specifically for this purpose [13], and neural network-based learning. The first and second approaches are based on the assumption that experts in the domain can accurately provide the causality coefficients in causal relationships. In contrast, this study used structured interviews with experts to determine causal relationships between the relevant KONEPS factors extracted from literature and then calculate causality coefficients among them, in a bid to construct an appropriate CM and simulate the expected performance of KONEPS that would most improve the three output variables C26, C27, C28. The causal relationships among the variables describing KONEPS (Figure 1) include the constraints that the MOCA-CAMS adheres to during the simulation. The next step would normally be to determine the causal coefficients; however, in our study, all causal coefficients are limited to +1 or -1 for the sake of simplicity. To induce more objective and unbiased causality coefficients, a questionnaire survey can be conducted as per the approach of Montazemi and Conrath [13], in which CM was used for information requirements analysis.

3.3 Experiment Results

Let us suppose that the five nodes C3, C4, C6, C8, and C24 are to be assessed relative to a competitor's software. It is uncertain precisely how a customer country will evaluate the five nodes, because each country has its own peculiar culture, policy, IT infrastructure, user characteristics, etc. Some nodes are quantitative and others are qualitative, but the five nodes have uncertainty in common, so the Monte Carlo simulation should be applied during the simulation. Whether the MOCA-CAMS simulation process ends or not depends on the results achieved using the 1/2 threshold described in Lee and Kwon[2].

Let us suppose that the customer country is considering using KONEPS as its e-government procurement management system. It is certain that the customer will want to evaluate the anticipated performance of KONEPS by setting the five nodes to high volatility. If the simulation results for the values of the three output nodes C26, C27, and C28 seem good despite the high volatility of other nodes, the customer country can safely conclude that KONEPS will perform as expected upon its adoption.

MOCA-CAMS performed the simulation using the adjacency matrix $\underline{E}$ and 1/2 threshold. In the meantime, we may assume probability distributions for the five uncertain nodes. If we want some nodes to remain highly volatile or uncertain, then a uniform distribution of minimum -0.5 and maximum 1 is applied. If we already have supplementary information about the nodes, then normal distribution with mean 0.2 and standard deviation 0.2 is used. For example, we found that evidential data existed for the three nodes C4, C6, and C8, so the normal distribution was applied to them, while the uniform distribution was applied to the rest of the nodes.

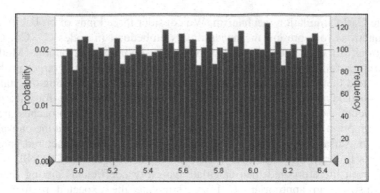

Fig. 2. Graphical Display of Output variables : User satisfaction (C28)

To compare the effects of number of random sampling trials of the Monte Carlo simulation, we performed two kinds of trials - (1) 5,000 trials and (2) 10,000 trials. For the scenario 1, as shown in Table 1, the MOCA-CAMS simulation yielded results for the three output nodes C26, C27, and C28. What should be noted here is that Table 1 shows almost equivalent statistical results between 5,000 trials and 10,000 trials. For example, for the case of 5,000 trials, C26 (Benefit for cost) had a minimum of 1.80 and a mean value of 3.0; C27 (Increase in per capita productivity) had a minimum of 0.7 and a mean value of 1.4; and C28 (User satisfaction) had a minimum of 4.9 and a mean value of 5.7. In all three cases, these results equate to a rating of "Very Good" on the 5-point Likert scale, indicating that customer countries will experience overall satisfaction upon adoption of KONEPS. These results of the 5,000 trials are almost same in the case of 10,000 trials, which proves that our proposed MOCA-CAMS can yield statistically significant results for the target problem of this study- predicting and measuring the performance of KONEPS. The results of Scenario 1 are summarized in Table 1. Meanwhile, Figure 2 depicts the output values (C28) of Scenario 1 for the 5,000 trials.

Table 1. MOCA-CAMS Simulation Results for Scenario 1

Node Name	Statistics	Number of Trial	
		5,000	10,000
Benefit for cost	Mean	3.0	3.1
(C26)	S.D	0.5	0.5
Increase in per capita	Mean	1.4	1.4
productivity (C27)	S.D	0.2	0.2
User satisfaction	Mean	5.7	5.7
(C28)	S.D	0.4	0.4

We also changed the probability distributions for the five nodes C3, C4, C6, C8, and C24 in order to determine whether the three output values would change significantly. In Scenario 2, beta distribution was used for C3 and C24, while normal distribution was used for C4, C6, and C8. In Scenario 3, only uniform distribution was used for all five nodes. For Scenario 4, uniform distribution was applied to C4, C6, and C8, and beta distribution was used for C3 and C24. Table 2 summarizes the MOCA-CAMS simulation results for all four scenarios using the 5,000 trials and 10,000 trials. In case of the 5,000 trials, all mean values for the three output nodes are statistically stable, under 95% confidence level. In average terms, the benefit for cost (C26) of KONEPS can be expected to range from 3.0 to 3.2, with an increase in per capita productivity (C27) of between 1.4 and 1.5. This indicates that the mean value is very stable, and the likelihood that adopting KONEPS will cause significant increase in per capita productivity is "Very Good". Meanwhile, the mean value for user satisfaction (C28) falls between 5.6 and 5.8, which is also very stable from a statistical perspective with 95% confidence level. When a customer country adopts KONEPS to manage the G2B procurement process, then, that country can expect that all mean values for the three output nodes (benefit for cost, increase in per capita productivity, user satisfaction) will be "Very Good" because all of their average values are greater than 1.0. As for the 10,000 trials, the results were almost the same with the 5,000 trials except infinitesimal changes in benefit for cost (C26) and user satisfaction (C28).

Table 2 reveals the statistical results from the four scenarios, comparing the 5,000 and 10,000 trials. All the results were same except minute changes in benefit for cost (C26) and user satisfaction (C28), showing that the proposed MOCA-CAMS can provide statistically significant and robust results under 95% confidence level.

Table 2. Summary Means of Four Scenarios

Node Name / Scenario	Benefit for cost (C26)		Increase in per capita productivity (C27)		User satisfaction (C28)	
	5,000	10,000	5,000	10,000	5,000	10,000
Scenario 1	3.0	3.1	1.4	1.4	5.7	5.7
Scenario 2	3.2	3.2	1.4	1.4	5.8	5.8
Scenario 3	3.1	3.1	1.5	1.5	5.6	5.6
Scenario 4	3.2	3.2	1.4	1.4	5.8	5.8

4 Conclusion

This study proposes a new method for predicting the performance of e-government software upon its adoption by a new country. Historically, this process has been problematic because it requires significant field knowledge and ample information about the customer country, and because uncertainty and volatility exist among the relevant variables that must be considered in any analysis. The target problem of this study was to predict the performance of KONEPS, an e-procurement system for G2B transactions which has been successfully implemented and operated by the Korean Public Procurement Service. In order to solve the target problem, we developed a new simulation approach called MOCA-CAMS in which the conventional causal map is modified to incorporate the Monte Carlo simulation mechanism. Experimental results showed that MOCA-CAMS could yield statistically significant and valid estimation results for predicting the performance of KONEPS when it is adopted for use in another country. The primary advantage of MOCA-CAMS is that it enables users to predict the performance of the KONEPS without spending a large amount of money or time. Another advantage is that MOCA-CAMS can be easily generalized and revised to incorporate additional variables and update the causal relationships depending on the characteristics of the KONEPS adoption situations. Most significantly, MOCA-CAMS can be used to predict the performance of other kinds of e-government systems, as well. We hope that this study will trigger more serious future studies in the field of e-government systems performance assessment in the future.

Acknowledgment. This research was supported by WCU(World Class University) program through the National Research Foundation of Korea funded by the Ministry of Education, Science and Technology (Grant No. R31-2008-000-10062-0).

References

1. Stratford, J.S., Stratford, J.: Computerized and networked government information. Journal of Government Information 27, 385–389 (2000)
2. Lee, K.C., Kwon, S.: The use of cognitive maps and case-based reasoning for B2B negotiation. Journal of Management Information Systems 22, 337–376 (2006)
3. Salling, K.B., Leleur, S., Jensen, A.V.: Modeling decision support and uncertainty for large transport infrastructure projects: The CLG-DSS model of the Øresund Fixed Link. Decision Support Systems 43, 1539–1547 (2007)
4. Parssian, A.: Managerial decision support with knowledge of accuracy completeness of the relational aggregate functions. Decision Support Systems 42, 1494–1502 (2006)
5. Caudle, L.S., Sharon, L., Gorr, L.W., Newcomer, E.K.: Key information systems management issues for the public sector. MIS Quarterly 15, 171–188 (1991)
6. Ein-Dor, P., Myers, M.D., Raman, K.S.: Information technology in three small developed countries. Journal of Management Information Systems 13, 61–90 (1997)
7. ChileCompra. Ministerio De Hacienda, Public Procurement System: Strategic Plan 2002–2004 (2002)
8. Public Procurement Service. The present and future of public procurement e-commerce. Republic of Korea, PPS (2002)
9. OECD. ICT Diffusion to Business: Peer Review, Country Report: Korea. Working Party on the Information Economy. DSTI/ICCP/IE(2003)9/FINAL (2004)
10. Tolman, E.C.: Cognitive Maps in Rats and Men. Psychological Review 55, 189–208 (1948)
11. Axelrod, R.: Structure of Decision: The Cognitive Maps of Political Elites. Princeton University Press, Princeton (1976)
12. Eden, C., Jones, C., Sims, D.: Thinking in Organizations. McMillan Press Ltd., London (1979)
13. Montazemi, A.R., Conrath, D.W.: The Use of Cognitive Mapping for Information Requirements Analysis. MIS Quarterly 10, 45–56 (1986)

Multi-agent Approach to Monitoring of Systems in SOA Architecture*

Dominik Ryżko and Aleksander Ihnatowicz

Warsaw University of Technology,
Intitute of Computer Science, Ul. Nowowiejska
15/19, 00-665 Warsaw, Poland

Abstract. The paper introduces a novel approach to monitoring of systems in Service-oriented architecture (SOA) with the use of multi-agent paradigms. Intelligent agents located across the system perform asynchronous, distributed measurements separately for each selected service or process. Compliance with required performance measures or SLAs can be defined and appropriate warnings are generated and reported. Alignment with definition of business processes together with proactive nature of agents allows early prediction of problems. The approach is flexible and scalable and naturally reflects the distributed nature of SOA.

1 Introduction

A Service Oriented Architecture (SOA) was introduced to overcome problems with integration of heterogeneous information systems and agile development of new solutions in the enterprise. It was thought to allow design and administration of flexible and scalable systems and facilitate communication between different company divisions responsible for its IT solutions and core operations.

Despite its advantages SOA architecture brings several challenges. Dynamic nature of collaborating services means several issues can be experienced at runtime. Therefore monitoring of such systems becomes a crucial task. Administrators need to be able to pinpoint quickly where the source of the business process failure lies. Obviously any information which can help to anticipate potential problems in SOA is of great value.

Traditional tools to monitoring of service archtiectures typically consists of a centralized module, which overlooks trafic on service bus and reports any anomalies. This approach has several limitations, including single point of failure, lack of scalability and distance from real problems occuring in varius places in the distributed environment. In such a setup its very difficult to predict problems early and react to them. The approach proposed in this paper shows how to

---

* The research was supported by the Polish National Budget Funds 2009-2011 for science under the grant N N516 3757 36.

N.T. Nguyen et al. (Eds.): New Challenges for Intelligent Information, SCI 351, pp. 309–318.
springerlink.com

overcome these limitations by the use of intelligent agents, which collaborate in order to deliver flexible, scalable and proactive monitoring capabilities.

2 Previous Work and Motivation

The idea of bringing together SOA and Multi-Agent Systems has already been proposed in previous research. In [7] an agent-oriented SOA programming model is proposed. It is argued that this approach allows natural introduction into SOA of concepts such as autonomy, uncoupling, data-oriented interaction and coordination. Another architecture presented by Sheremetov and Contreras [8] combines Web services and intelligent agent technologies orchestrated by a business process management system. Hahn et al. [4] presents a model-driven approach to design interoperable agents in SOA.

Monitoring of systems in SOA architecture has been studied extensively, since it is the key problem for implementing such solutions. Wang et al. [9] introduce a comprehensive QoS management framework for service level management (SLM) of SOA. The crucial aspect of managing SOA systems in the enterprise is compliance with Service Level Agreements (SLA), since any violations of SLA can cost money. Examples of managing this issue by SOA monitoring architectures can be found in [3], [1].

There already exist examples of application of agents for managing SOA performance. Some approaches try to attack the problem early at the design, development or testing phase. In [2] a framework for multi-agent based testing of Web Services is proposed. The system covers all phases of testing from their generation to execution and monitoring. In [6] it is argued that formal encoding of SLA goals within the business process development tools allows end-to-end management. Specificaly, runtime monitoring of services is described. In the proposed architecture light-weight agents collaborate to execute a larger process. However, with lots of existing legacy systems and possible changes in SLA, we need means of monitoring systems without any built-in facilities. Such approach is taken in [5], where agents collaborate on different levels in order to monitor SLA compliance and react in case of any deviations. Agents are grouped in clusters assigned to particular Web Services and act as proxies between WS and the client. The approach allows redirection of requests in case of problems, but creates significant overhead and, by becoming part of the flow, agents can generate problems by themselves.

The main goal of the approach presented in this paper was the introduction of a generic, platform independent approach to monitoring of SOA systems, which would allow for early problem detection. The envisaged monitoring architecture is flexible, scalable and adapting to the changing SOA topology. We argue that the multi-agent architecture is the most suitable for this task. Intelligent, autonomous and proactive agents are able to monitor particular services and processes, while communicating with each other in order to exchange information about encountered problems and gathered statistics. Means of limiting the overhead generated by the agents have been introduced, by allowing the administrator to switch on and off monitoring of selected services and processes.

3   Multi-Agent Architecture for SOA Monitoring

This section provides description of the Multi-agent architecture for monitoring of SOA systems. In general agents, with specific roles assigned to them, are located within a separate monitoring module. This multi-agent environment can be split over several physical servers. This allows for example agents monitoring a particular remote service to be placed on the same machine and gather relevant data localy. In such a setup only aggregated information and exception reports are passed over the network.

3.1   Agent Roles

Following roles, which correspond to abstract description of entity expected functions, were identified in the system: processes definition monitor, process instances monitor, service's monitor, queue's monitor, reporting tool.

At the lowest level data is gathered by service, queue and business process instance monitors. The multi-agent architecture allows to differentiate the number of details caputured for each element. By default only exceptions and aggregate statistics are captured. However, whenever we want we can ask the monitors to pass more details to study in detail process execution. On the higher level process definitions monitors and reporting agents take care of administration and data presentation tasks. Each of the roles will be now described in more detail.

Processes Definition Monitor. Description: Processes Definition Monitor role is responsible for loading processes' definitions, keeping track of changes in definitions. Other agents are able to get these definitions. Process Definition Monitor controls Service and Queue Monitor's life cycle, which allows turn off/on agents at user request.
Protocols and Activities: *GetDefinition*, *GetDefinitions*, CreateAgent and DestroyAgent.
Permissions: Read processes definitions.

Process Instances Monitor. Description: Process Instances Monitor gathers statistical data related to particular process instances within system. It generates messages about instances which take too long to complete.
Protocols and Activities: *LongInstance*, *NewInstance*, *SubscribeObserver*, *UnsubscribeObserver* and InstanceCheck.
Permissions: Read instance processing time in business process engine.

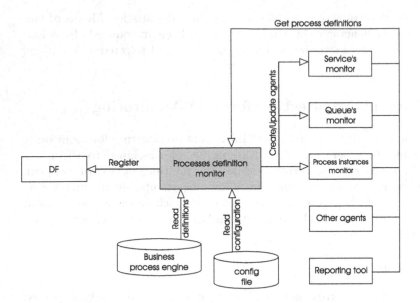

Fig. 1. Processes Definition Monitor

Service Monitor. Description: Service Monitor collects data: processing time (min/max/mean value), processed messages (all/failed) from service and transmit it to related roles.

Protocols and Activities: *LongService, MessageFailed, GetStatistices, Error, SubscribeObserver, UnsubscribeObserver* and ServiceMonitor.

Permissions: Read service processing time, messages count and service status.

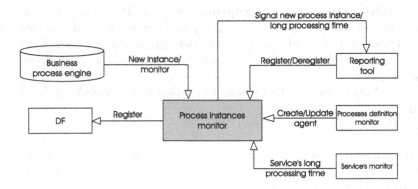

Fig. 2. Process Instances Monitor

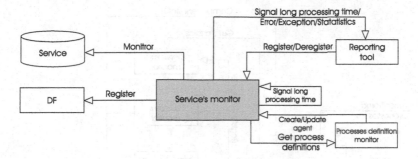

Fig. 3. Service Monitor

Queue Monitor. Description: Queue Monitor checks message count in the queue. It signal new message arrival and informs about threshold exceeded (too many messages in the queue).

Protocols and Activities: *ThresholdExceeded, NewMessageQueue, SubscribeObserver, UnsubscribeObserver* and QueueMonitor.

Permissions: Read queue message count, queue state.

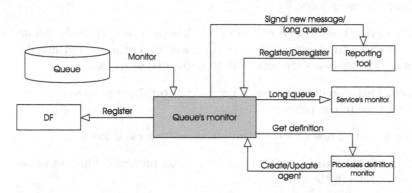

Fig. 4. Queue Monitor

Report. Description: Report is responsible for providing status reports and interaction with user.

Protocols and Activities: *LongInstance, NewInstance, GetDefinition, GetDefinitions, LongService, MessageFailed, GetStatistics, Error, ThresholdExceeded, NewMessageQueue, SubscribeObserver, UnsubscribeObserver, CreateAgent, DestroyAgent,* Register and Deregister.

Permissions: –

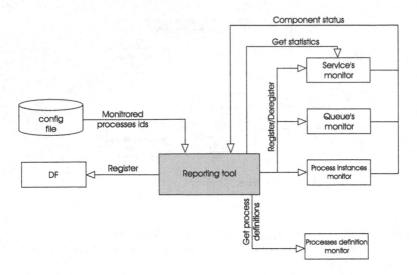

Fig. 5. Report

Interaction Model for the Proposed System. Interaction models are used to represent links between roles.

SubscribeObserver/UnsubscribeObserver. These two protocols allow observers to subscribe to the monitor (service, queue or instance) and unsubscribe from it. Event messages are sent only to registered observers.

LongInstance. This protocol represent is used to inform about process instance which took long time to complete.

NewInstance is sent to observer each time business process is invoked.

GetDefinition/GetDefinitions. This protocol allows process definitions to be accessed by other roles in the system.

LongService/MessageFailed/Error/GetStatistics. This protocol enables Service Monitor role to send messages related to state of the service.

ThresholdExceeded/NewMessageQueue. This protocol permits observer to get information about queue's state.

UpdateQueue/UpdateService. This protocol allows Process Definition Monitor role to inform other roles about changes in definitions of processes.

CreateAgent/DestroyAgent. This protocol allows Report role to ask Process Definition Monitor to turn on/off particular Service or Queue Monitor.

The attributes of each of the protocols are listed in Table 1.

Table 1. Attributes of the protocol definitions

Protocol	Initiator	Responder	Input	Output
SubscribeObserver/ UnsubscribeObserver	Report	Service, Queue and Instance Monitor (Mon.)	Observer Id	Accept/ Reject
LongInstance	Instance Mon.	Report	——	Instance Id
NewInstance	Instance Mon.	Report	——	Instance Id
GetDefinitions	Queue&Service Mon., Report	Definition Mon.	— / Definition Id	Process definition
LongService	Service Mon.	Service and Instance Mon., Report	——	Service Id
MessageFailed	Service Mon.	Service and Instance Mon., Report	——	Service Id
Error	Service and Queue Mon.	Report, Instance Mon.	——	Error message
GetStatistics	Report	Service Mon.	Service Id	Service Stat.
ThresholdExceeded	Queue Mon.	Report, Service Mon.	——	Queue Id
NewMessageQueue	Queue Mon.	Report	——	Queue Id
MessageFailed	Service Mon.	Report, Service and Instance Mon.	——	Service Id
UpdateService	Definition Mon.	Service Mon.	——	Service update
UpdateQueue	Definition Mon.	Queue Mon.	——	Queue update
Destroy/Create Agent	Report	Service and Queue Mon.	——	Acknowledge

3.2 Agents and Organisation of the System

Agents are based on the reactive model, which is accurate for the task. Complex actions are the effect of interaction between agents and organisation, which is modelled as community of experts. There are four types of agents in the system:

- monitoring agent for service and queue (which mapps Service Monitor and Queue Monitor roles);
- instance monitoring agent;
- process defintions monitoring agent;
- reporting agent.

4 Experimental Results

In order to test ideas presented in the paper a multi-agent system was built, which implements SOA monitoring capabilities. The scope of the tool includes identification and reporting of performance problems, errors and exceptions as well as non-compliance with SLA of services and business processes. The functions provided by this system are as follows:

- autonomous monitoring of services and business processes and broadcasting messages related to performance problems, errors and SLA violation;
- collecting statistical data associated with services (e.g. processing time);
- capturing failed messages;

- detecting failed services and processes which take too much time to complete;
- informing about events in the system – new message in the queue or new process instance;
- providing status reports to the user through a GUI based agent.

Elements of analysis and design methodology *Gaia* [10] were adopted to design the tool with the functionalities mentioned above. Apart from provided functionalities, it was crucial to design a tool in such a way that it does not overload SOA infrastructure and services. As a Enterprise Service Bus JBoss ESB 4.7 deployed on JBoss AS 5.1.0.GA was chosen with JADE 3.7 as the multi-agent platform. Agents use *Java Management Extension* interface and *Aspect-Oriented Programming* to communicate with the environment.

4.1   Test Scenarios

This section presents results of test scenarios. The tool was tested using different scenarios which simulate performance problems, exceptions in the system, queuing new messages or invocation of new process instance. Tests were based on a business process, consisting of sequential invocation of services and more complex one implementing an on-line shop (figure 6).

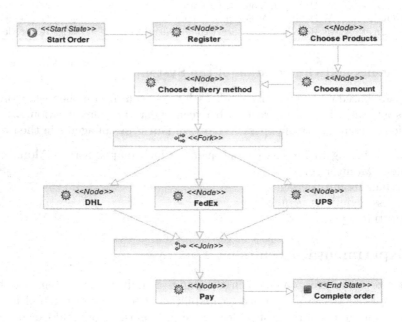

Fig. 6. Business process definition for on-line shop (Order Process)

4.2   Scenario 1 – Service's Long Processing Time

The *service level agreement* for the process is set as throughput of 100 process instances per hour. One of the service for choosing delivery method (UPS) has a performance issue which results in SLA violation of the whole process. Monitoring system presents following output to the user (only relevant messages):

```
...
22341-Instance: New instance: 22341
UPS Service:    new message on queue
...
UPS Service: long service
22341-Instance:  long service: UPS Service
...
UPS Service: long service

Order Process: SLA violation: suspected problem: UPS Service
```

New instance is started in the environment (`instance id 22341`). Message is arriving to the queue and processing starts. After processing completion service monitoring agent recognise that UPS Service takes too long to complete (based on statistical data) and informs instance monitoring agent. Finally, if no actions are taken, SLA is violated and this fact is also reported to the administrator.

4.3   Scenario 2 – Messages Queuing

The SLA set for the system is completion time less than 10 seconds. Due to unusually large queue to one the service (Register) processing time is becoming longer. The output from the system is as follows:

```
22156-Instance:  New instance: 22156
Register Service:     new message on queue
...
Register Service: many messages in the queue
...
22156-Instance: long service: Register Service
...
Order Process:  SLA violation: suspected problem: Register Service
```

New instance is started. Messages are arriving to Register service queue. *ThresholdExceeded* warning is issued as soon as there are more messages waiting then the threshold set by administrator based on historical data. Finally, information about SLA violation is sent to Reporting agent.

In situations as described above, the user recives warnings about performance issues before SLA is violated and can take actions after repeted warnings. Each warning includes information about specific problem class and source.

5   Conclusions

The paper presents how to use intelligent agents to monitor SOA systems, by overlooking particular services and processes in the system. Relevant statistics are gathers and sent to a single point of reporting, where administrator can review them. Any exceptional situations are reported both to the administrator and other agents related by the business process flow. This proactive behaviour allows to detect problems early and anticipate issues before their occurance.

Conducted tests proved feasibility of the agent-based approach to monitoring of Web Services in SOA systems. Cases such as service failure, overload with requests, long processing etc. have been simulated and reported successfuly. Monitoring of compliance with SLA was also demonstrated.

The prototype system is platform independent and can be used to monitor different systems built according to SOA principles and with the use of most common standards (e.g. BPEL). It is only required to develop specific plugins for gathering information for monitoring agents. The monitoring architecture is lightweight and introduces little overhead. It is independent from the SOA layer and it's failure does not interfere with the monitored functionality.

References

1. Ameller, D., Franch, X.: Service level agreement monitor (salmon). In: Proc. of the 7th Int. Conference on Composition-Based Software Systems (ICCBSS). IEEE, Los Alamitos (2008)
2. Bai, X., Dai, G., Xu, D., Tsai, W.: A Multi-Agent Based Framework for Collaborative Testing on Web Services. In: Proc. of the The 4th IEEE Workshop on Software Technologies for Future Embedded and Ubiquitous Systems, and the 2nd Int. Workshop on Collaborative Computing, Integration, and Assurance, SEUS-WCCIA 2006 (2006)
3. Berbner, R., et al.: An approach for the management of service-oriented architecture based application systems. In: Proceedings of the Workshop Enterprise Modelling and Information Systems Architectures (EMISA 2005), pp. 208–221 (2005)
4. Hahn, C., et al.: Meta-models, models, and model transformations: Towards interoperable agents. In: Fischer, K., Timm, I.J., André, E., Zhong, N., et al. (eds.) MATES 2006. LNCS (LNAI), vol. 4196, pp. 123–134. Springer, Heidelberg (2006)
5. Miede, A., et al.: Cooperation Mechanisms for Minitoring Agents in Service-oriented Architectures. In: Tagungsband der 9. internationalen Tagung Wirtschaftsinformatik, February 2009, vol. 1, pp. 749–758. sterreichische Computer Gesellschaft (2009)
6. Muthusamy, V., et al.: SLA-Driven Business Process Management in SOA. In: Proc. of the 2009 Conference of the Center for Advanced Studies on Collaborative Research (2009)
7. Ricci, A., Buda, C., Zaghini, N.: An Agent-Oriented Programming Model for SOA & Web Services. In: Proc. of 5th IEEE International Conference on Industrial Informatics (INDIN 2007), Vienna, Austria (July 2007)
8. Sheremetov, L., Contreras, M.: Industrial Application Integration Using Agent-Enabled Sematic SOA: Capnet Case Study. In: Information Technology For Balanced Manufacturing Systems. Springer, Heidelberg (2006)
9. Wang, G., et al.: Service Level Management using QoS Monitoring, Diagnostics, and Adaptation for Networked Enterprise Systems. In: Proceedings of the Ninth IEEE International EDOC Enterprise Computing Conference (2005)
10. Wooldridge, M., Jennings, N., Kinny, D.: The Gaia Methodology for Agent-Oriented Analysis and Design. Autonomous Agents and MAS 3 (2000)

Next Generation Mobile Convergence Service Analysis and Business Model Development Framework

Dongchun Shin[1], Jinbae Kim[2], Seungwan Ryu[1,*], and Sei-Kwon Park[1]

[1] Department of Information Systems, Chung-Ang University, Korea
[2] Department of Computer&Information, Anyang Technical College, Anyang, Korea
{dcshin,ryu,psk3193}@cau.ac.kr, jbkim@ianyang.ac.kr

Abstract. Next generation mobile communication will be developed by service-driven approach to realize human-centric mobile convergence services. As a preliminary research work on next generation mobile communication system, we have developed the scenario based service analysis process (2SAP) framework to derive core service technologies and functionalities. In this paper, we propose the next generation mobile convergence service business model creation methodology based on the 2SAP research framework. To achieve this goal, we first establish a service model which contains several components such as infrastructures, operations and service provisioning process. Then, after defining necessary components of business model including actors, their relationships, and roles, the next generation mobile services and business models can be created by including value flows to the developed service model.

1 Introduction

The next generation mobile communication system will be a service-driven developed system capable to realize the human-centric mobile convergence services [1,2,3]. As a preliminary research work on such a service-driven system development approach for the next generation mobile communication system, we have developed the scenario based service analysis process (2SAP) framework to derive core service technologies, service functionalities and network elements [4].

In this paper, we propose the next generation mobile convergence service business model creation methodology based on our previous research results (e.g., the 2SAP framework) on the next generation mobile services [3,4]. To achieve this goal, we first establish a service model contains several basic service elements such as infrastructures, operations, and service provisioning processes that are indispensable for providing next generation mobile services. Then, after defining necessary components of business models including actors, their relationships, and roles, the next generation mobile convergence service business models are created by adding service and value flows among participants to the developed service model.

---

* Corresponding author.

N.T. Nguyen et al. (Eds.): New Challenges for Intelligent Information, SCI 351, pp. 319–328.
springerlink.com

This paper is organized as follows. In next section, we briefly introduce the scenario-based service analysis process (2SAP) framework. In section 3, we propose next generation mobile service model development framework. We also propose how to establish the next generation mobile services infrastructure construction and operation model, in that participants of the model and their roles and activities are defined. In section 4, we propose next generation mobile convergence service business model development framework. Finally, we conclude this paper.

2   Scenario Based Service Analysis Process

In this section we propose the scenario based analysis method for next generation mobile services, called the *scenario-based service analysis process (2SAP)*. The proposed analysis method consists of four sequential steps, the scenario refinement process, the scenario decomposition process, the functional analysis process for each situation, and the service functionalities elicitation process. In the first process, various crude narrative scenarios are refined or better analysis. In the second process, based on several criteria, each scenario is decompose into a set of situations having common characteristics generated while using a system. In the third process, each situation obtained in previous process is also divided into several scenes to elicit service functionalities that are required to realize such scenes. Finally, various basic core technologies that are required to implement each service functionality are obtained in the last process.

In this paper, we use descriptive method that uses natural language for representation and graphical representation methods together for better analysis of next generation mobile services. In detail, initial and refined scenarios are described in natural language, and then expressed by scenario trees in the analysis step. Finally it is modelled as a *use case diagram* in both of functionality analysis and core technologies and service functionalities elicitation processes.

2.1   Scenario Refinement

In order to obtain core technologies and service functionalities of next generation mobile communication systems, various crude next generation mobile service scenarios are generated. These scenarios are written as a narrative story with lots of modifiers such as adjectives and adverbs. Although the "story" type crude service scenario is easy to understand, it is hard to get core technologies and service functionalities via analyzing such service scenarios. Thus those scenarios is required to be transformed into refined ones for better analysis.

In this process, initially generated story type scenario is refined to a concise one containing essential and necessary contents. In particular, all modifiers exists in a scenario are removed because they are used to make the scenario elegant and understandable for people. Then, the refined scenario is transformed into an enumerative type having list of different situations.

A possible initial story type service scenario and a refined and transformed enumerative scenario are shown in figure 1 and 2 respectively.

Tom living in LA, CA. is getting up in the morning with listening beautiful morning call music coming from his mobile phone put on the table. Today he is leaving to Toronto, Canada to meet a researcher work at Nortel Research Center to get useful information about network gaming business. He gets weather information about Toronto using his mobile phone. However, thanks to the ubiquitous computing and communication environment equipped in his house, he sees scenery of Toronto via display devices located in the wall and gets weather forecast via his audio system. Then various sensing devices having context information collection capabilities checks his physical status, and report his physical status to his mobile phone which is the gateway of the personal area network around him. ....

Fig. 1. An example narrative story type scenario.

1. Tome get alarming service from his mobile phone.
2. Tom gets weather information about place which he is planning to visit.
3. Physical status of him is sent to his private doctor, and discuss with him via chatting service. Prescription is sent to pharmacy and medicine will be delivered to the airport.
4. Download photos of his daughter on his mobile phone.
5. Starts up his car using his mobile phone.
6. .....

Fig. 2. An example of refined and transformed scenario into an enumerative one.

2.2 Scenario Decomposition

In previous process a scenario is refined and transformed into an enumerative type having list of different situations. Although it is possible to analyze all the listed situations together, a better and efficient approach is to analyze it after decomposing the whole scenario into a number of small sub-problems. Therefore, in 2SAP methods, the refined scenario is decomposed into a number of groups of situations based on coherency of them using grouping criteria.

The fundamental grouping criteria are location of a user, devices, and time. For example, in most cases, a new service is initiated when location and/or devices are changed. In addition, when, time especially date, is changed, it is highly probable to begin a new situation. When there are groups having larger numbers of list than other groups, they are decomposed into small sub-groups by applying the secondary decomposition criteria such as the number of actors and objectives of each situation. Figure 3 shows an example of scenario decomposition and grouping process.

2.3 Functional Analysis of Each Situation

In order to get service functionalities from situations described at the end nodes of the scenario tree, it is required to decompose each situation into smaller units called *scene*. A scene is defined as a target service moment that we are

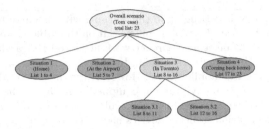

Fig. 3. An example of scenario tree representation based on decomposition and grouping approach.

interested in. A scene is thus corresponds service functionality that is essential basis service technology felt directly by users. The decomposition criteria used for getting several scenes from a situation are the initiation and/or releasing moment of services for a user gets it through his(or her) five senses. By means of intensive analysis of these scenes, we can obtain functionalities that make these scenes possible.

In this study, we first establish a pool of possible service functionalities required to realize next generation mobile services. Then, the pool of service functionalities is used to functional analysis of each situation. Detailed functional analysis procedures for each situation are described below.

- Establish a pool of service functionalities. In this case, the pool can be established by assistant of expert groups or investigation of related technologies.
- derives scenes from a situation, and obtains service functionalities required to realize each scene using the pool of service functionalities.
- Gets a list of required functionalities for each scene.
- Finally, represents each scene (or a situation) as a use case diagram which describes relations between a user and service functions.

Table 1 gives example lists of scenes and required functionalities for the situation 1.

Table 1. Example lists of scenes and required functionalities for the situation 1.

Scenes Description	Functionalities
Scene 1 Gets alarming service via mobile phone	• Mobile phone alarming
Scene 2 Checking weather of the place to visit	• web browsing
Scene 3 Health check & data transferring to doctor	• Remote Control • data messaging & file transfer
Scene 4 Chatting request from a doctor	• Chatting / Messenger
Scene 5 prescription is delivered to a pharmacy	• File transfer
Scene 6 Download daughter's photos	• File transfer

2.4   Elicitation of Service Technologies

Several types of core technologies such as platform technologies, mobile terminal technologies, and service application technologies are necessary to implement service functionalities. Similar to the previous functional analysis process for each situation, a procedure for elicitation of service technologies can be obtained as follow

– Establish a pool of service technologies with assistant of expert groups or investigation of related reference books or technical documents.
– Derives service technologies for each service functionalities using use case diagrams obtained in previous process. If there is a service technology that is required in realizing a service functionality but does not exist in the pool, creates a new service technology and makes a detailed technological list.
– Gets a list of required service technologies for each service functionalities.
– Finally, for each service functionality, draw a use case diagram describes relations between a user and service technologies.

3   Next Generation Mobile Services Model

3.1   Next Generation Mobile Service Business Model Creation Framework

As shown in figure 4, various next generation customized mobile services will be provided as an impromptu combination of service functionalities and unit service technology components selected from the functionality-based service pool and the service technology component pool respectively. Here, we call the service functionalities and service technology components are *the unit service components*. When a new customized mobile service is requested, it is first analyzed into situations, scenes, service functionalities and unit technologies by means of 2SAP procedure. Then, proper unit service components, i.e., service functionalities and unit service technology components, are selected from the pools and combined to create the new customized services. Since pools of service functionalities and service technology components do not cover all possible next generation service functionalities and technologies, newly discovered and elicited ones from the 2SAP based analysis are added to those pools. Then, they could be utilized to create another next generation mobile services.

In order to provide new mobile services by combining existing unit service components, it is required to establish an effective way to express the unit service component and how to combine those service components. To address this problem, we introduce a service model consisting of the service infrastructure construction and operation model and the service provisioning model. A unit mobile service can be composed using several unit service component extracted from the pool, and then provided through the service infrastructure. The service infrastructure construction and operation model defines participants and their roles that make composition and provision of the unit mobile services possible. The service provisioning model defines service and commodity flows among

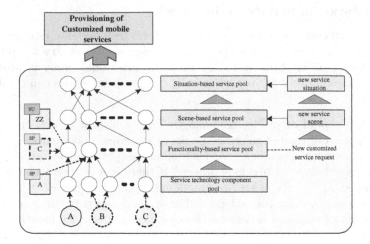

Fig. 4. Scenario analysis based next generation service provisioning procedure.

service participants over the service infrastructure. In addition, if it is possible to describe revenue sharing among service participants, next generation mobile service business model can be established. The established service model and business model for each unit mobile service will be used to establish new service and business models for the newly requested mobile services.

3.2   Mobile Service Infrastructure Construction and Operation Methodologies

Next generation mobile service infrastructure could be classified into software and hardware infrastructures, and components of these infrastructures used to provide unit mobile services can be established in various ways. Hence, in this section, we define detailed classification and representation diagram of several types of infrastructure construction and operation methods, several types of operators, and groups of participants existing in the mobile service provisioning process.

Based on the central operating body of the mobile service infrastructure we propose three types of infrastructure construction and operation methods, *a self constructor (SC)*, *a outsourced constructor (OC)* and *a leased constructor (LC)*, as shown in figure 5.

Fig. 5. Representation diagrams of three types of mobile service infrastructure construction and operation methods.

After establishing the service infrastructure, the central operation body and operation methodology should be determined. The central operation body, i.e., the service platform operator, can be classified into three types, *a self operator(SO)*, *a leased operator(LO)* and *a residential operator(RO)* as shown in figure 6.

SO: Self-   LO: Leased-   RO: Residential-
Operator   Operator   Operator

Fig. 6. Representation diagram of the service platform operators.

Actors participating in provisioning next generation mobile services using the above infrastructure, e.g., mobile services providers or service users, can be classified into three groups based on their role in this service provisioning process: *a service provider (SP)* group, *a contents provider (CP)* group, and *a service user (SU)* group. Each of participant group also can be classified into detailed subgroups as follow. The service provider group can be divided into groups of direct service providers (DSP), indirect service providers (ISP), and aggregate service providers (ASP), and the content provider group can be divided into groups of information contents providers (ICP) and advertisement contents providers (ACP). However, with appearance of new mobile services or changes of business approaches, the above sub-grouping could be changed to include newly created mobile services or evolved services. Representation diagram of the detailed groups of participants are shown in figure 7.

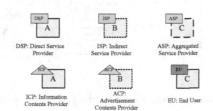

DSP: Direct Service   ISP: Indirect   ASP: Aggregated
Provider   Service Provider   Service Provider

ICP: Information   ACP: EU: End User
Contents Provider   Advertisement
                        Contents Provider

Fig. 7. Representation diagram of the detailed groups of participants

3.3 Mobile Service Model

In this section, we give readers an example use case diagram for the next generation service model established using the proposed infrastructure construction and operation methodologies. Most of mobile services including current and future next generation mobile services could be analyzed by 2SAP analysis process, and represented as a use case diagram as shown in figure 4 [4]. Furthermore, by using the proposed infrastructure construction and operation methods with the resulting use case diagram obtained from 2SAP, most mobile services can be modeled into a service model in which roles, activities and relation between participants of the service can be clearly and efficiently defined and represented.

Figure 8 shows an example representation diagram of next generation mobile service provisioning process for the case of service situation (figure 4) obtained from the 2SAP analysis process. In this example, it is assumed that company A is responsible for providing unit services and contents. However, for case of web browsing services, as an ASP, the company A is responsible for providing web browsing service by integrating service, information and/or advertisement contents provided by company N, P, U and K. For example, company A provides weather forecasting service to Tom (an end user(EU)) via web browsing service by aggregating weather information contents provided from company P with other contents. Similar to the weather service, a online banner advertisement is provided to Tom via web browsing by aggregating such advertisement contents supplied by company K with other contents. Besides, the doctor and the pharmacist are also assumed to be served by a DSP company.

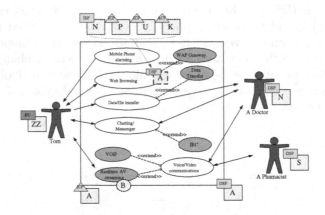

Fig. 8. Representation diagram of the detailed groups of participants.

4   Next Generation Mobile Service Business Model

The next generation mobile service model is established based on the service infrastructure construction and operation model, in that participants of the model and their roles and activities are defined. In other words, the next generation mobile convergence service business model can be developed as an architecture that is capable to support mobile communication services and information flows among various business actors participating the service provisioning process. The proposed mobile services business model also describes potential benefits for the various business actors and sources of revenues. Therefore, mobile services business models should be able to answer the following questions.

– which services or products are provided in the business model?
– who are the participants?
– what are roles and activities of participants, and how these are related among participants?
– what value will be generated from this model, and how it is distributed to participants?

When roles and activities of participants are appeared in concrete shapes, value that each participant want to get will be revealed clearly. Such value could be either a real object such as product and money or an abstract object such as knowledge, reliability, and satisfaction. In any case, such value can be distributed to each participant depending on contribution of roles or activities of them in the value creation process. Since the representation diagram for each service situation shows how mobile services are provided for each cohesive situation through interaction among participants, the corresponding business model can be obtained via finding merchant (e.g., services) and monetary flows among participants.

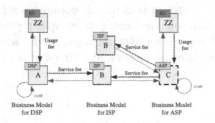

Fig. 9. Business models for DSP, ISP and ASP cases.

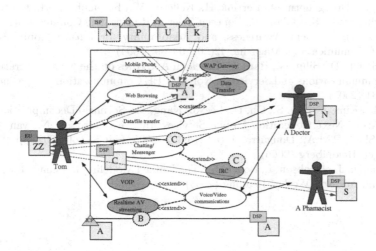

Fig. 10. mobile service business model for the example case.

Figure 9 shows business models for DSP, ISP and ASP cases that represents service flows (red solid line), service or usage fee (blue dashed line) and the cost (red dashed line). After identifying these flows between participant and applying them to the representation diagram show in figure 8, we can establish mobile service business model for the example case as shown in figure 10.

5 Conclusion

In this paper, we propose the next generation mobile convergence services business model development framework. We first introduce the scenario-based service analysis process (2SAP) framework developed in our previous research in order for analyze next generation mobile convergence services. Then, we propose a generic mobile services business mode development framework from the service scenario analysis process. In this framework, service models are established by utilizing the result of the 2SAP based next generation mobile convergence service analysis. The next generation mobile service model is established based on the service infrastructure construction and operation model, in that participants of the model and their roles and activities are defined. In other words, the next generation mobile convergence service business model can be developed as an architecture that is capable to support mobile communication services and information flows among various business actors participating the service provisioning process. The proposed mobile services business model also describes potential benefits for the various business actors and sources of revenues.

References

1. Ballon, P., Arbanowski, S.: Busness models in the future wirless world. WWRF White Paper (August 2002)
2. David, K., Droegehorn, O., Eertink, H., Kellerer, W., Kranenburg, H., Raatikainen, K., Arbanowski, S., Ballon, P., Popescu-Zeletin, R.: I-centric Communications: Personalization, Ambient Awareness, and Adaptability for Future Mobile Services. IEEE Communications Magazine 42(10), 63–69 (2004)
3. Ryu, S., Oh, D., Sihn, G., Han, K.: Research Activities on the Next Generation Mobile Communications and Services in Korea. IEEE Communications Magazine 43(9), 122–131 (2005)
4. Shin, D., Kim, J., Ryu, S., Oh, D., Lee, J., Kang, M.: Scenario Decomposition based Analysis of Next Generation Mobile Services. In: Alexandrov, V.N., van Albada, G.D., Sloot, P.M.A., Dongarra, J. (eds.) ICCS 2006. LNCS, vol. 3992, pp. 977–984. Springer, Heidelberg (2006)
5. Rumbaugh, J., Jacobson, I., Booch, G.: The Unified Modeling Language User Guide, 2nd edn. Addison-Wesley, Reading (2005)

Optimizing of Artificial Neural Network Based on Immune Genetic Algorithm in Power Load Forecasting

Yongli Wang

School of Economics and Management,
North China Electric Power University, Beijing, China
wyl_2001_ren@163.com

Abstract. Accurate forecasting of electricity load has been one of the most important issues in the electricity industry. Recently, along with power system privatization and deregulation, accurate forecast of electricity load has received increasing attention. There are many difficulties in the application of the method of BP neural network, for example, it is difficult to define the network structure and the network is easy to fall into local solution. To overcome these, in this paper, at first, by giving the undefined relation between learning ability and generalization ability of BP neural network, the hidden notes are obtained. Secondly, it poses to optimize the neural network structure and connection weights and defines the original weights and bias by means of genetic algorithm. Meanwhile, it reserves the best individual in evolution process, so that to build up a genetic algorithm neural networks model. This new model has high convergent speed and qualification. In order to prove the rationality of the improving GA-BP model, it analyses the network load with an area. Compare with BP neural network, it can be found that the new model has higher accuracy for power load forecasting.

Keywords: power load forecasting, the hidden notes, genetic algorithm, BP neural network.

1 Introduction

Load forecasting is a very crucial issue for the operational planning of electric power systems. Short-term load forecasting (STLF) aims at predicting electric loads for a period of minutes, hours, days, or weeks. Short-term load forecasting is absolutely necessary in the real-time control and the daily operation of electric power system[1]. However, the load change of electric power system is very complicated. It involves many factors, which are not only related to the load characters of the grid itself and the structure of power consumption, but also the gap of regional economic development and the population. Short-term load is a periodic non-stationary random process, including the relative stationary normal load and the floating load disturbed by many stochastic factors. Considerable research efforts have been devoted to enhance the accuracy and practicality of the short-term load forecasting. Many models and methods

---

[1] This research has been supported by Natural Science Foundation of China(71071052).

N.T. Nguyen et al. (Eds.): New Challenges for Intelligent Information, SCI 351, pp. 329–338.
springerlink.com © Springer-Verlag Berlin Heidelberg 2011

have been developed, such as time-series prediction, regression methods and so on. In fact, the relationship between the load and weather is not static, depending on the space and time factors, therefore there are many disadvantages in those methods. While expert system limited by the "knowledge bottle-neck" is not that successful currently. The artificial neural network-based models have been found to be the most popular for load forecasting applications. Their advantage lies in the fact that ANN with strong robustness can learn new knowledge and handle the complicated non-linear relationship.

Considerable simulation tests and theory studies have proved that BP algorithm with the character of dealing with the non-linear problems is an effective learning algorithm for neural network. Despite of the wide application in practice, the BP neural network has its own drawbacks, such as poor convergence, and easy to fall in local optima. Besides lacking of evidence the choosing of the initial weights and the network architecture has the character of randomness, therefore it's hard to choose the best initial point and get the best results, which limits its practical application a lot. Genetic algorithm is a global optimization algorithm simulating the genetic selection theory of Darwin and the biological evolution process. It has adaptability, global searching ability and parallel nature, showing its strong problem solving skills.

2 Determining the Number of the Nodes in Hidden Layer

The number of the nodes in hidden layer is depended on the number of the training samples and the complication of the rules in them. No scientific and fixation methods are found in present researches. Some use the trial-and-error method, the calculation in which is not easy; some adopt the threshold method, setting threshold value for the nodes in hidden layer, which is usually 0.8-2.0 times as much as the input nodes. However, those methods are based on the subjective judgment and lack of scientific evidence. In this paper, the number of the hidden layer nodes is determined though the uncertain relationship.

The multiple correlation coefficient R describing the close relationship between the whole of independent variables and the dependent variable can be computed though single correlation coefficient.

$$R = \sqrt{1 - R^* / R_{yy}^*} \tag{1}$$

$$R_{(n+1)\times(n+1)} = \begin{bmatrix} r_{11} & r_{12} & \cdots & r_{1n} & r_{1y} \\ r_{21} & r_{22} & \cdots & r_{2n} & r_{2y} \\ \vdots & \vdots & \vdots & \vdots & \vdots \\ r_{n1} & r_{n2} & \cdots & r_{nn} & r_{ny} \\ r_{y1} & r_{y2} & \cdots & r_{yn} & r_{yy} \end{bmatrix} \tag{2}$$

The formula for the average training relative error of the training samples ε_1 is as follow:

$$\varepsilon_1 = \frac{1}{m} \sum \frac{O_i - T_i}{T_i} \tag{3}$$

We expect the average training relative error ε_1 is 0.05. Though the formula $R = \sqrt{1 - R^* / R_{yy}^*}$, we can calculate the multiple correlation coefficient R.

According to the uncertain relationship, the formula for the hidden layer nodes is as follow:

$$H^* = [\frac{30.8}{R} \varepsilon_1^{\frac{3}{2}} \log_2 (1 + \frac{1}{\varepsilon_1})] \tag{4}$$

Put in ε_1 and R, and get the rounded H^*, which is the number of the hidden layer nodes in the network.

3 Optimizing the Initial Weights Using Genetic Algorithm

BP algorithm has high accuracy, but it has some disadvantages: it converges slowly, gets local minimization too easily and courses oscillation. local minimization can be solved by adjusting the initial weights, while slow convergence and the oscillation are usually coursed by getting into local minimization in the later period of the network training. Considering this, if the weight value can be searched efficiently and take it as the initial weight in BP algorithm, the problem above can be solved. Due to the genetic algorithm has the characters of strong global searching ability, simplicity, robustness and parallel nature, it can be used in earlier research to overcome the disadvantages of BP algorithm.

The process of the weights optimization is as follows:

(1) Coding method

Proper coding method is necessary in genetic algorithm. Learning weights is a complicated continuous parameter optimal problem. Adapting binary code will make the codes too long, besides need to recode it back to real number, which will influence the accuracy of network learning. Therefore, this paper utilizes the real number coding. Because the forecasting model in this paper is BP network with single hidden layer, we can connect the weights and threshold values together first, then the BP network here has only one hidden layer, we can connect the weight with the threshold at first, as a result, a vector consist of real-number network weights and threshold is obtained as below:

$$X = (w_{11}, w_{12}, \cdots, w_{1n}, \theta_1, \cdots, w_{m1}, w_{m2}, \cdots, w_{mn}, \theta_m, \theta_{m+1}) \tag{5}$$

Where, m presents the number of the nodes in hidden layer; n is the number of the weight connect with the hidden layer; w is the connection weight; θ is the weight from hidden layer to output layer.

(2) Original colony
Since the network built, distribute t groups of random weights to t networks at first, and the network will been initialized. The t groups of random weights construct the original colony of the weights, which become the basis of optimal operation in the genetic algorithm of network weights. A great number of practices shows that if the individual number of original colony t is too small, it will lead to the degradation of reproduction, as a result, the solution will converge to several feasible solutions too early and lose the global searching ability. So it's necessary to make sure the population has certain scale. The value of network weights is determined by probability distribution $e^{-|r|}$ randomly, which make sure the genetic algorithm can search all the range of the feasible solution. This method is got though many experiments. It's found that once the network is converged, the absolute value of the weights are usually small, few of them are big. Using $e^{-|r|}$ probability distribution rule can make the most network weights small.

(3)Fitness function
Since the adaptive value is the only deterministic index for the individual's chance to live, the fitness function determines s the population's evolution directly. In this paper the function is described as below:

$$f = 1/E \tag{6}$$

Where, E presents the error of output in the neural network. This function shows that the bigger the load forecasting error is, the worse the adaptability of the neural network is, which is what we expect.

(4)Genetic Operation
① Choose excellent individuals to enter into the next generation. N is the number of the individuals, f_i is the adaptability of the individual i. the function for each individual's duplication probability is as follows:

$$P_i = f_i \bigg/ \sum_{i=1}^{N} f_i \tag{7}$$

The copy probability is $NP_i = N \bullet P_i$, the number of copied individuals is $MP_i = \text{int}(NP_i)$, save the remained partial part as $SP_i = NP_i - MP_i$。

If $MP_i = 0$, the individual will be cleaned out, otherwise, it will copy MP_i times. After the copy, only N_1 copied individuals of the N individual are filled, the left are free $N_2 = N - N_1$. According to the ascending order of the SP_i, put the individual with the greatest decimal part in N_2 to the solution space.

② Crossover

Every two copied colonies as a group randomly, according to crossover probability $P_0 = 0.8$, have crossover operator. When it comes to the real-number coded individual, the new individual is got by the linear composite of every two of them, suppose after k times iterations, the two individuals become X_k and Y_k, and the cross functions are described as:

$$\begin{cases} X_{k+1} = \alpha \bullet Y_k + (1+\alpha) \bullet X_k \\ Y_{k+1} = \alpha \bullet X_k + (1+\alpha) \bullet Y_k \end{cases} \tag{8}$$

Where, $\alpha = \min(0.2, 1/(1+k))$, using variable length, a large range of new individuals are produced in the early research, so that it wouldn't destroy good individuals excessively later.

③variation

According variation operation P_m, operate the mutation for the individuals in population. Here self-fitted mutation probability is adopted, the individual with higher fitness use lower P_m, the ones with lower fitness use higher P_m, which enhance the chance for mutation before entering into the next generation. The function is described as below:

$$P_m = \begin{cases} \dfrac{k_m (f_{\max} - f_m)^2}{(f_{\max} - \overline{f})^2}, f_m \le \overline{f} \\ k_m, f_m \le \overline{f} \end{cases} \tag{9}$$

Where, k_m is the constant less than 1 (in this paper it's 0.05), f_m is the fitness of the mutation individual, $f_{\max}$ and f is the greatest and the average of the fitness separately.

(5)optimal guaranteed

Calculate the error of the new individuals using BP algorithm, find the greatest and the smallest ones. Then compare the smallest one with that of last generation. If the smallest error in new population is smaller, store the error and weight of the best individual to Min and w, otherwise instead the best individual in new population with the original one.

4   Load Forecasting Study Based on the GA-BP Neural Network

The main methods for encoding in genetic algorithm are binary code and real-number code. In recent years, the latter is widely applied in the optimal of weight and threshold gradually.

GA-BP algorithm is from the modified BP algorithm. First get the initial value though genetic algorithm. Then take it as the initial weight for BP algorithm. This is the principle of the new model.

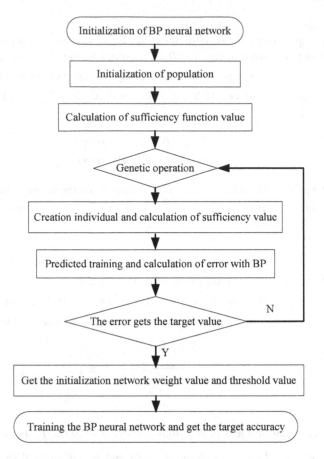

Fig. 1. The flow diagram of GA-BP model

The steps of GA-BP algorithm is as follows:

①Initialize the population P , including crossover probability Pc 、 mutation probability Pm and WIH_{ij}, WHO_{ji} ; And code with the real number.

②Compute the evaluation function and sort them in decreasing order; Choose the individual according to the probability function below:

$$p_s = f_i / \sum_{i=1}^{N} f_i \tag{10}$$

f_i is the fitness of individual i, which is got from under equation(14) and (15):

$$f(i) = 1 / E(i) \tag{11}$$

$$E(i) = \sum_{P} \sum_{k} (V_k - T_k)^2 \tag{12}$$

Where, $i = 1, \cdots, N$ is the number of chromosome; $k = 1, \cdots, 4$ is the nodes in out layer ; $p = 1, \cdots, 5$ is the learning samples: T_k is the teaching signs。

③Crossover the G_i and G_{i+1} with Pc, and produce the new individuals G'_i and G'_{i+1}. The left individuals copy themselves directly; and produce the new individual G'_j from the mutation of G_j with Pm;

④Insert the new individuals to the population P, then calculate the evaluation function of the new individual;

⑤Calculate the sum of squares of errors, if reach the predetermined ε_{GA}, turn to ⑦, otherwise, turn to ③ and carry on;

⑥Take the optimal value in genetic algorithm as the initial weigh, and train the network in BP algorithm until reach the predetermined ε_{BP} ($\varepsilon_{BP} < \varepsilon_{GA}$)

5 Empirical Analysis

5.1 Choose Sample

Use the new method in this paper to forecast the load in some region as empirical study. Choose the load database of this power grid from 00:00 April 2, 2007 to 12:00 June 23, 2008 as training sample, and the period from 13:00 to 24:00 on June 23rd, 2008 as testing sample.

Using the hourly load data of a certain power grid, we built the short-term load forecasting model with three-layer BP neural network. Since the rule of hourly load is diverse, each hour has a model separately. Hence, the input-output relationship for the ith hour is described as follow:

$$L_{i,t} = f(L_{i-1,t-1}, L_{i,t-1}, L_{i+1,t-1}, T_{max,t-1}, T_{min,t-1}, L_{i-1,t-7}, L_{i+1,t-7},$$
$$T_{max,t-7}, T_{min,t-7}, T_{max,t}, T_{min,t}, T_{average}, R_t) \qquad (13)$$

Where $L_{i,t}$ is the load of the i_{th} hour on the t_{th} day. $T_{max,t}$, $T_{min,t}$ and $T_{average}$ present the highest ,lowest and average temperature on the t_{th} day separately. R_t is the rainfall on the t_{th} day. This function considers both daily and weekly periodic characters of load.

From Eq (1), The multiple correlation coefficient R is computed $R = 0.4223$. According to ε_1, R and Eq (4), rounded H^* is 6, Therefore, the number of the nodes in hidden layer is 6 in this paper.

In data training, we assume predetermined target precision is 0.002, the predetermined iteration times is 10000.The target precision is achieved at the 8532th Epochs, using the initial weights and threshold under GA algorithm.

Seen from Fig.3, the error is stable when it comes to the last generation 100. The fitness is increasing continuously in the process, while the Sum-Squared error is decreasing.

5.2 Results and the Evaluation Index

Take the relative error and the relative error of RMS as the final evaluation indexes:

$$E_r = \frac{x_t - y_t}{x_t} \times 100\% \tag{14}$$

$$RMSRE = \sqrt{\frac{1}{n}\sum_{t=n}^{n}\left(\frac{x_t - y_t}{x_t}\right)^2} \tag{15}$$

Table 1. The analysis of forecasting error MW

Time point	Actual load	GA-BP		BP	
		Forecasting load	E_r	Forecasting load	E_r
13:00	1012.1662	1000.7287	-1.13%	1030.6888	1.83%
14:00	1015.1629	1028.5631	1.32%	990.6975	-2.41%
15:00	1021.9054	1000.9563	-2.05%	1058.2852	3.56%
16:00	994.2215	972.1498	-2.22%	1023.8493	2.98%
17:00	1093.0934	1112.0039	1.73%	1051.3372	-3.82%
18:00	1484.8892	1463.9523	-1.41%	1529.5844	3.01%
19:00	1631.3091	1682.2059	3.12%	1557.0845	-4.55%
20:00	1749.9822	1725.8324	-1.38%	1810.1816	3.44%
21:00	1796.4101	1835.0329	2.15%	1915.5121	6.63%
22:00	1422.3833	1403.0389	-1.36%	1357.8071	-4.54%
23:00	1039.2594	1052.5619	1.28%	1069.1901	2.88%
24:00	855.6952	827.7995	-3.26%	823.5211	-3.76%
RMSRE			1.99%		3.81%

It can be seen from table 1 that the greatest gap between relative error of GA-BP model are 3.26% and -2.22%, and the gap between them is 5.48% (3.26%～-2.22%) ; while in the traditional BP network model, the greatest gap between relative error is 6.63% and -4.55%, the gap is 11.18% (6.63%～-4.55%) . Besides, the relative error of RMS in traditional BP network is3.81%, on the contrast, that of new GA-BP model is only 1.99%, which means the prediction precision is improved by1.82% Thus the new forecasting model built in this paper can express the mapping relationship among the loads well and has great accuracy.

It is said that if the relative absolute error is less than 3%, the model has reasonable accuracy. Take this as standard here, there is 10 points out of 12 points with relative absolute error less than 3% in GA-BP model. And the error of the most of the 11 points is very close to 0, the smallest one is -1.13%, the only two which are higher than 3% is just 3.12% and 3.26%. In contrast, there are only 4 points with the relative absolute error under 3%, and the value is all close to 3%. Therefore, from the both respects of forecasting error and precision, the GA-BP model performs much better.

6 Conclusion

The practical data analysis has improved the good performance of the modified model in short-term load forecasting, at the same time, we can come to the conclusions below:

(1) Analyze the relationship among the network learning ability, generalization ability and other influence factors, introduce the multiple correlation coefficient to describe the complexity of the samples; According to the uncertain relationship, derive the multiple correlation coefficient , calculate the number of the nodes in hidden layer, satisfying the predetermined error.

(2) Optimize the connection weight and the network structure with genetic algorithm, keep the best individual in the evolution. It decreases the randomness in choosing the initial weight of network and the oscillation in the decision of network structure, also solves to fall in local optima too easily. The practical study indicates this new method overcomes the difficulty in determining the network depended on experience, what's more, the advantages of GA-BP algorithm over the regression method is that it has special learning ability and the forecasting stability, which can be used widely. This provides new thoughts and methods in load forecasting under neural network.

(3) By empirical analysis, we found the modified new algorithm performs much better than the BP network in respected both forecasting error and the accurate number.

(4) The determining of the solution space is very important in the GA-BP model. If the solution space is too big, adopting the short binary code will affect the accuracy of the weight and threshold; To improve the accuracy, it's necessary to lengthen the code, as a result, the solution space will become large, which means more time consumed. Therefore how to decide the solution space properly is worthy for future study.

References

1. Niu, D.-x., Cao, S.-h., Lu, J.-c., Zhao, L.: Technology and application of power load forecasting, 2nd edn. China Power Press, Beijing (2009)
2. Fatemi-Ghomi, N.: Performance Measures for Wavelet-Based Segmentation Algorithms. Ph.D. Thesis, Surrey University (1997)
3. Rumelhart, D.E., Hinton, G.E., Williams, R.J.: Learning international representations by error propagation, parallel distributed processing: Explorations in the microstructures of cognition. MIT Press, Cambridge (1986)

4. Sivakumar, B., Jayawardena, A.W., Fernando, T.M.K.G.: River flow forecasting: Use of phase-space reconstruction and artificial neural networks approaches. Journal of Hydrology 265(1–4), 225–245 (2002)
5. Zealand, C.M., Burn, D.H., Simonovic, S.P.: Short term streamflow forecasting using artificial neural networks. Journal of Hydrology 214, 32–48 (1999)
6. Riad, S., Mania, J., Bouchaou, L., Najjar, Y.: Rainfall–runoff model using an artificial neural network approach. Mathematical and Computer Modelling 40(7-8), 839–846 (2004)
7. Dawson, C.W., Wilby, R.L.: A comparison of artificial neural networks used for rainfall–runoff modelling. Hydrology and Earth Systems Sciences 3, 529–540 (2000)
8. Kang, K.W., Park, C.Y., Kim, J.H.: Neural network and its application to rainfall–runoff forecasting. Korean Journal of Hydroscienee 4, 1–9 (1993)
9. Coulibaly, P., Anctilb, F., Bobec, B.: Daily reservoir inflow forecasting using artificial neural networks with stopped training approach. Journal of Hydrology 230(3-4), 244–257 (2000)
10. Jain, S.K., Das, D., Srivastava, D.K.: Application of ANN for reservoir inflow prediction and operation. Journal of Water Resources Planning and Management – ASCE 125(5), 263–271 (1999)
11. Jin, L., Wu, J.-s., Lin, K.-p., et al.: Short-term climate prediction model of neural network based on genetic algorithms. Plateau Meteorology 24(6), 981–987 (2005)
12. Shang, J., Zhu, Z.-l.: Application of genetic algorithms neural networks in predicting corrosion rate of carbon steel. Corrosion Science and Protection Technology 19(3), 225–228 (2007)
13. Miller, G.F., Todd, P.M., Hegde, S.U.: Designing neural networks using genetic algorithms. In: Schaffer, J.D. (ed.) Proceedings of the Third International Conference on Genetic Algorithms and their Applications, pp. 379–384. Morgan Kaufmann, San Mateo (1989)
14. Yao, X.: Evolving artificial neural networks. IEEE Trans. Neural Networks 87(9), 1423–1447 (1999)
15. Niu, D.-x., Wang, Y.-l., Ma, X.: Optimization of support vector machine power load forecasting model based on data mining and Lyapunov exponents. J. Cent. South Univ. Technol. 17(2), 406–412 (2010)
16. Balakrishnan, K., Honavar, V.: Evolutionary design of neural architectures— a preliminary taxonomy and guide to literature. Technical Report, AI Research Group, CS-TR 95-01 (January 1995)
17. Bebis, G., Georgiopoulos, M., Kaspairs, T.: Coupling weight elimination with genetic algorithms to reduce network size and preserve generalization. Neurocomputing 17(3-4), 167–194 (1997)
18. Li, T.S., Su, C.T., Chiang, T.L.: Applying robust multi-response quality engineering for parameter selection using a novel neural-genetic algorithm. Computers in Industry 50(1), 113–122 (2003)
19. Niu, D.-x., Wang, Y., Wu, D.D.: Power load forecasting using support vector machine and ant colony optimization. Expert Systems with Applications 37(3), 2531–2539 (2010)
20. Niu, D.-x., Wang, Y.-l., Duan, C., Xing, M.: A new short-term power load forecasting model based on chaotic time series and SVM. Journal Of Universal Computer Science 15(13), 2726–2745 (2009)

What Factors Do Really Influence the Level of Technostress in Organizations?: An Empirical Study

Chulmo Koo and Yulia Wati[*]

College of Business, Chosun University
375 Seosukdong Donggu Gwangju South Korea
{helmetgu,yuliawati}@gmail.com

Abstract. Technostress is important fallout of the inevitable use of ICTs in organizations and illustrates the bivalent nature of their organization influence. In this research, we studied the influence of innovation culture, self-efficacy, and task complexity on technostress. In addition, we examined moderating effect of literacy facilitation in influencing the relationship between task complexity and technostress. The data for this research were collected from 98 employees of Korean companies. The results indicated that innovation culture influenced technostress negatively while task complexity had positive effect on technostress. However, we found an insignificant relationship between self-efficacy and technostress. The result also showed a negative moderating effect of literacy facilitation in influencing the relationship between task complexity and technostress. Additional discussion and implications of the empirical findings are discussed in this paper.

Keywords: Technostress, innovation culture, self-efficacy, task complexity, literacy facilitation.

1 Introduction

Information Communication Technologies (ICTs) are becoming increasingly indispensable in many aspects of business and everyday life [1], [2]. ICTs have brought a significant impact on the processes and outcomes of organizational life. Organizations have enthusiastically absorb the benefits of ICT, including speed, replicability, responsiveness, and accuracy. On the other hand, ICTs often force employees to try and accomplish more tasks in less time, result in eliminating of manual jobs, and affect relationships with colleagues [3]. In the organizational setting, it is common for employees to handle multiple task from different sources such as internet, emails, cellphones, short messages and faxes, aided by devices such as laptop, smart-phones, etc. [4]. They feel overwhelmed with information and are forced to work faster to cope with increased processing requirements. This situation may create stress and leave users feeling frustrated [4]. These negative reactions toward the technology have been mentioned as technostress, the condition when the complexity and changeability of IT

---

[*] Corresponding author.

N.T. Nguyen et al. (Eds.): New Challenges for Intelligent Information, SCI 351, pp. 339–348.
springerlink.com   © Springer-Verlag Berlin Heidelberg 2011

require more work, lead to excessive multitasking, and are accompanied by technological problems and errors [5].

Technostress is important fallout of the inevitable use of ICTs in organizations and illustrates the bivalent nature of their organizational influence [5]. To understand this phenomenon, stress literature is being increasingly recognized as a potential basis for recognizing user attitude towards ICTs in workplace [6]. However, this area received less attention from IS scholars. Thus, it is important to investigate the impact of technostress on organization conceptually and empirically [5]. Several studies have summarized the various causes of technostress. For organizational perspective, Wang et al. [2] pointed out the magnitude of organizational environment in influencing the level of employee technostress. From individual perspective, Bloom [7] stated that the lack of computer ability and experience are the major causes of computer related technostress. Likewise, regarding task characteristics, Goodhue and Thompson [8] argued that task characteristics might move a user to rely more heavily on certain aspects of the information technology. From these perspectives, we conclude that an empirical research is needed to investigate the role of organizational culture, individual trait, and task characteristics in influencing technostress.

Research by Ragu-Nathan et al. [5] has proven that technostress has negative impact on job satisfaction, commitment to organization, and intention to stay. They also recommended two second order constructs: technostress creator (factors that create stress from the use of ICTs) and technostress inhibitors (organizational mechanisms that reduce stress from the use of ICTs). However, they also called for further research to create a formative modeling to identify factor affecting technostress and to examine whether the relationship is moderated by the coping strategies provided by organization. Thus, our research questions are: (1) What is the effect of innovation culture of a company on technostress creation?; (2) How does individual train (i.e. self-efficacy) influence technostress among individuals?; (3) How does task complexity influence technostress in organizational setting?; and (4) Does organizational mechanism as a coping strategy moderate the effect of task complexity on technostress?

2 Literature Review

2.1 Technostress and Its Coping Strategies

The term of technostress first appeared in Craig [9] which is he defined as a modern disease of adaptation caused by an inability to cope with new computer technologies in a healthy manner [10]. Technostress is stress experienced by individuals due to the use of ICTs and has been defined as a modern disease of adaptation caused by an inability to cope with new computer technologies in a healthy manner [11] and as a state of arousal observed in certain employees who are heavily dependent on computers in their work [12]. Important variables that affect the probability of developing technostress include the age of the user, past experience with technology, perceived control over new tasks, and organizational climate [11], [13]. Components of technostress [4], [14] include: Techno-overload (the ICTs pushes employees to work faster); Techno-invasion (the pervasive ICTs invades personal life); Techno-complexity (the complexity of new ICTs makes employees feel incompetent);

Techno-insecurity (the job security of employees threatened by fast changing ICTs); and Techno-uncertainly (the constant changes, upgrades and bug fixes in ICT hardware and software impose stress on the end-users).

Generally, coping strategies or mechanisms were divided into two categories: psychological and behavioral [15]. In technostress context, these strategies have been classified into two major categories: emotion focused strategies (psychological) and problem focused strategies (behavior) [2]. Problem focuses coping strategies are viewed as direct approaches, while emotional focused coping strategies are viewed as indirect approaches. In this research, we will focus on behavior problem focus strategies rather than emotional focuses strategies. We adopted coping strategies recommended by Ragu-Nathan et al. [5], whereas they called these strategies as technostress inhibitors. We limited the research on literacy facilitation as moderating variable under organizational environment.

3 Research Model and Hypothesis

Our proposed research model can be seen in Figure 1. In this model, we measured the effect of innovation culture, self-efficacy, and task complexity on technostress. Moreover, we also measured the interaction effect of literacy facilitation as a coping strategy and task complexity on technostress.

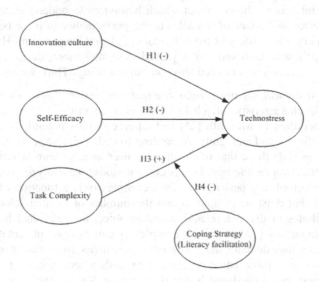

Fig. 1. Research Model

3.1 Hypotheses Development

Culture is a complex system of norms and values that is shaped over time and affects the types and variance of organizational processes and behaviors [16], [17]. Innovation is defined as an idea, practice, or object that is perceived as new by an individual or another unit of adoption [18]. An innovation presents an individual or an organization with a new alternative or with new means of solving problem. Innovation at the organizational level presents potential adopters with new means for solving problems and exploiting opportunities [19]. Thus, organizations whose cultures emphasize innovation when resources are available tend to implement more innovations and develop competitive advantage [17]. In this study, we refer to the innovativeness construct (as organizational culture). This innovativeness can be measured through the level of learning and development, support and collaboration, and participative decision making [17]. Previous literature has reported that innovation culture would influence individual belief [16]. That is, the study has indicated that a firm with an innovation orientation is likely to focus on learning, participative decision making, support and collaboration, and thus, affect the level of technology adoption in the organization [17] and reduce the technostress level among employees. Thus, we hypothesize:

H1: Innovation culture will negatively influence technostress level of an employee.

Hendrix et al. [20] argued that individual characteristic is one of the causes of stress. An individual may experience computer anxiety which manifest in several ways: temporary confusion as to how to use the technology and feel of being rushed by technology [11]. Bandura and Cervone [21] suggested that perceived self-efficacy affects the level of self-set goals, strength of goal commitment, and the level of cognitive performance. Social learning theory distinguishes the effects between strength of perceived self-efficacy on effort during learning and during established skills [22]. Self efficacy influences choices about which behaviors to undertake, the effort and persistence exerted in the face of obstacles to the performance of those behaviors, and in turn, ultimately, the mastery of the behaviors [23]. In Compeau and Higgins's [23] study, individuals with high self-efficacy used computer more, derived more enjoyment from their use, and experienced less computer anxiety. Thus, we argued that:

H2: Self-Efficacy will negatively influence technostress level of an employee.

The intrinsic characteristic of job has also been identified as one of sources of stress [24]. According to McGrath [25] task characteristics potentially create stress, such as task difficulty and ambiguity. According to task technology fit theory [8], task characteristics include those that might move a user to rely more heavily on certain aspects of information technology. In this current study, we analyze task characteristics by using a complexity perspective. Task complexity is a function of the number of distinct acts that must be completed and the number of distinct information cues about the attributes of the task-related stimulus object an individual has to process when performing a task [26]. High task complexity can increase information processing requirements and demand more cognitive resources from task executors [26], [27]. If the amount of information processing exceeds a certain limit, people's attention to tasks may get diluted and hence they cannot learn effectively, leading to a higher stress [26]. Thus, following these prior studies, we argued that task complexity will drive the technostress level among individuals.

H3:Higher task complexity level will increase the level of technostress of an employee

Coping strategies refers to behavior that protects people from being psychologically harmed by problematic social experience, a behavior that importantly mediates the impact that societies have on their members [28]. Ragu-Nathan et al. [5] identified literacy facilitator as one of technostress inhibitors, which may reduce the impact of technostress. Moreover, study by Mick and Fournier [15] indicated that selecting the appropriate coping strategies might decrease the level of technostress. Coping strategies may result in improving user effectiveness and efficiency and minimizing the perceived threats of the technology [29]. Thus, following the coping strategy theory, we hypothesize:

H4: Coping strategies (literacy facilitator) moderates the relationship between task complexity and technostress.

4 Research Methodology

4.1 Data Collection

For this preliminary proposed, we collected data from 100 employees of several companies in South Korea. Respondents were chosen randomly and they were asked to filled out the questionnaire indicating their company's condition and their attitude toward ICTs. Items for the questionnaire were adopted from previous studies with revisions to meet our objectives. Total of 98 valid questionnaires were used for further analysis. Among them, 58 respondents (59.2%) were male, and 40 respondents (40.8%) were female. Around 63.3% respondents (62 respondents) were from 26-35 years old, while 30.6% respondents were from 36-50 years old. More than 80% (80 respondents) have undergraduate degree, and more than 60% respondents are IT staff/technician/clerical.

5 Analysis and Result

5.1 Reliability and Validity of Measurement Items

Analysis was conducted using Partial Least Square (PLSGraph 3.0). PLS was used for this preliminary research because it does not require as large of a sample and it does not make the same distributional assumptions about the data [30]. The measurement model was tested with respect to individual item reliability, internal consistency, and discriminant validity. Each item was modeled as a reflective indicator of its latent construct. A common rule of thumb is that item loadings should exceed .5 and the average variance extracted for each construct should exceed .50 (see table 1). Composite reliabilities ranged from 0.802 to 0.916, while AVE (average variance extract) ranged from 0.671 to 0.756 (table 2). In assessing discriminant validity, the square root of the average variance extracted for the construct should exceed the correlations with the other constructs. The diagonal elements in table 2 represent the square roots of variance extracted (AVE) of latent variables, while off-diagonal elements are the correlations between latent variables. All scales exhibited discriminant validity, with the square root of the variance extracted exceeding the correlations for all scales.

Table 1. Confirmatory Factor Analysis (PLS result)

Item	IN	SE	TC	TS	CS
IN2	**0.75**	0.16	0.03	-0.11	0.41
IN3	**0.81**	0.26	0.02	-0.06	0.31
IN4	**0.93**	0.24	-0.02	-0.21	0.26
SE1	0.22	**0.78**	0.14	-0.05	0.36
SE2	0.23	**0.95**	-0.02	-0.10	0.26
TC1	0.07	0.10	**0.77**	0.19	0.19
TC2	0.14	0.12	**0.82**	0.26	0.13
TC3	-0.02	-0.04	**0.87**	0.35	0.03
TC4	-0.08	0.01	**0.88**	0.48	0.04
TS1	-0.07	0.02	0.35	**0.85**	-0.03
TS2	-0.09	-0.03	0.32	**0.86**	-0.07
TS3	-0.23	-0.15	0.38	**0.87**	-0.09
TS4	0.01	0.07	0.21	**0.72**	-0.02
TS5	-0.24	-0.17	0.42	**0.84**	-0.15
CS1	0.38	0.20	0.04	-0.09	**0.89**
CS2	0.18	0.37	0.13	-0.07	**0.75**

Table 2. Discriminant Validity

Var.	CR	AVE	IN	SE	TC	TS	CS
IN	0.871	0.695	**0.834**				
SE	0.860	0.756	0.255	**0.869**			
TC	0.902	0.699	0.002	0.035	**0.836**		
TS	0.916	0.685	-1.180	-0.095	0.423	**0.828**	
CS	0.802	0.671	0.359	0.325	0.093	-0.098	**0.819**

5.2 Structural Model

The result of the tests of the main structural model and associated path coefficients are found in figure 2. Consistent with previous research and recommendations [30], bootstrapping (100 subsamples) was performed to determine the statistical significance of each path coefficient using t-tests. Among 3 hypotheses, there was only 1 unsupported hypothesis. Results indicated that self efficacy has no significant effect on technostress, rejecting H2. As we expected, innovation culture negatively influenced technostress. Thus, H1 was accepted. Moreover, the path from task complexity to technostress was also positive and significant, supporting H3.

After we tested the main effect, we measured the moderating effect of literacy facilitation. The tests for the moderating effects were conducted by following Chin [31].

Firstly, Cohen's f^2 was performed to calculate the change in R-square values between main and interaction effects.

Similar to Cohen [32], f^2 of 0.02 may be regarded as weak, effect size from 0.15 as moderate, and effect size above 0.35 as large effect at the structural level. Chin et al. [33] state that a low effect size does not specify that the underlying moderator effect is negligible. The moderating effect of literacy facilitation was negatively significant, supporting H4 *(F-value = 1.81, p<0.05, Cohen f²=0.03)*.

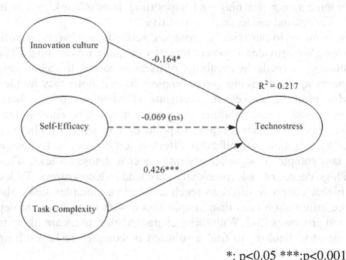

*: p<0.05 ***:p<0.001

Fig. 2. Analysis Result (Main Effect)

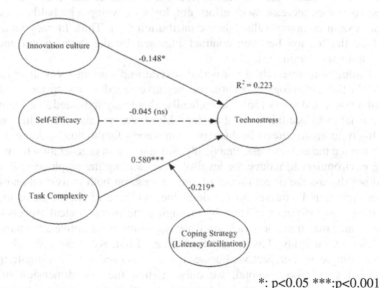

*: p<0.05 ***:p<0.001

Fig. 3. Analysis Result (Moderating Effect)

6 Discussion and Implications

The main purpose of this research is to investigate the influence of innovation culture, self-efficacy, and task complexity on technostress. Firstly, the results revealed that innovation culture negatively influences technostress level. This finding suggests that innovation culture may reduce the technostress level among employees. The adoption of technology always involves the degree of uncertainty. Thus, by providing an innovative environment (e.g. learning and supporting atmosphere), potential users are encouraged to adopt and use technology positively.

Secondly, contrary to our initial hypothesis, self-efficacy had no significant effect on technostress. We provided some arguments to explain this finding. First, the overload of technology exceeds the capability of users to cope with technology. Although most technology applications are quite complex, for example, they frequently should be modified to meet the business requirements. However, even after these modifications have been made, the capabilities of systems may not be enough, leading to another problems such as computer crash or application are slow [4]. These uncontrolled situations may explain the insignificance effect of self efficacy on technostress.

Thirdly, task complexity significantly influences technostress level. This result implied that a high degree of task complexity may lead to technostress. Task complexity requires a higher cognitive effort to reach a solution. Complex tasks also typically contain more information cues than simple tasks, with these cues are dependent on other cues and processes [34]. With those characteristics, users are likely to use more than one type of technology to find a solution of complex task, resulting in higher level of technostress.

Lastly, we also found the significant effect of literacy facilitation as a moderating variable. To meet socio-emotional needs and to determine the organization's readiness to reward increase work effort, employees develop a global belief concerning to what extent company values their contribution [10]. Thus, literacy facilitation would reduce the tension between contract imposed by the organization and employee's psychological strain [35].

Through this research, we provided several implications for both research and practice. In the technology literature, the negative reaction toward the technologies may lead anxiety and stress [36]. Theoretically, this study indicated that innovation culture is crucial to reduce this level of anxiety. Moreover, we also found that task complexity is likely to create stress. In this case, company's facilitation in doing a complex task can reduce the technostress. Practically, we suggest an organization to create a supporting environment to reduce the level of stress among the employees. Even though we did not discuss the direct impact of technostress on both individual and firm performance, previous literature has provided the evidence of the negative impact of technostress on performance. Thus, by managing the psychological aspects of employees, companies may transform their innovativeness into competitive advantage.

This research also has several limitations. First, we measured task characteristics from complexity perspective. Further research is needed to investigate the other characteristics of tasks. Second, we only studied the one dimension of technostress (techno-overload). A research examining the other dimensions of technostress would provide us with a better understanding of technostress. Lastly, we also encourage the IS scholars to explore the roles of coping strategies in the case of technostress.

References

1. Hoffman, D.L., Novak, T.P., Venkatesh, A.: Has the internet become indispensable? Communications of the ACM 47(7), 37–42 (2004)
2. Wang, K., Shu, Q., Tu, Q.: Technostress under different Organizational Environment: An Empirical Investigation. Computer in Human Behavior 24, 3002–3013 (2008)
3. Burke, R.: Organizational-level interventions to reduce occupational stressors. Work and Stress, 777–787 (1993)
4. Tu, Q., Tarafdar, M., Ragu-Nathan, T.S., Ragu-Nathan, B.S.: Improving end-user satisfaction through techno-stress prevention: some empirical evidences. In: Proceedings of the Fourteenth Americas Conference on Information Systems, Toronto, ON, Canada, August 14-17 (2008)
5. Ragu-Nathan, T.S., Tarafdar, M., Ragu-Nathan, B.S.: The consequences of Technostress for End-user in Organizations: Conceptual Development and Empirical Validation. Information System Research 19(4), 417–433 (2008)
6. Ahuja, M.K., Thatcher, J.B.: Moving beyond intentions and towards the theory of trying. MIS Quarterly 29, 427–459 (2005)
7. Bloom, A.J.: An anxiety management approach to computer-phobia. Training and Development Journal 39(1), 90–94 (1985)
8. Goodhue, D.L., Thompson, R.L.: Task-technology fit and individual performance. MIS Quarterly (1995)
9. Craig, B.: Technostress: The Human Cost of the Computer Revolution, pp. 1–3. Addison-Wesley, Readings (1984)
10. Wang, K., Shu, Q.: The moderating impact of perceived organizational support on the relationship between technostress and role stress. In: Bhowmick, S.S., Küng, J., Wagner, R. (eds.) DEXA 2008. LNCS, vol. 5181, Springer, Heidelberg (2008)
11. Brod, C.: Technostress: The Human Cost of the Computer Revolution. Addison-Wesley, Reading (1984)
12. Arnetz, B.B., Wiholm, C.: Technological Stress: Psychophysiological Symptom in Modern Offices. Journal of Psychometric Research 43(1), 35–42 (1997)
13. Burke, M.K.: The incidence of technological stress among baccalaureate nurse educators using technology during nurse preparation and delivery. New Education Today 29, 57–64 (2008)
14. William, S., Cooper, L.: Managing Workplace Stress: A Best Practice Blueprint. John Wiley & Sons, West Sussex (2002)
15. Mick, D.G., Fournier, S.: Paradoxes of technology: customer cognizance, emotions, and coping strategies. Journal of Customer Research 25 (1998)
16. Barney, J.B.: Organizational culture: can it be a source of sustained competitive advantage? Academy of Management Review 11(3), 656–665 (1986)
17. Hurley, R.F., Hult, T.M.: Innovation, market orientation, and organizational learning: an integration and empirical examination. Journal of Marketing 62, 42–54 (1998)
18. Rogers, E.M.: Diffusion of Innovations, 3rd edn. The Free Press, New York (1983)
19. Brancheau, J.C., Wetherbe, J.C.: The adoption of spreadsheets software: testing innovation diffusion theory in the context of end-user computing. Information System Research 1(2), 115–143 (1990)
20. Hendrix, W.H., Summers, T.P., Leap, T.L., Steel, R.P.: Antecedents and organizational effectiveness outcomes of employee stress and health. In: Rick, C., Pamela, L.P. (eds.) Occupational Stress: A Handbook, pp. 75–86 (1995)

21. Bandura, A., Cervone, D.: Self-evaluative and self-efficacy mechanisms governing the motivational effects of goal systems. Journal of Personality and Social Psychology 45(5), 1017–1028 (1983)
22. Bandura, A.: Self-efficacy mechanism in human agency. American Psychologist (37), 122–147 (1982)
23. Compeau, D.R., Higgins, C.A.: Computer self-efficacy: development of a measure and initial test. MIS Quarterly, 189–211 (1995)
24. Cooper, C.L., Sloan, S.J., Williams, S.: Occupational Stress Indicator Management Guide. NFER-Nelson, Oxford (1988)
25. McGrath, J.E.: Stress and behavior in organizations. In: Dunnette, M.D. (ed.) Handbook of Industrial and Organizational Psychology, pp. 1351–1395. Rand-McNally, Chicago (1976)
26. Jiang, Z.J., Benbasat, I.: The effect of presentation formats and task complexity on online consumers' product understanding. MIS Quarterly 31(3), 475–500 (2007)
27. Klemz, B.R., Gruca, T.S.: Dueling or the battle royale? The impact of task complexity on the evaluation of entry threat. Psychology & Marketing 20(11), 999–1016 (2003)
28. Pearlin, L.I., Schooler, C.: The structure of coping. Journal of Health and Social Behavior 19, 2–21 (1978)
29. Beaudry, A., Pinsonneault, A.: Understanding user responses to information technology: A coping model of user adaptation. MIS Quarterly 29(3), 459–524 (2005)
30. Chin, W.W.: Issues and opinions on structural equation modeling. MIS Quarterly 22(1), vii–xvi (1998)
31. Chin, W.W.: How to Write Up and Report PLS Analysis. In: Esposito Vinzi, V., et al. (eds.) Handbook of Partial Least Squares, pp. 655–690. Springer, Heidelberg (2010)
32. Cohen, J.: Statistical power analysis for the behavioral sciences. L. Erlbaum Associates, Hillside (1988)
33. Chin, W.W., Marcolin, B.L., Newsted, P.R.: A Partial Least Squares Latent Variable Modeling Approach for Measuring Interaction Effects: Results from a Monte Carlo Simulation Study and an Electronic-Mail Emotion / Adoption Study. Information Systems Research 14(2), 189–217 (2003)
34. Wood, R.E.: Task complexity: Definition of the construct. Organizational Behavior and Human Decision Processes 37, 60–82 (1986)
35. Aselage, J., Eisenberger, R.: Perceived organizational support and psychological contracts: a theoretical integration. Journal of Organizational Behavior (24), 491–509 (2003)
36. Mick, D.G., Fournier, S.: Paradoxes of technology: consumer cognizance, emotions, and coping strategies. Journal of Consumer Research 25, 123–143 (1998)

Why Do Customers Keep or Switch
Their Mobile Telecommunication Service Provider?

KyoungAe Kim[1], JaeSung Park[1], JaeJon Kim[2], and Joon Koh[2]

[1] Business Incubating Center, Chonnam National University, Korea
skyae@cibi.or.kr, pamto@daum.net
[2] Department of Business Administration, Chonnam National University, Korea
jaejon@jnu.ac.kr, kjoon@jnu.ac.kr

Abstract. Severe competition among firms is a hallmark of the Korean mobile telecommunications industry, largely because so many similar services are being provided in a saturated market. There are no significant differences among mobile communication providers in terms of 3G quality criteria in the Korean mobile telecommunications market. Further, the switching behaviors of frequent mobile telecommunication services users have been a consequence of the aggressive marketing tactics of each of these mobile telecommunication firms, which make great efforts to lure away their competitor's customers; the costs associated with these marketing activities amount to at least seven billion dollars, which come directly out of the pockets of the mobile telecommunication service providers. Therefore, it is clearly necessary to assess the efficacy of these marketing activities in luring away competitors' customers.

Keywords: Mobile Telecommunication Service, Alternative Attractiveness, Bonds.

1 Introduction

Recent turbulence in the business environment has compelled firms to search out new sources of competitive advantages. Firms who must cope with such environments have experienced major difficulties in securing sustainability management. Competition among firms in the Korean mobile telecommunication industry has been remarkably robust, owing to the presence of so many similar services in the saturated market. Thus, customers have become increasingly likely to switch their mobile telecommunication service provider when the customer feels that the service quality is low, and customers' decisions as to whether or not to maintain current relationships with their mobile telecommunication service provider are a matter of critical import to mobile providers (Thibaut and Kelley, 1959).

The mobile communication service in Korea began with KMT (Korea Mobile Telecommunications Corp.), the predecessor of SK Telecom, which first provided an analog car-phone service in 1984. The size of the mobile telecommunications market in Korea grew at a rate of 100% per year from 1995 to 1999, and thus, as of 2009, forty-seven million users (this exceeds the population of the Republic of Korea, because it includes repeat and overlap in the customer base) have used mobile telecommunication services (KCC, 2009b). Since mobile communication service firms must be permitted by the Korean government to participate in the mobile communications industry, only three companies (SK Telecom, KT, LG U+) currently provide mobile communication services in Korea.

According to the Korea Communication Commission (2009a), no significant differences could be detected among the three mobile communication providers in terms of the 3G quality criteria of mobile communication, which include voice telephony quality (success rate of access, drop call rate, voice error rate), wireless data quality (success rate of access, transmission complete rate, transmission speed), and video telephony quality (success rate of access, drop call rate, picture quality). Mobile communications service providers in Korea focus on strategies to retain their customers and to lure competitors' customers, as the mobile telecommunications market in Korea has reached a growth plateau. Thus, the mobile telecommunication service providers in Korea have developed a variety of legal relationships, including contract period systems. Meanwhile, they have also implemented a host of inducements to retain their customer base, including mobile equipment subsidies, in addition to their strategies to lure other companies' customers. These activities are largely responsible for the fierce competition that exists among mobile telecommunication service providing firms in Korea.

Statistics from the KCC (2010a) show that 300 million mobile telecommunication users changed their service provider in 2004 (the year mobile number portability was implemented); this number increases every year. The number of mobile number portability users reached 10 million users in 2009; that is to say, a full quarter of the Korean population has changed their provider. The size of the entire mobile communications in Korea has been estimated to be as high as thirty billion dollars in 2010 (KCC, 2010b). Therefore, we must assess whether these marketing activities effectively lure away competitors' customers, and if so, how. In marketing strategy, an offensive and defensive strategy toward the target market may be followed simultaneously. The offensive strategy, in this case, involves the capture of market share and increasing market size, whereas the defensive strategy involves minimizing customer exits and switching behaviors. The principal objective of the defensive strategy is to minimize customer turnover (that is, maximize customer retention), considering certain cost constraints, by protecting products and markets from competitive inroads (Fornell and Wernerfelt, 1987; 1988). In general, defensive strategy is more effective than offensive strategy (Fornell, 1992).

Whether to maintain or to break off the relationship with the existing service provider can be determined via comparisons between the strength of the bond to the existing service provider and the level of attractiveness to the alternative service provider in the fiercely competitive mobile telecommunications market of Korea, in which other service providers constantly attempt to take their competitors' customers (Anderson and Narus, 1990). Thus, for a mobile telecommunication firm, it is a matter of critical importance to understand the criteria of firm's resource allocation strategies in order to implement both customer retention and customer attraction strategies simultaneously, thereby allowing for survival in the face of fierce competition. To accomplish this, we must first attempt to evaluate the factors that influence continuous usage intentions in the Korean mobile telecommunications industry. Thus, the principal objective of this research is to identify and assess the principal factors that may affect customers' decisions to reuse or switch their existing mobile telecommunication service.

2 Theoretical Background

2.1 Standards of Exit or Contiguous Use

The decision as to whether customers continuously employ their existing mobile telecommunication service or change their service can be calculated by comparing cost and rewards deriving from switching activity based on the exchange theory (Homans, 1961, Blau, 1964). Homans (1961) previously asserted that performers should evaluate the values associated with giving and receiving from interactions. These include not only physical profits, but also psychic profits in the economics dimension.

Thibaut and Kelley (1959) proposed that profit was not an absolute, but rather a relative concept. When performers decide on something, they consider all alternatives as much as possible, and outcomes from the exchange relationship become a major standard of judgment for the decision to continue a relationship. They also suggested the comparison level, which can be regarded as a standard representing the quality of outcomes the service provider has come to anticipate from a given type of relationship, based on present and past experience with similar relationships, and knowledge of other service provider's similar relationships. By way of contrast, the comparison level for alternatives is a standard representing the average quality of outcomes available from the best alternative exchange relationship (Anderson and Narus, 1984).

When customers elect to continue the relationship with their service provider, they evaluate sunk costs from the existing channel, as well as relationships with alternative channels. If no alternative that can provide better outcomes exists, customers will elect to maintain the existing relationship. However, if not, customers have sufficient motivation to break off the existing relationship and engage in a new alternative relationship.

Thus, the decision to maintain or change the existing channel can be influenced by both bond levels, which were the sum of cost and benefits deriving from the maintenance of the existing channel, and the anticipated profit level due to moving to the best alternative. Customers should maintain their relationship with the existing channel, if they feel that the the level of the two bonds to the existing channel is greater than the attractiveness of the best alternative. On the other hand, customers may consider switching the existing partner, if the attractiveness of the best alternative is higher than the bonds of the existing channel.

2.2 Alternative Attractiveness

The alternative attractiveness is defined as the best expected alternative service level relative to the existing service provider (Ping 1993). Alternative attractiveness performs a function in the offsetting of bonds to the existing channel, and customers should attempt to break their relationship with the existing channel, in cases in which they recognized that the level of alternative attractiveness was higher than the level of bonds to the existing channel (Ping, 1993). When the alternative attractiveness increases, customers frequently find problems in their relationship with the existing service provider and begin to complain aggressively about their problems (Hirschman, 1970; Rusbult et al., 1982; Rusbult et al., 1988; Park et al., 2010).

On the contrary, when the alternative attractiveness is insufficient from the perspective of the service providers, a relatively positive situation exists for the retention of their customers, even though their core services were below average. If the customers perceive attractive alternatives, including better service, geographic proximity, full range of service affordability, and low cost or high economic contribution, they are generally likely to switch their existing service provider (Sharma and Patterson, 2000).

When the customers' alternative attractiveness is low, the perceived profit or benefit deriving from the switching of one's existing service provider may be low. In this case, customers recognize the relatively high binding force of the existing provider. This empirical evidence has been detected in the area of relationships among individuals and the job market (Rusbult, 1980; Farrell and Rusbult, 1981) or relationships with channels (Ping, 1993). This logic is quite reminiscent of resource dependence theory. Resource dependence theory is predicated on the notion that environments are a source of scarce resources, and organizations are dependent on these finite resources for survival. Thus, a lack of control over these resources creates uncertainty for firms that operate in that environment. Organizations must develop methods to exploit these resources, which are also being sought by other firms, in order to assure their own survival (Pfeffer and Salancik, 1978). According to resource dependence theory, the high power of providers relative to the customers may prevent their customers from searching for alternative providers. At this moment, the level of customers' expected benefits from alternatives are similar to the alternative attractiveness (Jones et al., 2000).

Economic alternative attractiveness is the expected scale of profit from the best alternatives, which has some aspects in common with the anticipated financial or economical elements from the best alternatives (Olsen and Ellram, 1997), or the anticipated value deriving from the best alternatives (Sirohi et al., 1998). Thus, economic alternative attractiveness can be defined as the level of economic value from the best alternatives. If the customers perceive high economic alternative attractiveness, they may build a new relationship with the best alternatives rather than maintaining their existing service provider.

Social alternative attractiveness is associated with the anticipated fair business relationship from the transactional partner from the best alternatives (Ping, 1993) and the perceived social reputation and image associated with the social culture (Olsen and

Ellram, 1997; Kim et al., 2004). If the customers perceive a high level of social attractiveness from the best alternative, a fraudulent act of the existing service provider might cause them to reconsider switching their service provider.

Technical alternative attractiveness can be described as the anticipated level of attractiveness associated with the product or service technology from the best alternatives (Ping, 1993; Jones et al., 2000), or the anticipated attractiveness associated with the service range or level from the best alternatives (Sharma and Patterson, 2000; Patterson and Smith, 2003). This attractiveness can be regarded as the quality attractiveness based on the technology. Thus, when the customers perceive a high level of quality attractiveness in the best alternative, they may decide to move to the best alternatives, rather than maintaining the existing service provider.

Thus, alternative attractiveness can be classified into three categories: economic attractiveness associated with economic profit, social attractiveness associated with social reputation, relationship fairness and business image, and quality attractiveness associated with the service or product range or differentiation level.

2.3 Bonds

Bonds are psychological, emotional, economic, or physical attachments in a relationship that are fostered by association and interaction and serve to bind parties together under a relational exchange paradigm (McCall, 1970; Turner, 1970). Storbacka et al. (1994) previously suggested bonds as the factor that involuntarily induced loyalty, and defined bonds as a type of restriction in which customers were subordinate to the service provider for political, economical, social, and cultural reasons. For example, if an automaker limited his customers' guarantee to specific auto-repair centers, customers could not readily switch to another auto-repair center for the aforementioned reasons, and bonds performed a pivotal role in restricting customers' switching behavior (Liljander and Roos, 2002). Even in cases in which customer satisfaction is extremely low, if customers maintain strong bonds with the service provider, the possibility of maintaining the relationship still increases (Gronhaug and Gilly, 1991). Thus, bonds can be regarded as a type of negative concept (Bendapudi and Berry, 1997; Liljander and Strandvik, 1995).

However, we do not consider bonds as solely negative, since bonds may afford the customers convenience--for example, providing a knowledge base about their product (like Dell), thereby creating an easy process for purchasing music files from a vast music database such as Apple's iTunes service, operating communities, and offering new services such as LG U+'s blog. Therefore, some bonds can be perceived as both positive and negative.

Generally, social bonds created by the community and associated with products and services, knowledge bonds created to promote the product or service convenience, and psychological bonds such as brand loyalty are all recognized as positive by customers. On the contrary, compulsory bonds--such as legal bonds--can be regarded as negative (Liljander and Strandvik, 1995). Thus, we can predict that the customers' perceptions of bonds and the effect thereof will differ according to the forms of bonds.

In this context, Berry and Parasuraman (1991) attempted to divide bonds into three types: exit barrier such as financial bonds, social bonds, and structural bonds. They reported that the establishment of each bond strategy should restrain customers from exiting. Berry (1995) previously suggested the roles of three types of bonds in relationship marketing. Peltier and Westfall (2000) evaluated the relationships between the three types of bonds and overall satisfaction in health maintenance organizations.

Financial bonds provide a special low price to loyal customers or strengthen customer relations via other financial incentives (Berry, 1995; Strauss and Frost, 2001). When the accumulated points reach a certain level, the customers may even employ their points to pay a charge. Further, these firms have provided ratable fare discounts to their best customers and have offered discount services relative to communication services, in addition to their mobile telecommunication services.

Social bonds refer to personal linkages that pertain to service dimensions offering interactions between individuals (Beatty, Mayer, Coleman, Reynolds and Lee, 1996; Wilson, 1995). Thus, social bonds push customers toward self-disclosure, listening, caring and helping improve mutual understanding, openness between relationship partners, and degree of closeness (Thorbjornsen et al. 2002; Hsieh et al, 2005). For example, the sharing of know-how regarding mobile telecommunication services promote strong social bonds among customers in communities. Korean mobile telecommunication service-providing firms induce customers to participate in their on-line communities and blogs. However, the majority of power users joined communities that had been voluntarily created without any support from firms. Although mobile telecommunication service-providing firms did not invest in their communities, spontaneous customers and their bonds appear to function as a deterrent against switching to alternatives.

Structural bonds can be defined as knowledge and information related to the customization of products and industries, which require a very heavy price upon termination of the relationship (Hsieh et al., 2005). Expert knowledge and information reduces risks and results in competitive advantage generated through greater asymmetrical information in the structural bonds. For instance, customers who employ mobile telecommunication service possess their own schedules, lists of phone numbers, mp3 music files which are categorized based on their needs, and online game software in their phone. However, when customers elect to switch their mobile telecommunication service provider, in Korea, it has proven difficult to transfer these contents to a new provider. In terms of customers' schedules and phone number databases, music files, and purchased software, for example, phone platforms and file formats can all differ among mobile telecommunication service providers. If the customers wish to use their contents, they should exploit third-party utilities or exert greater efforts to reuse their contents. Thus, know-how, information, and knowledge created through long-term customer usage may be a sunk cost when a customer switches service providers. This would obviously constitute a barrier against moving to other mobile service provider companies in Korea.

In addition to the three kinds of bonds proposed by Berry and Parasuraman (1991), Korea mobile telecommunications service firms maintain relationships with customers via legal contracts, including handset subsidies and contract discounts. Cannon and Parreault (1999) have defined such bonds as contracts with concrete legal binding force that applies to both sides. They evaluated the effects of legal bonds on customer evaluations of suppliers. Thus, legal bonds go beyond the basic obligations and protections that regulate commercial exchange, whether or not the parties sign a formal document (Uniform Commercial Code, 1978). The profit derived from legal bonds to the service or product provider can be found on both sides; one derives from maintaining the relationship via the legal system, and the other involves routinizing the transactional relationships by offering a plan for the future (Macneil, 1980). Actually, mobile telecommunication service providers in Korea have frequently implemented contracts for long-term phone service usage. When their customers cancel the contract before the stipulated time period, they request a cancellation fee from their customers.

Financial bonds might be the weakest bond among the four dimensions of bonds, and from them can be derived the formal binding relationship between service providing firms and customers. Social bonds form comparatively strong binding relationships because customers express their needs and inconvenience regarding the specific service to their service provider or other customers in the communities (Berry, 2000). Moreover, structural bonds induce inimitable relationships because customers produce product knowledge and information via continuous usage.

3 Research Model and Hypotheses

3.1 Research Model

This study suggests a model that explains the effects of antecedents such as alternative attractiveness and bonds on continuous usage intentions, depending on the social exchange theory in the Korean mobile telecommunication industry.

Alternative attractiveness is the anticipated profit deriving from switching one's existing service provider (Ping, 1993; Sharma and Patterson,2000; Rusbult,1980; Anderson and Narus,1990; Olsen and Ellaram,1997; Kim et al., 2004; Jones et al., 2000; Petterson and Smith, 2003), and bonds represent the expected cost for switching the existing service provider (Berry and Parasuraman,1991; Berry, 1995; Strauss and Frost, 2001; Wilson,1995; Hsieh et al., 2005; Cannon and Parreault,1999, Lijander and Strandvik, 1995). Thus, we regard alternative attractiveness as a negative factor for continuous usage intentions in mobile telecommunication services, whereas bonds are perceived as a positive factor in continuous usage intentions.

Additionally, alternative attractiveness can be classified into three categories: economical attractiveness, social attractiveness, and quality attractiveness. These factors can influence continuous usage intentions. Bonds can be classified into four types-- financial bonds, social bonds, structural bonds, and legal bonds. Among the four types of bonds, the three types of bonds--financial bonds, social bonds, and structural bonds--are voluntary binding forces, which operate as positive factors affecting continuous usage intentions, whereas legal bonds are coercive binding forces strengthened by contracts, and function as negative factors influencing continuous usage intentions.

3.2 Hypotheses

Alternative Attractiveness and Continuous Usage Intention. Jones and Sasser (1995) previously argued that customers intend to change their existing service provider when they are dissatisfied with the existing service provider and also have many alternatives. Therefore, the absence of alternative attractiveness would be expected to be advantageous for the existing service providers (Ping, 1993). Sharma and Patterson (2000) previously determined, through their in-depth interviews, that the existence of competitors increases the possibility that customers will quit their existing relationship. According to Frazier (1983), alternative attractiveness is associated closely with dependence, which means it is necessary to continue the existing relationship (Anderson and Narus, 1990). Furthermore, Ping (1993) reported that alternative attractiveness affects the decision to change transaction partners. Thus, when outcomes from the existing partner are worse than the outcomes deriving from the alternatives, it would be difficult to continue the relationship with the existing partner. When the profit expected from the alternatives is larger than the profit from the existing partner, the possibility of changing the existing partner will be relatively high (Kumar et al., 1996). Consequently, these degrees of economical, social, and quality alternative attractiveness may affect customers' decisions as to whether or not they could maintain their existing service provider.

H1: Economical alternative attractiveness is negatively associated with the customer's continuous usage intention.

H2: Social alternative attractiveness is negatively associated with the customer's continuous usage intention.

H3: Quality alternative attractiveness is negatively associated with the customer's continuous usage intention.

Bonds and Continuous Usage Intention. Financial bonds derive from the benefits of low prices and financial incentives to loyal customers (Berry, 1995; Strauss and Frost, 2001). Social bonds refer to personal linkages that offer interactions between individuals (Beatty, Mayer, Coleman, Reynolds and Lee, 1996; Wilson, 1995), and structural bonds can be defined as knowledge and information associated with the customization of products and industries, and require a very heavy price upon termination of the relationship (Hsieh et al., 2005). If customers receive financial benefits from the existing service provider, harbor good feelings toward the existing service provider, and obtain knowledge and information from their transactions with the existing service provider, than they will harbor favorable feelings toward their service provider. Hence, financial bonds, social bonds, and structural bonds, all of which are voluntary binding forces, may positively influence customers' continuous usage intentions in the mobile telecommunication industry.

H4: Financial bonds are positively associated with customers' continuous usage intentions.

H5: Social bonds are positively associated with customers' continuous usage intentions.

H6: Structural bonds are positively associated with customers' continuous usage intentions.

A contract might keep customers silent even when they are dissatisfied with their service provider. These bonds should be regarded as negative by customers (Bendapudi and Berry, 1997; Lijander and Strandvik, 1995). For example, phone users engaged in a two-year contract with their mobile telecommunications service provider would incur a cancellation charge fee if they canceled their contract. Thus, these users cannot readily change their service provider, even if they are dissatisfied with their service provider during the stipulated time. That is to say, legal bonds deprive customers of the freedom to switch their service provider. Legal bonds can be regarded as a negative factor that can ameliorate the continuous usage intentions in the Korean mobile telecommunications industry. Therefore, the following hypothesis can be proposed:

H7: Legal bonds are associated negatively with customers' continuous usage intentions.

4 Analysis of Research Results

4.1 Test of Validity and Reliability

We mailed the survey to all members that had employed internet communities, an invitation to participate in our web-based survey over a two-week period. In an attempt to prevent double participations in the survey, we checked and confirmed each respondent's IP address and e-mail address. A total of 359 users participated in the web survey. However, we excluded 15 cases owing to a lack of consistency, using reversed items in the answers of the respondents. Thus, 344 cases were employed for the analysis. However, 53.2% of the respondents were male, and 60.8% were younger than 30. Further, 71.2% had switching experience with moving to another mobile telecommunications service. Additionally, 50.3% of the respondents used SK Telecom, 33.4% were KT customers, and the other 16.3% were LG U+ customers; these results are very similar to the real market share in the Korean mobile telecommunication industry.

Each variable was measured using multiple items. We conducted an exploratory factor analysis to evaluate their uni-dimensionality. In order to assess the construct validity of the instruments, a principal-component factor analysis with Varimax rotation was conducted. Furthermore, the majority of the factor loadings for the items appeared to be above 0.5. Thus, the items corresponded well to each singular factor, evidencing a high degree of convergent validity. As the factor loadings for a variable (or factor) were higher than the factor loadings for the other variables, the instrument's discriminant validity was shown to be clear (Hair et al, 1998). Eight factors were extracted as anticipated, which explained 75.07% of the total variance, with eigen values in excess of 1.000.

The internal consistency reliability of the variables was evaluated by computing Cronbach's alphas. The Cronbach's alpha values of the variables ranged from 0.728 to 0.928, but in all cases over 0.700 (Nunally, 1978). Both the reliability and validity of the variables are, therefore, acceptable.

4.2 Tests of Hypotheses

The effects of alternative attractiveness and bonds on continuous usage intentions in the mobile telecommunication service industry were evaluated via multiple regression analysis. The Pearson correlations were calculated for the variables, and measured by the interval or ratio scales. Potential multicollinearity among the antecedents was evaluated prior to the multiple regression analysis, as some of the variables were correlated significantly with others. Although several variables exhibited significant correlations, their tolerance values ranged between 0.521 and 0.979, thus demonstrating that multicollinearity is not a likely threat to the parameter estimates (Hair et al., 1998). After checking some basic assumptions for the regression analysis, we elected to conduct multiple regression analysis and moderated regression analysis to evaluate the given hypotheses.

Table 1 shows the results of multiple regression analysis. Our results demonstrate that the regression models were significant at $p<0.001$ (F=32.280), and that the predictors of each model explain 40.2% of the total variance. Hypotheses 1 to 3 evaluate the relationships between economical alternative attractiveness and continuous usage intentions, between social alternative attractiveness and continuous usage intentions, and between quality alternative attractiveness and continuous usage intentions. Social alternative attractiveness was found to negatively affect continuous usage intention to a significant degree ($\beta=-0.175$, $p<0.01$). Therefore, our findings support Hypothesis 2. We interpret this to indicate a negative relationship existing between social alternative attractiveness and continuous usage intentions.

Table 1. The Results of Multiple Regression Analyses.

Model	R^2	adj. R^2	F	β	Results
Continuous Usage Intention (CUI)					
CUI=EAA+SAA+QAA+FB+SB+StB+LB+errors	0.402	0.390	32.280***		
EAA (Economical Alternative Attraction)				0.039	H-1 Not accepted
SAA (Social Alternative Attraction)				-0.175**	**H-2 Accepted**
QAA (Quality Alternative Attraction)				-0.073	H-3 Not accepted
FB (Financial Bonds)				0.080	H-4 Not accepted
SB (Social Bonds)				0.360***	**H-5 Accepted**
StB (Structural Bonds)				0.203***	**H-6 Accepted**
LB (Legal Bonds)				-0.120**	**H-7 Accepted**

*$p<0.05$, **$p<0.01$, ***$p<0.001$.

Hypotheses 4 to 7 assesses four relationships: namely, between financial, social, structural, and legal bonds, and continuous usage intentions. Social bonds and structural bonds were found to significantly influence continuous usage intentions (β=0.360, p<0.001; β=0.203, p<0.001). Therefore, our findings support Hypotheses 5 and 6, indicating the existence of positive relationships between social bonds/structural bonds and continuous usage intentions. Additionally, legal bonds negatively affected continuous usage intentions to a significant degree, of 0.01 (β= -0.120). Thus, a negative relationship appears to exist between legal bonds and continuous usage intentions, thereby supporting Hypothesis 7.

5 Discussion and Implications

This study evaluated the principal factors influencing continuous usage intentions based on the social exchange theory in the mobile telecommunications industry. The study results demonstrated that only social alternative attractiveness significantly affected continuous usage intentions among the three dimensions of alternative attractiveness. Most customers appear to recognize charge, point, and fare discounts from Korean mobile telecommunication service-providing firms as economically attractive. However, as the economic benefit is quite similar among several mobile telecommunication service-providing firms, it may prove difficult for customers to distinguish the differences in the benefits of alternatives in the Korean mobile telecommunication industry. Furthermore, based on the statistics from KCC (2009a), there have been no significant differences between the three mobile communication providers in terms of the 3G quality criteria of mobile communications, including voice telephony quality, wireless data quality, and video telephony quality. Therefore, quality alternative attractiveness may be an insignificant factor affecting continuous usage intentions.

On the other hand, social alternative attractiveness has been associated with the perceived social reputation and image associated with social culture (Olsen and Ellram, 1997; Kim et al., 2004). Although fierce competition has resulted in a profound similarity among mobile telecommunication services, levels of social attractiveness are not so easily equalized. Social alternative attractiveness forms over a long period via the implementation of consistent customer policies and corporate image due to social contribution. Thus, when customers recognize the best alternatives as socially attractive, the continuous usage intentions of customers of existing mobile telecommunication service providers will tend to be relatively low.

Financial bonds were found to be insignificant in affecting continuous usage intentions. However, structural bonds and social bonds increased continuous usage intentions, whereas legal bonds reduced continuous usage intentions in the mobile telecommunication industry. Financial bonds refer to the provision of special low prices to loyal customers or the strengthening of customer relations via other financial incentives (Berry, 1995; Strauss and Frost, 2001). However, owing to the fierce competition inherent to the mobile telecommunications service industry, customers may not perceive any gap in financial benefits between service providers. We interpret our findings to mean that financial bonds were not associated with continuous usage intentions.

Additionally, because structural bonds consist of knowledge and information related to the customization of products, which requires a very heavy price upon termination of the relationship (Hsieh et al., 2005), and social bonds allow customers to tend toward self-disclosure, listening, caring, and helping to improve mutual understanding, openness between relationship partners, and degree of closeness, these social bonds and structural bonds exert positive effects on continuous usage intentions.

On the contrary, as legal bonds go beyond the basic obligations and protections that regulate commercial exchange whether or not the parties sign a formal document, their customers may tend to perceive a strong binding force to their service provider during the stipulated time period. Legal bonds might elicit resistance from customers during the stipulated time. Although legal bonds clearly contribute to the short term business strategies of service providers, they may prove injurious to the long-term business strategy to lure their customers.

Therefore, we interpret our results to mean that legal bonds undermine customer's satisfaction with mobile telecommunication services, and increase the possibility of customer intentions to switch providers due to of compulsory restrictions such as contract discounts. This result indicates that mobile telecommunication service-providing firms should focus on financial bonds, social bonds, and structural bonds for the voluntary continuous usage intentions of Korean customers.

Despite the implications of this study, the study design was limited in several regards. First, since we conducted a web-based survey, the sample was restricted to those who could use a computer. Thus, it will eventually be necessary to conduct a more comprehensive study. Second, the cross-sectional data collation was limited to our findings: For example, alternative attraction and bonds generally accumulate over time. Therefore, future research, in the form of a longitude study, is clearly necessary. Third, with the advent of smart phones, such as the iPhone and Android Phone, the mobile telecommunication industry has been facing sudden changes. Therefore, the development of relevant models reflecting the innovation and changes in the mobile telecommunication service is important.

References

1. Anderson, J., Narus, J.: A Model of Distributor Firm and Manufacturer Firm Working Partnerships. The Journal of Marketing 54(1), 42–58 (1990)
2. Bendapudi, N., Berry, L.: Customers' Motivations for Maintaining Relationships with Service Providers. Journal of Retailing 73(1), 15–37 (1997)
3. Berry, L.: Relationship Marketing of Services-Growing Interest, Emerging Perspectives. Journal of the Academy of Marketing Science 23(4), 236–245 (1995)
4. Berry, L.: Cultivating Service Brand Equity. Journal of the Academy of Marketing Science 28(1), 128–137 (2000)
5. Beatty, S., Mayer, M., Coleman, J., Reynolds, K., Lee, J.: Customer-sales Associate Retail Relationships. Journal of Retailing 72(3), 223–247 (1996)
6. Fornell, C., Wernerfelt, B.: Defensive Marketing Strategy by Customer Complaint Management: A Theoretical Analysis. Journal of Marketing Research 24(4), 337–346 (1987)
7. Fornell, C., Wernerfelt, B.: A Model for Customer Complaint Management. Marketing Science 7(3), 287–298 (1988)

8. Fornell, C.: A National Customer Satisfaction Barometer: the Swedish experience. The Journal of Marketing 56(1), 6–21 (1992)
9. Grönhaug, K., Gilly, M.: A Transaction Cost Approach to Consumer Dissatisfaction and Complaint Actions. Journal of Economic Psychology 12(1), 165–183 (1991)
10. Hair, J., Anderson, R., Tatham, R., Black, W.: Multivariate Data Analysis. Prentice Hall, Upper Saddle River (1998)
11. Hirschman, A.: Exit, Voice, and Loyalty. Harvard University Press, Cambridge (1970)
12. Hsieh, Y., Chiu, H., Chiang, M.: Maintaining a Committed Online Customer(A Study Across Search-Experience-Credence Products. Journal of Retailing 81(1), 75–82 (2005)
13. Jones, M., Mothersbaugh, D., Beatty, S.: Switching Barriers and Repurchase Intentions in Services. Journal of Retailing 76(2), 259–274 (2000)
14. Korea Communication Commission. The Result of Evaluation for Mobile Telecommunication Service, Korea Communication Commission (2009a), http://www.kcc.go.kr
15. Korea Communication Commission. Statistics of Subscribers to Communication Service in September 2009, Korea Communication Commission (2009b), http://www.kcc.go.kr
16. Korea Communication Commission. Statistics of Subscribers to Communication Service in June 2010, Korea Communication Commission (2010a), http://www.kcc.go.kr
17. Korea Communication Commission. Result of Marketing and Investments Through the First Half of 2010, Korea Communication Commission (2010b), http://www.kcc.go.kr
18. Kim, M., Park, M., Jeong, D.: The Effects of Customer Satisfaction and Switching Barrier on Customer Loyalty in Korean Mobile Telecommunication Services. Telecommunications Policy 28(2), 145–160 (2004)
19. Liljander, V., Roos, I.: Customer-Relationship Levels-from Spurious to True Relationships. Journal of Services Marketing 16(7), 593–614 (2002)
20. Liljander, V., Strandvik, T.: The Nature of Customer Relationships in Services. Advances in Services Marketing and Management 4, 141–167 (1995)
21. Macneil, I.: The New Social Contract. Yale University Press, New Haven (1980)
22. McCall, G.: The Social Organization of Relationships. Social Relationships, Aldine (1970)
23. Nunnally, J., Bernstein, I.: Psychometric theory (1978)
24. Olsen, R., Ellram, L.: Buyer-Supplier Relationships: Alternative Research Approaches. European Journal of Purchasing and Supply Management 3(4), 221–231 (1997)
25. Park, J., Kim, J., Koh, J.: Determinants of Continuous Usage Intention in Web Analytics Services. Electronic Commerce Research and Applications 9(1), 61–72 (2010)
26. Patterson, P., Smith, T.: A Cross-Cultural Study of Switching Barriers and Propensity to Stay with Service Providers. Journal of Retailing 79(2), 107–120 (2003)
27. Pfeffer, J., Salancik, G.: The External Control of Organizations, New York (263)(1978)
28. Ping, R.: The Effects of Satisfaction and Structural Constraints of Retailer Exiting, Voice, Loyalty, Opportunism, and Neglect. Journal of Retailing 69(3), 320–352 (1993)
29. Rusbult, C., Farrell, D., Rogers, G., Mainous III, A.: Impact of Exchange Variables on Exit, Voice, Loyalty, and Neglect: An Integrative Model of Responses to Declining Job Satisfaction. Academy of Management Journal 31(3), 599–627 (1988)
30. Rusbult, C., Zembrodt, I., Gunn, L.: Exit, Voice, Loyalty, and Neglect: Responses to Dissatisfaction in Romantic Involvements. Journal of Personality and Social Psychology 43(6), 230–242 (1982)
31. Sharma, N., Patterson, P.: Switching Costs, Alternative Attractiveness and Experience as Moderators of Relationship Commitment in Professional, Consumer Services. International Journal of Service Industry Management 11(5), 470–490 (2000)

32. Sirohi, N., McLaughlin, E., Wittink, D.: A Model of Consumer Perceptions and Store Loyalty Intentions for a Supermarket Retailer. Journal of Retailing 74(2), 223–245 (1998)
33. Straus, J., Frost, R.: E-marketing. Prentice-Hall, Inc., Englewood Cliffs (2001)
34. Thibaut, J., Kelley, H.: The Social Psychology of Groups. Wiley, Chichester (1959)
35. Thorbjørnsen, H., Supphellen, M., Nysveen, H., Egil, P.: Building Brand Relationships Online: A Comparison of Two Interactive Applications. Journal of Direct Marketing 16(3), 17–34 (2002)
36. Turner, R.: Family interaction. John Wiley, New York (1970)
37. Wilson, D.: An Integrated Model of Buyer-Seller Relationships. Journal of the Academy of Marketing Science 23(4), 335–345 (1995)
38. Uniform Commercial Code, American Law Institute and the National Conference of Commissioners on Uniform State Laws, Washington. DC (1978)

Author Index